COMBINATORIAL TOPOLOGY
VOLUME 2
THE BETTI GROUPS

OTHER *GRAYLOCK* PUBLICATIONS

KHINCHIN: *Three Pearls of Number Theory*

PONTRYAGIN: *Foundations of Combinatorial Topology*

NOVOZHILOV: *Foundations of the Nonlinear Theory of Elasticity*

ALEKSANDROV: *Combinatorial Topology, Vol.* 1

PETROVSKIĬ: *Lectures on the Theory of Integral Equations*

KOLMOGOROV and FOMIN: *Elements of the Theory of Functions and Functional Analysis, Vol.* 1. *Metric and Normed Spaces*

COMBINATORIAL TOPOLOGY

VOLUME 2

THE BETTI GROUPS

BY

P. S. ALEKSANDROV

GRAYLOCK PRESS

ROCHESTER, N. Y.

1957

TRANSLATED FROM THE FIRST (1947) RUSSIAN EDITION

BY

HORACE KOMM

Library of Congress Catalog Card Number 56-13930

Manufactured in the United States of America

CONTENTS

PART THREE

THE BETTI GROUPS

Chapter

Chapter

Chapter

Part Three

THE BETTI GROUPS

The Betti groups (defined in Chapter VIII) are the central concept of combinatorial topology. All of Part Three, as well as much of the sequel, is devoted to their study.

The underlying algebraic theory of combinatorial topology is developed in Chapter VII. It is used, in particular, to define and investigate the Betti groups themselves. The algebraic apparatus is made up of the two concepts: a *chain* and a *boundary operator* Δ. These two ideas are fundamental to Chapter VII. These in turn depend on the notions of orientation (discussed at the beginning of Chapter VII) and of an a-complex. The latter is a natural generalization of the set of all oriented simplexes of a triangulation.

Having introduced all the auxiliary algebraic concepts in Chapter VII, we deal in Chapter VIII with the definition and elementary theory of the "lower" Betti groups (homology groups) or, as we shall call them here, the Δ-groups of triangulations (and, in general, of a-complexes). Chapter VIII concludes with an investigation of orientable and nonorientable pseudo-manifolds which, in addition to other examples, serve to illustrate the theory.

Chapter IX deals with more complicated problems of the theory of Betti groups. First, the "upper" Betti groups or ∇-groups (cohomology groups) are introduced. Then the Betti groups are studied by means of canonical bases which, in particular, enable us to derive a relation among the Betti groups over various coefficient domains.

In Chapter X we prove the invariance of the Betti groups, that is, that all the (topological) triangulations of a polyhedron (or of homeomorphic polyhedra) are isomorphic.

In Chapters XI and XII the concept of Δ-group is extended from polyhedra to arbitrary compacta. It should be noted, however, that this generalization can be effected in a completely adequate fashion only by means of the theory of topological groups, which is beyond the scope of this book. I have succeeded in avoiding this difficulty, but only by defining the ∇-groups, and not the Δ-groups, of arbitrary compacta. This is done, but considerably later, in Chapter XIV.

Chapter XII, among other things, contains an account of the local Δ-groups. These are used in XIII, 1.1, to give a simple invariant definition of h-manifolds.

Chapter VII

CHAINS. THE OPERATOR Δ

§1. Orientation

§1.1. Orientation of the space R^n. The concept of an oriented or directed segment will already be familiar to the reader who has had a course in elementary algebra. In this section the notion of an oriented segment will be generalized to n dimensions.

We shall call a collection of $n + 1$ linearly independent points of R^n written in a definite order an *ordered skeleton* of R^n. According to this definition two different ordered skeletons may consist of the same points, differing from each other only in the order of these points. We recall (Appendix 1, 1.5) that there is precisely one affine mapping of R^n onto itself which carries a given ordered skeleton $| e_0, e_1, \cdots, e_n |$ into a preassigned ordered skeleton $| e'_0, e'_1, \cdots, e'_n |$, that is, which maps the points

$$e_0, e_1, \cdots, e_n$$

into the points $e'_0, e'_1, \cdots, e'_n$.

We shall say that two ordered skeletons $| e_0, e_1, \cdots, e_n |$ and

$$| e'_0, e'_1, \cdots, e'_n |$$

are *equivalent* if this mapping is positive (Appendix 1, 1.5). It follows from the properties of affine mappings that this equivalence relation partitions the set of all ordered skeletons of a given Euclidean n-space into two classes.

A Euclidean space is said to be oriented if the skeletons of one of these classes are designated as positive and those of the other as negative. The same idea can be expressed in the following way:

Definition 1.1. *An orientation of Euclidean n-space is a function*

$$R^n | e_0, e_1, \cdots, e_n | = \pm 1,$$

defined on all the ordered skeletons $| e_0, e_1, \cdots, e_n |$ of R^n, which assumes the value $+1$ on all the skeletons of one class and the value -1 on all the skeletons of the other class. A space with an orientation $R^n | e_0, e_1, \cdots, e_n |$ is called an oriented space.

Remark 1. In this section R^n will denote an n-dimensional oriented space and $| R^n |$ a space without orientation. In the sequel, after the elementary theorems on orientation have been established, R^n will denote both an oriented and a nonoriented space wherever this does not lead to ambiguity.

Remark 2. Every Euclidean space obviously has two opposite and mutually exclusive orientations. If one of these orientations is denoted by R^n, the other will be denoted by $-R^n$.

Remark 2_0. If $n = 0$, that is, the space consists of a single point o, there is but one uniquely ordered skeleton in the whole space. Nevertheless, we retain Def. 1.1 and in this case the orientation R^0, by definition, assumes the value $+1$ at the point o, while the orientation $-R^0$ has the value -1 at this point.

A zero-dimensional space o with orientation $R^0 \mid o \mid = +1$ ($R^0 \mid o \mid = -1$) is usually referred to as a point o with coefficient $+1$ (-1).

Remark 2_1. An orientation R^1 of the straight line $\mid R^1 \mid$ has the same value on the ordered skeletons $\mid e_0, e_1 \mid$ and $\mid e'_0, e'_1 \mid$ if the vectors e_0e_1 and $e'_0e'_1$ have the same direction.

Remark 2_2. An orientation of the plane assumes the same value on the ordered skeletons $\mid e_0, e_1, e_2 \mid$ and $\mid e'_0, e'_1, e'_2 \mid$ if both circuits $\mid e_0, e_1, e_2 \mid$ and $\mid e'_0, e'_1, e'_2 \mid$ of the triangles $e_0e_1e_2$ and $e'_0e'_1e'_2$ are described in the same sense (i.e., both counterclockwise or both clockwise).

The following remark is very important:

Remark 3. Let $(e_1 \cdots e_n)$ be an $(n - 1)$-simplex in $\mid R^n \mid$ and let e_0', e_0'' be two points in the exterior of the plane $\mid R^{n-1} \mid$ of the simplex. Then an arbitrary orientation R^n of $\mid R^n \mid$ assumes the same value on the ordered skeletons $\mid e_0', e_1, \cdots, e_n \mid$ and $\mid e_0'', e_1, \cdots, e_n \mid$ if e_0' and e_0'' lie on the same side of the plane $\mid R^{n-1} \mid$, and opposite values if e_0' and e_0'' lie on different sides of $\mid R^{n-1} \mid$ (see Appendix 1, 1.1).

Indeed, the affine mapping carrying $\mid e_0', e_1, \cdots, e_n \mid$ into

$$\mid e_0'', e_1, \cdots, e_n \mid$$

is positive if e_0' and e_0'' are on the same side of $\mid R^{n-1} \mid$ and negative if they are on different sides (see Appendix 1, 1.52).

§1.2. Orientation of a simplex and of a skeleton. Let us consider the $(n + 1)!$ distinct ordered skeletons which can be obtained from a given n-skeleton; in particular, from the skeleton of a given n-simplex. (An n-skeleton is an arbitrary set of $n + 1$ elements; in particular, the set of vertices of an n-simplex or of a degenerate n-simplex in Euclidean space is an n-skeleton.) These ordered skeletons are called the set of ordered skeletons associated with the given skeleton or simplex.

The set of all these ordered skeletons can be divided into two classes: two ordered skeletons are in the same class, by definition, if one is an even permutation of the other.

Remark 1. If an n-skeleton is the skeleton of a simplex $T^n \subset \mid R^n \mid$, then two of its ordered skeletons are in the same class if, and only if, they are equivalent in the sense of 1.1; that is, if one is mapped into the other by a positive affine mapping (see Appendix 1, Theorem 1.53).

A skeleton or simplex is said to be oriented if all its ordered skeletons of one class are assigned the sign +, and all ordered skeletons of the other class are assigned the sign −. Otherwise stated:

DEFINITION 1.2. *An orientation of a skeleton or simplex* $T^n = (e_0 \cdots e_n)$ *is an odd function* $t^n \mid e_{i(0)}, \cdots, e_{i(n)} \mid$, *defined on all the ordered skeletons* $\mid e_{i(0)}, \cdots, e_{i(n)} \mid$ *of* $(e_0 \cdots e_n)$, *which assumes the values* ± 1. (An odd function of $n + 1$ arguments is a function which changes sign, but remains the same in absolute value, for every odd permutation of its arguments. It follows from this definition that an odd function preserves its sign for an even permutation of its arguments.)

REMARK 2. To define the orientation of a simplex it is sufficient to assign its value on any one of its ordered skeletons, for example, on the ordered skeleton $\mid e_0, \cdots, e_n \mid$; then the orientation will have the same value on all the ordered skeletons of the same class, and the opposite value on all the ordered skeletons of the other class. The orientation assuming the value $+1$ on the ordered skeleton $\mid e_0, \cdots, e_n \mid$ will be denoted by $\mid e_0 \cdots e_n \mid$; the orientation assuming the value -1 on the ordered skeleton $\mid e_0, \cdots, e_n \mid$ will be denoted by $- \mid e_0 \cdots e_n \mid$.

1.20. A pair consisting of a simplex (skeleton) and an orientation of the simplex (skeleton) is called an *oriented simplex* (*skeleton*).

Hence,

1.21. *For arbitrary* $n = 0, 1, 2, \cdots$ *every n-simplex (skeleton)* $T^n = (e_0 \cdots e_n)$ *has two orientations or induces two oriented simplexes (skeletons):* $\mid e_0 \cdots e_n \mid$ *and* $- \mid e_0 \cdots e_n \mid$.

One of these oriented simplexes is usually denoted by t^n, the other by $-t^n$. If t^n is an orientation of the simplex T^n, we shall write $\mid t^n \mid = T^n$. Then also $\mid -t^n \mid = T^n$.

REMARK 3. In the sequel we shall identify the orientation of a simplex with the oriented simplex itself, since each uniquely defines the other.

REMARK 4_0. If $n = 0$, the simplex $T^0 = (e_0)$ has only one vertex and consequently only one ordered skeleton $\mid e_0 \mid$. Nevertheless, Def. 1.2 remains in force: if $\mid e_0 \mid$ is the orientation whose value at the point e_0 is $+1$, then $- \mid e_0 \mid$ is the orientation whose value at e_0 is -1. We shall refer to these orientations of a 0-simplex as a point e_0 with coefficient $+1$ or -1, respectively.

REMARK 4. If $n \geq 1$, the two orientations of an n-simplex correspond (1−1) to its two classes of ordered skeletons: to a given orientation corresponds that class on which the orientation assumes the value $+1$.

Hence if $n \geq 1$, it is possible (and in most cases customary) to identify an orientation of a simplex with a class of ordered skeletons of the simplex, it being understood that the oriented simplex $\mid e_0 \cdots e_n \mid$ is the class containing the ordered skeleton $\mid e_0, \cdots, e_n \mid$. However, it is necessary to remember that for $n = 0$ this identification is impossible, since in this case there is only one ordered skeleton, but as always two orientations.

REMARK 5. All the definitions of 1.2 hold without qualification also for degenerate simplexes (see IV, 0.1, Remark).

§1.3. The body of an oriented simplex. Let t^n be an oriented simplex in $| R^n |$. The corresponding closed simplex, that is, the closed convex hull of the skeleton of the simplex $| t^n |$, is called the *body of the oriented simplex* t^n.

REMARK. We shall use the same definition for oriented degenerate simplexes: the body of an oriented degenerate simplex is the closed convex hull of its skeleton [we recall that degenerate simplexes coincide with their skeletons (see IV, 0.1, Remark)].

The body of an oriented simplex t^n is denoted by $\bar{t}^n$.

§1.4. Extension of an orientation t^n to an orientation R^n. The product orientations t^nR^n and $t_1^nt_2^n$. Let $T^n = (e_0^0 \cdots e_n^0)$ be a simplex in $| R^n |$ and let t^n be any orientation of T^n. The function t^n is defined on all the ordered skeletons of T^n; in virtue of 1.2, Remark 1, t^n can be extended to the set of all ordered n-skeletons $| e_0 , \cdots , e_n |$ of $| R^n |$ by setting

$$t^n \,| e_0 , \cdots , e_n | = t^n \,| e_0^0, \cdots , e_n^0 |$$

if the affine mapping which transforms $| e_0^0, \cdots , e_n^0 |$ into $| e_0 , \cdots , e_n |$ is positive, and by putting

$$t^n \,| e_0 , \cdots , e_n | = -t^n \,| e_0^0 , \cdots , e_n^0 |$$

if this affine mapping is negative. (The reader will recall that t^n is a function, but that $t^n \,| e_0 , \cdots , e_n |$ is the value of this function on the given skeleton.)

Hence by extending the orientation t^n to all of $| R^n |$ (in the sequel we shall use this phrase instead of the more precise but longer phrase "to the set of all ordered n-skeletons of $| R^n |$") we obtain an orientation R^n of $| R^n |$; this orientation R^n is said to be coherent with t^n: if t^n has been extended to $| R^n |$, we may simply write

$$R^n \,| e_0 , \cdots , e_n | = t^n \,| e_0 , \cdots , e_n |$$

for any ordered skeleton $| e_0 , \cdots , e_n |$ of $| R^n |$. If the orientations R^n and t^n are coherent, the orientations $-R^n$ and t^n are said to be noncoherent.

Now, let t^n be any orientation of a simplex $T^n = | t^n | \subset | R^n |$ and let R^n be an orientation of the space $| R^n |$; we shall think of the function t^n as extended to all of $| R^n |$. Then the product

$$t^n \,| e_0 , \cdots , e_n | \cdot R^n \,| e_0 , \cdots , e_n |$$

is defined for any ordered skeleton $| e_0 , \cdots , e_n |$ of $| R^n |$. This product is equal to 1 if t^n and R^n are coherent, and is equal to -1 if they are noncoherent.

Hence

1.41. *The product of two orientations* t^n *and* R^n (that is, of two functions $t^n \mid e_0, \cdots, e_n \mid$ and $R^n \mid e_0, \cdots, e_n \mid$ defined on all the ordered skeletons of the space $\mid R^n \mid$) *is a constant, equal to* 1 *if* t^n *and* R^n *are coherent and equal to* -1 *if they are noncoherent.*

Finally, let $T_1^n = (e_{10} \cdots e_{1n})$, $T_2^n = (e_{20} \cdots e_{2n})$ be two simplexes in $\mid R^n \mid$ with orientations t_1^n and t_2^n. If these orientations are thought of as being extended to all of $\mid R^n \mid$, we may without further explanation speak of their coherence, noncoherence, and product. As before, the product of t_1^n and t_2^n is a constant, equal to 1 if the orientations are coherent, and equal to -1 if they are not.

Remark. Orientation of Domains of $\mid R^n \mid$. Since there is a $(1-1)$ correspondence between the orientations of a simplex T^n and those of the space $\mid R^n \mid$ which carries T^n (see Appendix 1, 2.41), we can redefine the concept of an oriented simplex: an oriented simplex t^n is a pair consisting of the simplex $\mid t^n \mid$ and an orientation of the space $\mid R^n \mid$ which carries the simplex. The convenience of this definition lies in the fact that it can immediately be extended to arbitrary domains of $\mid R^n \mid$; in particular, to convex polyhedral domains, spheres, half-spaces, etc.

Definition 1.42. An *oriented domain* of $\mid R^n \mid$ is a pair consisting of the given domain and an orientation R^n of the space $\mid R^n \mid$ which carries the domain.

Clearly, every domain $T^n \subseteq \mid R^n \mid$ has two orientations; if one of these is denoted by t^n, the other will be denoted by $-t^n$.

§1.5. The orientation $(e_0 t^{n-1})$. Let the simplex $T^{n-1} = (e_1 \cdots e_n)$ be a face of the simplex $T^n = (e_0 e_1 \cdots e_n)$ and let t^{n-1} be an orientation of T^{n-1}. We shall define the orientation $(e_0 t^{n-1})$ of T^n by setting

$$(e_0 t^{n-1}) \mid e_0, e_{i(1)}, \cdots, e_{i(n)} \mid = t^{n-1} \mid e_{i(1)}, \cdots, e_{i(n)} \mid,$$

where $\mid e_0, e_{i(1)}, \cdots, e_{i(n)} \mid$ is any ordered skeleton of the simplex T^n with initial vertex e_0. In other words, the orientation $(e_0 t^{n-1})$ is assigned on $\mid e_0, e_{i(1)}, \cdots, e_{i(n)} \mid$, by definition, that value which t^{n-1} assumes on $\mid e_{i(1)}, \cdots, e_{i(n)} \mid$. Hence (1.2, Remark 2) the orientation $(e_0 t^{n-1})$ is defined on all ordered skeletons of T^n.

If $\mid e_0, e_{i(1)}, \cdots, e_{i(n)} \mid$ and $\mid e_0, e_{j(1)}, \cdots, e_{j(n)} \mid$ are two ordered skeletons whose initial vertex is e_0, the parity of the permutation

$$\begin{pmatrix} e_0 e_{i(1)} & \cdots & e_{i(n)} \\ e_0 e_{j(1)} & \cdots & e_{j(n)} \end{pmatrix}$$

is the same as that of the permutation

$$\begin{pmatrix} e_{i(1)} & \cdots & e_{i(n)} \\ e_{j(1)} & \cdots & e_{j(n)} \end{pmatrix}.$$

Consequently, the orientation whose value on $|e_0, e_{i(1)}, \cdots, e_{i(n)}|$ is $t^{n-1} | e_{i(1)}, \cdots, e_{i(n)} |$ assumes the value $t^{n-1} | e_{j(1)}, \cdots, e_{j(n)} |$ on $| e_0, e_{j(1)}, \cdots, e_{j(n)} |$. In other words, the orientation $(e_0 t^{n-1})$ is independent of the choice of the ordered skeleton $| e_0, e_{i(1)}, \cdots, e_{i(n)} |$ which is used to define it, but depends only on t^{n-1}; this justifies the notation $(e_0 t^{n-1})$.

REMARK. *For $n > 1$, the definition of the orientation $(e_0 t^{n-1})$ can be simplified: if $t^{n-1} = | e_1 \cdots e_n |$, define $(e_0 t^{n-1})$ to be $| e_0 e_1 \cdots e_n |$.*

The following important proposition follows from Appendix 1, Theorem 1.521:

1.5. *Let $| R^{n-1} |$ be a plane in $| R^n |$ and let e', e'' be two points exterior to $| R^{n-1} |$. If t_1^{n-1} and t_2^{n-1} are two coherently oriented simplexes in $| R^{n-1} |$, then the orientations $(e' t_1^{n-1})$, $(e'' t_2^{n-1})$ are coherent if e', e'' are on the same side of $| R^{n-1} |$ and noncoherent if these points are on opposite sides of $| R^{n-1} |$.*

The following special case of Proposition 1.5 is particularly important:

1.51. *If t^{n-1} is an oriented simplex in the plane $| R^{n-1} |$ and e', e'' are two points of $| R^n |$ exterior to $| R^{n-1} |$, then the orientations $(e' t^{n-1})$ and $(e'' t^{n-1})$ are coherent if e', e'' are on the same side of $|R^{n-1}|$ and noncoherent if the two points are on opposite sides of $| R^{n-1} |$.*

§1.6. Affine images of orientations. Let S_α^β be an affine mapping of $| R_\beta^n |$ onto $| R_\alpha^n |$. If R_β^n is an orientation of $| R_\beta^n |$, let us define the orientation $S_\alpha^\beta R_\beta^n$ of $| R_\alpha^n |$ by requiring that it assume on the ordered skeleton $| S_\alpha^\beta e_0^\beta, \cdots, S_\alpha^\beta e_n^\beta |$ the value of R_β^n on $| e_0^\beta, \cdots, e_n^\beta |$:

$$(1.6) \qquad S_\alpha^\beta R_\beta^n | S_\alpha^\beta e_0^\beta, \cdots, S_\alpha^\beta e_n^\beta | = R_\beta^n | e_0^\beta, \cdots, e_n^\beta |,$$

or, what is the same,

$$(1.6^{-1}) \qquad S_\alpha^\beta R_\beta^n | e_0^\alpha, \cdots, e_n^\alpha | = R_\beta^n | (S_\alpha^\beta)^{-1} e_0^\alpha, \cdots, (S_\alpha^\beta)^{-1} e_n^\alpha |$$

for every ordered skeleton $| e_0^\alpha, \cdots, e_n^\alpha |$ of $| R_\alpha^n |$.

Since each orientation of a simplex corresponds to a unique orientation of the carrying plane of the simplex, we may use (1.6^{-1}) to define the image $S_\alpha^\beta t_\beta^n$ of an orientation t_β^n of an arbitrary simplex (or in general of a convex polyhedral domain T_β^n) under an affine mapping S_α^β.

Hence

1.61. The image of an oriented simplex $t_\beta^n = | e_0^\beta \cdots e_n^\beta |$ under a (non-degenerate) affine mapping S_α^β is the oriented simplex

$$S_\alpha^\beta t_\beta^n = | S_\alpha^\beta e_0^\beta \cdots S_\alpha^\beta e_n^\beta |.$$

An immediate result of this definition is that

$$S_\alpha^\beta t_\beta^n \cdot S_\alpha^\beta R_\beta^n = t_\beta^n \cdot R_\beta^n,$$
$$S_\alpha^\beta t_{\beta 1}^n \cdot S_\alpha^\beta t_{\beta 2}^n = t_{\beta 1}^n \cdot t_{\beta 2}^n,$$

where t_β^n, $t_{\beta 1}^n$, $t_{\beta 2}^n$ are any orientations of any three n-simplexes of the space $| R_\beta^n |$.

Now let $| R_\beta^n | = | R_\alpha^n | = | R^n |$ and let S be an affine mapping of $| R^n |$ onto itself. Then $R^n \mid Se_0 , \cdots , Se_n \mid$ is equal to $R^n \mid e_0 , \cdots , e_n \mid$ or $-R^n \mid e_0 , \cdots , e_n \mid$ depending on whether S is a positive or negative mapping. Since $SR^n \mid Se_0 , \cdots , Se_n \mid$ is, by definition, equal to

$$R^n \mid e_0 , \cdots , e_n \mid$$

and $\mid Se_0 , \cdots , Se_n \mid$ is an arbitrary ordered skeleton of $| R^n |$, it follows that

$$SR^n = \text{sign } S \cdot R^n, \tag{1.62}$$

where sign S is the sign of the affine mapping.

In the same way, for an arbitrary orientation t^n of a simplex T^n extended to all of $| R^n |$ we have

$$St^n = \text{sign } S \cdot t^n. \tag{1.63}$$

REMARK. It goes without saying that (1.63) has meaning only on the assumption that t^n (and hence St^n) has been extended to all of $| R^n |$.

§2. Intersection number of planes and simplexes

[This section is not needed until Chapter XV.]

§2.1. Intersection number of planes. Let X^p and Y^q be two oriented planes of the oriented space R^n, with $p + q = n$ and in general position, that is, intersecting in a single point o. We shall define the intersection number $(X^p \times Y^q; R^n)$ of the oriented planes X^p and Y^q in the oriented space R^n. To this end, we choose in the planes $| X^p |$ and $| Y^q |$ two simplexes $T_1^p = (oa_1 \cdots a_p)$ and $T_2^q = (ob_1 \cdots b_q)$ with common vertex o. Then the points $o, a_1 , \cdots , a_p , b_1 , \cdots , b_q$ are linearly independent in $| R^n |$ and $T^n = (oa_1 \cdots a_p b_1 \cdots b_q)$ is an n-simplex of $| R^n |$ (Fig. 99).

Let us choose arbitrary orientations t_1^p and t_2^q of the simplexes T_1^p and T_2^q and let us assume that the notation has been so chosen that $t_1^p = | oa_1 \cdots a_p |$ and $t_2^q = | ob_1 \cdots b_q |$; then the orientation

$$t^n = | oa_1 \cdots a_p b_1 \cdots b_q |$$

of the simplex T^n and the products $t_1^p X^p$, $t_2^q Y^q$, and $t^n R^n$ are defined.

Let us put

$$(2.1) \qquad (X^p \times Y^q; R^n) = t_1^p X^p \cdot t_2^q Y^q \cdot t^n R^n$$

and show that (2.1) depends only on the given orientations X^p, Y^q, and R^n and is independent of the choice of the simplexes T_1^p, T_2^q and their orientations t_1^p, t_2^q.

Let $T_{10}^p = (oa_1^0 \cdots a_p^0)$ and $T_{20}^q = (ob_1^0 \cdots b_q^0)$ be two other simplexes chosen in the same way as T_1^p and T_2^q; let $t_{10}^p = | oa_1^0 \cdots a_p^0 |$ and $t_{20}^q = | ob_1^0 \cdots b_q^0 |$ be arbitrary orientations of these simplexes; finally, let $t_0^n = | oa_1^0 \cdots a_p^0 b_1^0 \cdots b_q^0 |$. It is required to prove that:

$$(2.10) \qquad t_{10}^p X^p \cdot t_{20}^q Y^q \cdot t_0^n R^n = t_1^p X^p \cdot t_2^q Y^q \cdot t^n R^n.$$

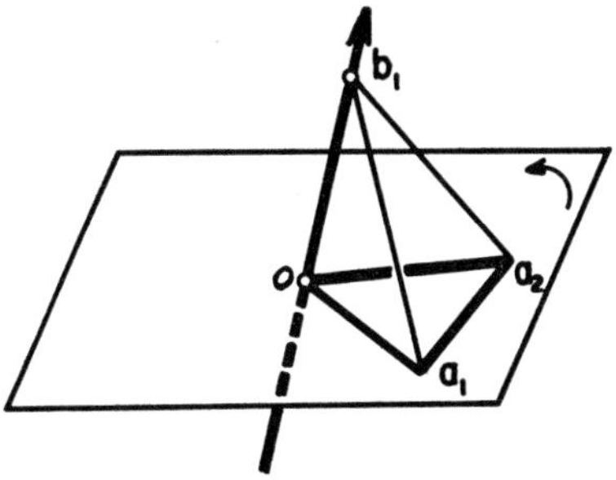

FIG. 99

Let us denote by S_1 (S_2) the affine mapping of $| X^p |$ ($| Y^q |$) onto itself which takes the ordered skeleton $| o, a_1, \cdots, a_p |$ ($| o, b_1, \cdots, b_q |$) into $| o, a_1^0, \cdots, a_p^0 |$ ($| o, b_1^0, \cdots, b_q^0 |$).

These affine mappings induce an afine mapping S of $| R^n |$ onto itself such that $S(| o, a_1, \cdots, a_p, b_1, \cdots, b_q |) = | o, a_1^0, \cdots, a_p^0, b_1^0, \cdots, b_q^0 |$ and $\operatorname{sign} S = \operatorname{sign} S_1 \cdot \operatorname{sign} S_2$.

By (1.63),

$$t_{10}^p = \operatorname{sign} S_1 \cdot t_1^p; \qquad t_{20}^p = \operatorname{sign} S_2 \cdot t_2^q; \qquad t_0^n = \operatorname{sign} S \cdot t^n,$$

or

$$t_{10}^p X^p = \operatorname{sign} S_1 \cdot t_1^p X^p; \qquad t_{20}^q Y^q = \operatorname{sign} S_2 \cdot t_2^q Y^q; \qquad t_0^n R^n = \operatorname{sign} S \cdot t^n R^n.$$

Hence

$$t_{10}^p X^p \cdot t_{20}^q Y^q \cdot t_0^n R^n = \operatorname{sign} S_1 \cdot \operatorname{sign} S_2 \cdot \operatorname{sign} S \cdot t_1^p X^p \cdot t_2^q Y^q \cdot t^n R^n.$$

This proves (2.10) since

$$\operatorname{sign} S = \operatorname{sign} S_1 \cdot \operatorname{sign} S_2, \qquad \text{i. e., } \operatorname{sign} S_1 \cdot \operatorname{sign} S_2 \cdot \operatorname{sign} S = 1.$$

Obviously,

$$(2.11) \quad (-X^p \times Y^q; R^n) = (X^p \times -Y^q; R^n)$$
$$= (X^p \times Y^q; -R^n) = -(X^p \times Y^q; R^n).$$

THEOREM 2.1. $(X^p \times Y^q) = (-1)^{pq}(Y^q \times X^p)$. [Wherever this can be done without ambiguity, we shall write $(X^p \times Y^q)$ in place of $(X^p \times Y^q; R^n)$.]

Indeed, setting $t_1^{p-1} = | a_1 \cdots a_p |$, $t_2^{q-1} = | b_1 \cdots b_q |$, we have

$$(X^p \times Y^q) = (ot_1^{p-1})X^p \cdot (ot_2^{q-1})Y^q \cdot (ot_1^{p-1}t_2^{q-1})R^n,$$
$$(Y^q \times X^p) = (ot_2^{q-1})Y^q \cdot (ot_1^{p-1})X^p \cdot (ot_2^{q-1}t_1^{p-1})R^n;$$

where

$$(ot_1^{p-1}t_2^{q-1}) = | oa_1 \cdots a_pb_1 \cdots b_q |,$$
$$(ot_2^{q-1}t_1^{p-1}) = | ob_1 \cdots b_qa_1 \cdots a_p |.$$

The permutation taking $| o, a_1, \cdots, a_p, b_1, \cdots, b_q |$ into

$$| o, b_1, \cdots, b_q, a_1, \cdots, a_p |$$

consists of pq transpositions. Hence

$$(ot_2^{q-1}t_1^{p-1}) = (-1)^{pq}(ot_1^{p-1}t_2^{q-1}).$$

This proves the theorem.

§2.2. The intersection number of simplexes. Let t_1^p and t_2^q be two oriented (perhaps degenerate) simplexes in the space R^n, with $p + q = n$. We shall assume that the two simplexes satisfy at least one of the following two conditions:

1°. The bodies (see 1.3) $\bar{t}_1^p$ and $\bar{t}_2^q$ of these simplexes are disjoint:

$$\bar{t}_1^p \cap \bar{t}_2^q = 0.$$

2°. The vertices of t_1^p and t_2^q are in general position (see Appendix 1, 1.4) in R^n. In this case then the simplexes are nondegenerate and $\bar{t}_1^p \cap \bar{t}_2^q$ is either empty or consists of a single point o which is an interior point of both simplexes $| t_1^p |$ and $| t_2^q |$.

We define the intersection number $(t_1^p \times t_2^q)$ of t_1^p and t_2^q as follows:

1°. If $\bar{t}_1^p \cap \bar{t}_2^q = 0$, then $(t_1^p \times t_2^q) = 0$.

2°. If $\bar{t}_1^p \cap \bar{t}_2^q \neq 0$, then t_1^p and t_2^q are nondegenerate and their planes $| X^p |$ and $| Y^q |$ intersect in a single point o. Let us denote by X^p and Y^q the orientations of these planes which are coherent with the orientations t_1^p and t_2^q, and let us set

$$(t_1^p \times t_2^q; R^n) = (X^p \times Y^q; R^n).$$

REMARK. The intersection number of two oriented convex polyhedral domains t_1^p and t_2^q in an oriented space R^n is defined in exactly the same way.

The requirement that t_1^p and t_2^q be in general position can be met by satisfying at least one of the following two conditions: 1) $\bar{t}_1^p \cap \bar{t}_2^q = 0$ or 2) the planes of the polyhedral domains and their faces are in general position.

§2.3. Intersections and simplicial mappings.

THEOREM 2.3. *Let R_β^n and R_α^n be oriented n-dimensional spaces and let*

$$X_\beta^p \text{ and } Y_\beta^q \text{ in } R_\beta^n, \qquad |X_\beta^p| \cap |Y_\beta^q| = o^\beta,$$

$$X_\alpha^p \text{ and } Y_\alpha^q \text{ in } R_\alpha^n, \qquad |X_\alpha^p| \cap |Y_\alpha^q| = o^\alpha,$$

be two pairs of intersecting planes. Let S_α^β be an affine mapping of $|R_\beta^n|$ onto $|R_\alpha^n|$ which maps $|X_\beta^p|$ onto $|X_\alpha^p|$ and $|Y_\beta^q|$ onto $|Y_\alpha^q|$. Finally, let $S_\alpha^\beta R_\beta^n = \varepsilon R_\alpha^n$, $\varepsilon = \pm 1$. Then

$$(S_\alpha^\beta X_\beta^p \times S_\alpha^\beta Y_\beta^q; R_\alpha^n) = \varepsilon(X_\beta^p \times Y_\beta^q; R_\beta^n). \tag{2.3}$$

This proposition, in essence, needs no proof; indeed, it is easily seen that

$$(S_\alpha^\beta X_\beta^p \times S_\alpha^\beta Y_\beta^q; S_\alpha^\beta R_\beta^n) = (X_\beta^p \times Y_\beta^q; R_\beta^n),$$

whence (2.3) follows by (2.11).

§3. Incidence numbers

§3.1. Definition of the incidence numbers. Let t^r and t^{r-1} be two oriented simplexes (of a given R^n or of a given simplicial complex). We shall define the *incidence number* $(t^r : t^{r-1})$ of the oriented simplexes t^r and t^{r-1} in the following way.

1°. If $|t^{r-1}|$ is not a face of $|t^r|$, then $(t^r : t^{r-1}) = 0$.

2°. Let $|t^{r-1}|$ be a face of $|t^r|$ and let e be the vertex of $|t^r|$ opposite $|t^{r-1}|$. Then the orientation (et^{r-1}) of $|t^r|$ is uniquely determined. If $(et^{r-1}) = t^r$, we shall set $(t^r : t^{r-1}) = 1$; if $(et^{r-1}) = -t^r$, we put $(t^r : t^{r-1}) = -1$. Hence $(t^r : t^{r-1})$ is $\varepsilon = \pm 1$ and is defined by the equation

$$(et^{r-1}) = \varepsilon t^r.$$

REMARK. The symbol "$\wedge$" placed over a vertex e_i in expressions of the form $(e_0 \cdots e_r)$, $|e_0 \cdots e_r|$, means that this vertex is to be omitted, i.e., $(e_0 \cdots \hat{e}_k \cdots e_r) = (e_0 \cdots e_{k-1}e_{k+1} \cdots e_r)$.

For $t^r = |e_0 \cdots e_r|$ and $t_k^{r-1} = |e_0 \cdots \hat{e}_k \cdots e_r|$ we note the important formula

$$(t^r : t_k^{r-1}) = (-1)^k, \tag{3.11}$$

which follows immediately from the definition of incidence number and from the identity

$$t^r = |\, e_0 \cdots e_r \,| = (-1)^k \,|\, e_k e_0 \cdots \hat{e}_k \cdots e_r \,|.$$

In particular,

$$(3.110) \qquad (\,|\, e_0 e_1 e_2 \cdots e_r \,| : |\, e_1 e_2 \cdots e_r \,|\,) = 1,$$

$$(3.111) \qquad (\,|\, e_0 e_1 e_2 \cdots e_r \,| : |\, e_0 e_2 \cdots e_r \,|\,) = -1.$$

EXAMPLES OF INCIDENCE NUMBERS.

1°. Let $t^1 = |\, e_0 e_1 \,|$. We have

$$(\,|\, e_0 e_1 \,| : |\, e_0 \,|\,) = -1, \qquad (\,|\, e_0 e_1 \,| : |\, e_1 \,|\,) = 1,$$

where $|\, e_0 \,|$ denotes the vertex e_0 with coefficient $+1$ (see 1.2, Remark 4_0).

2°. Let $t^2 = |\, e_0 e_1 e_2 \,|$, $t^1 = |\, e_0 e_2 \,|$. Then $(t^2 : t^1) = -1$.

3°. For $t^3 = |\, e_0 e_1 e_2 e_3 \,|$ and $t^2 = |\, e_0 e_1 e_2 \,|$ we have $(t^3 : t^2) = -1$; on the other hand, for $t^2 = |\, e_1 e_0 e_2 \,|$ we have $(t^3 : t^2) = 1$.

§3.2. Properties of the incidence numbers.

Property 1:

$$(3.2_1) \qquad (-t^r : t^{r-1}) = (t^r : -t^{r-1}) = -(t^r : t^{r-1}).$$

The first equality follows from the fact that if $t^r = \varepsilon(e t^{r-1})$, then $-t^r = -\varepsilon(e t^{r-1})$; the second from the fact that $(e(-t^{r-1})) = -(e t^{r-1})$.

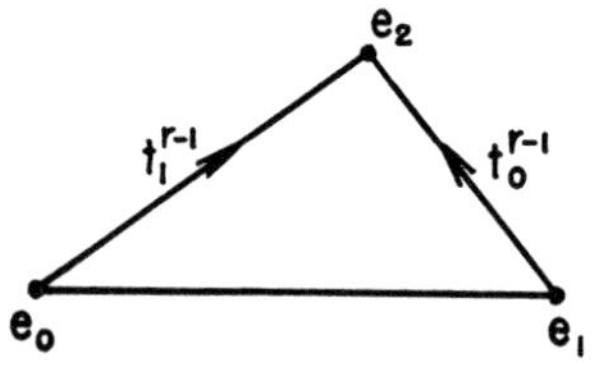

FIG. 100

Property 2. Let $T^r = (e_0 e_1 e_2 \cdots e_r)$ be a simplex and let

$$(3.20) \qquad t_0^{r-1} = |\, e_1 e_2 \cdots e_r \,|, \qquad t_1^{r-1} = |\, e_0 e_2 \cdots e_r |$$

be two of its $(r-1)$-faces with the indicated orientations. Then, for any orientation t^r of T^r,

$$(3.21) \qquad (t^r : t_1^{r-1}) = -(t^r : t_0^{r-1})$$

(see Fig. 100 for the case $r = 2$).

Proof. In virtue of the first basic property,

$$(-t^r : t_1^{r-1}) = -(t^r : t_1^{r-1}), \qquad (-t^r : t_0^{r-1}) = -(t^r : t_0^{r-1});$$

consequently, it is enough to prove (3.21) for any one orientation of T^r, e.g., for $t^r = |\, e_0 e_1 e_2 \cdots e_r \,|$. But, by (3.11), we have for this orientation:

$$(t^r : t_0^{r-1}) = 1, \qquad (t^r : t_1^{r-1}) = -1.$$

We shall give Property 2 another formulation which will be needed for further generalizations.

Let $|\, t^r \,| = (e_0 e_1 e_2 \cdots e_r)$ be a simplex and let $t^{r-2} = |\, e_2 \cdots e_r \,|$ be an arbitrarily oriented $(r-2)$-face of $|\, t^r \,|$. Let the two $(r-1)$-faces $|\, t_0^{r-1} \,| =$

$(e_1e_2 \cdots e_r)$ and $| t_1^{r-1} | = (e_0e_2 \cdots e_r)$ incident with $| t^{r-2} |$ be given orientations t_0^{r-1} and t_0^{r-1} such that

$$(3.22) \qquad (t_0^{r-1}:t^{r-2}) = 1, \qquad (t_1^{r-1}:t^{r-2}) = 1.$$

Then

$$(3.23) \qquad (t^r:t_0^{r-1}) + (t^r:t_1^{r-1}) = 0.$$

Indeed, since $(| e_0e_2 \cdots e_r |:| e_2 \cdots e_r |) = (| e_1e_2 \cdots e_r |:| e_2 \cdots e_r |) = 1$, the orientations t_0^{r-1}, t_1^{r-1} satisfying (3.22) are

$$t_1^{r-1} = | e_0e_2 \cdots e_r |, \qquad t_0^{r-1} = | e_1e_2 \cdots e_r |.$$

This and (3.21) imply (3.23).

Formula (3.23) can be generalized further if (3.22) is not imposed. We then obtain the following general formulation:

3.2$_{11}$. Let $| t_0^{r-1} |$ and $| t_1^{r-1} |$ be the two faces of a simplex $| t^r |$ which are incident with the face $| t^{r-2} |$. For arbitrary orientations t^r, t_0^{r-1}, t_1^{r-1}, and t^{r-2}, we have

$$(3.24) \qquad (t^r:t_0^{r-1})(t_0^{r-1}:t^{r-2}) + (t^r:t_1^{r-1})(t_1^{r-1}:t^{r-2}) = 0.$$

This formula is valid if t_0^{r-1} and t_1^{r-1} satisfy (3.22), since then it becomes (3.23) which has already been proved. Its validity in the general case follows from the fact that if either or both of the orientations t_0^{r-1}, t_1^{r-1} are replaced by their opposites, both terms of (3.24) retain their value.

Now let K be an unrestricted simplicial complex. Let t^r and t^{r-2} be arbitrary orientations of an r-simplex and an $(r - 2)$-simplex of K. We choose a definite orientation for each $(r - 1)$-simplex of K and denote it by t_i^{r-1}. In these conditions (3.24) can be rewritten in the form

$$(3.25) \qquad \sum (t^r:t_i^{r-1})(t_i^{r-1}:t^{r-2}) = 0,$$

where the sum is extended over all t_i^{r-1}.

To prove (3.25) it is enough to note that for fixed t^r and t^{r-2} there exist precisely two t_i^{r-1}, say t_0^{r-1} and t_1^{r-1}, for which the conditions $(t^r:t_i^{r-1}) \neq 0$ and $(t_i^{r-1}:t^{r-2}) \neq 0$ are satisfied. Because of this, (3.25) is identical with (3.24), which was proved above.

§4. Cell complexes; a-complexes

The reader may omit this section on a first reading if in the sequel he thinks of a cell complex $\mathfrak{K}$ (or an a-complex) as the set of all oriented simplexes of a triangulation K and of a closed (open) subcomplex of a cell complex $\mathfrak{K}$ as the set of all oriented simplexes of a closed (open) subcomplex of K. With this in mind, ρ^r denotes the number of r-simplexes of K; the number of oriented r-simplexes of K or of r-cells of $\mathfrak{K}$ is then, obviously,

$2\rho^r$. However, the general concept of a cell complex is required in Chapter X.

In a paper (see Aleksandrov [*f*], Bibliography, Vol. 1) which supplements this book, but was written after the book was completed, the term "cell complex" was used instead of the term "*a*-complex"; cell complexes (in the sense introduced in this section) were called "cell spaces".

§4.1. Definition of *a*-complexes and cell complexes. The properties of incidence numbers investigated in the preceding section lead to a natural definition of cell complexes and *a*-complexes; these appear as an immediate generalization of the set of all oriented simplexes of a simplicial complex (or of an unrestricted simplicial complex).

DEFINITION 4.11. Let $\mathfrak{K}$ be a set consisting of elements called *cells*. $\mathfrak{K}$ is said to be a *cell complex* if the following conditions are satisfied:

4.111. Every cell is assigned a nonnegative integer called the dimension of the cell.

4.112. With every r-dimensional cell (r-cell) $t^r \in \mathfrak{K}$ there is associated a unique r-cell $-t^r \in \mathfrak{K}$, with

$$-(-t^r) = t^r.$$

The cells t^r and $-t^r$ are said to be opposites.

4.113. Every pair of cells with first element an r-cell t^r and second element an $(r-1)$-cell t^{r-1} is assigned an integer $(t^r:t^{r-1})$ called the incidence number of the cells t^r and t^{r-1}.

4.114. For every cell t^r the set of t^{r-1} for which $(t^r:t^{r-1}) \neq 0$ is finite.

4.115. The incidence numbers satisfy the condition

$$(-t^r:t^{r-1}) = (t^r:-t^{r-1}) = -(t^r:t^{r-1}).$$

A cell complex is said to be an a-complex if it satisfies the following condition:

4.116. Let t^r and t^{r-2} be an arbitrary r-cell and $(r-2)$-cell, respectively; denote any one of the cells in each pair of $(r-1)$-dimensional opposites by t_i^{r-1}. Then

$$\sum (t^r:t_i^{r-1})(t_i^{r-1}:t^{r-2}) = 0,$$

where the sum is extended over all t_i^{r-1}.

DEFINITION 4.117. A $(1-1)$ mapping F of a cell complex $\mathfrak{K}_1$ onto a cell complex $\mathfrak{K}_2$ is called an *isomorphism* provided the following conditions are satisfied:

a) The image $F(t) \in \mathfrak{K}_2$ of an arbitrary element $t \in \mathfrak{K}_1$ has the same dimension as t.

b) If the elements t and $-t$ of $\mathfrak{K}_1$ are opposites, then $F(t)$ and $F(-t)$ are opposites in $\mathfrak{K}_2$.

c) $(F(t^r):F(t^{r-1})) = (t^r:t^{r-1})$.

Two cell complexes are said to be *isomorphic* if there is an isomorphism of one onto the other.

It is clear that a cell complex isomorphic to an a-complex is an a-complex.

DEFINITION 4.118. If the elements of a cell complex $\mathfrak{K}$ have a maximum dimension n, then n is said to be the *dimension of the cell complex* $\mathfrak{K}$.

REMARK 1. From the properties of the incidence numbers in an unrestricted simplicial complex it obviously follows that *the set of all oriented simplexes of an unrestricted simplicial complex K is an a-complex.* The following example shows that the set of all oriented simplexes of a simplicial complex K which is not unrestricted, while it is obviously a cell complex, may not be an a-complex. The complex K consists of a triangle $(e_0e_1e_2)$, a side (e_0e_1), and a vertex e_1. Setting $t^2 = |\, e_0e_1e_2 \,|$, $t^1 = |\, e_0e_1 \,|$, $t^0 = |\, e_1 \,|$, we have

$$\sum (t^r : t_i^{r-1})(t_i^{r-1} : t^{r-2}) = (t^2 : t^1)(t^1 : t^0) = 1 \neq 0.$$

We shall prove (in 7.2), however, that the set of all oriented simplexes of every *open* subcomplex of an unrestricted simplicial complex is an a-complex. The same assertion for closed subcomplexes follows from the fact that every closed subcomplex of an unrestricted simplicial complex is an unrestricted simplicial complex.

DEFINITION 4.12. If $(t^r : t^{r-1}) \neq 0$, the cells t^r and t^{r-1} are said to be *incident* (it is proper to use both phrases: "t^r is incident with t^{r-1}" and "t^{r-1} is incident with t^r").

DEFINITION 4.13. A cell t^{r-p} precedes or is less than a cell t^r:

$$t^r > t^{r-p} \qquad (p \geq 1)$$

in a cell complex $\mathfrak{K}$ if there are cells

$$t_0^r = t^r,\ t_1^{r-1},\ \cdots,\ t_p^{r-p} = t^{r-p}$$

in $\mathfrak{K}$ such that

$$(t_i : t_{i+1}) \neq 0 \qquad \text{for } i = 0, 1, \cdots, p-1.$$

REMARK 2. This ordering of the cells of $\mathfrak{K}$ and the assignment of a dimension to every cell enable us to consider a cell complex as a complex in the sense of IV, 1.7.

REMARK 3. Def. 4.13 implies that if

$$t > t',$$

then

$$-t > t', \qquad t > -t', \qquad -t > -t'.$$

DEFINITION 4.14. A cell complex $\mathfrak{K}'$ is said to be a *cell subcomplex* of a cell complex $\mathfrak{K}$ if every element of $\mathfrak{K}'$ is an element of $\mathfrak{K}$ and if, in addition, the following conditions are satisfied:

1°. Every element of $\mathfrak{K}'$ is assigned the same dimension in $\mathfrak{K}$ as it is in $\mathfrak{K}'$.

2°. Every pair of opposites in $\mathfrak{K}'$ is a pair of opposites in $\mathfrak{K}$.

3°. Every two elements t^r and t^{r-1} of $\mathfrak{K}'$ whose dimensions differ by 1 have the same incidence number in $\mathfrak{K}$ as they do in $\mathfrak{K}'$.

REMARK 4. Def. 4.14 implies that if an element t of a cell complex $\mathfrak{K}$ is an element of a cell subcomplex $\mathfrak{K}'$ of $\mathfrak{K}$, then $-t$ is also an element of $\mathfrak{K}'$.

DEFINITION 4.15. A cell subcomplex $\mathfrak{K}_0$ of a cell complex $\mathfrak{K}$ is said to be a *closed* (*open*) subcomplex of $\mathfrak{K}$ if every element of $\mathfrak{K}$ which precedes an element of $\mathfrak{K}_0$ (is preceded by some element of $\mathfrak{K}_0$) is an element of $\mathfrak{K}_0$, i.e., if $t \in \mathfrak{K}$, $t_0 \in \mathfrak{K}_0$ and $t_0 > t$ $(t > t_0)$, then $t \in \mathfrak{K}_0$.

REMARK 5. In the sequel we shall never consider any subcomplexes of a cell complex except cell subcomplexes. *We shall therefore in the sequel always refer to the cell subcomplexes of a cell complex $\mathfrak{K}$ simply as the subcomplexes of $\mathfrak{K}$.*

REMARK 6. In view of Remarks 2 and 3, the theory of connectedness of complexes (including the notion and properties of components) developed in IV, 7 can be applied verbatim to cell complexes.

REMARK 7. In most of the cell complexes considered in topology the incidence numbers assume only the values 1, -1, and 0. At the end of the next article we shall give examples of cell complexes in which the incidence numbers assume other values also (see 4.2, Examples 3 and 4).

§4.2. The incidence matrices of a cell complex. Let $\mathfrak{K}$ be a finite n-dimensional cell complex. We shall denote by $2\rho^r$ $(r = 0, 1, \cdots, n)$ the number of r-cells in $\mathfrak{K}$. Let us suppose that the cells of $\mathfrak{K}$ of each dimension are numbered in a definite way. Then for $r = 0, 1, 2, \cdots, n - 1$ it is natural to consider the matrix consisting of $2\rho^{r+1}$ rows and $2\rho^r$ columns in which the element in the ith row and jth column is the incidence number of the ith $(r + 1)$-cell and the jth r-cell. All the properties of the complex $\mathfrak{K}$ can be deduced from these matrices. However this can be effected more economically by denoting one cell of each pair of opposites by t_i^r and the other by $-t_i^r$, $i = 1, 2, \cdots, \rho^r$, and considering the matrices $\mathfrak{E}^r$ in which the element in the ith row and jth column $(i = 1, 2, \cdots, \rho^{r+1}; j = 1, 2, \cdots, \rho^r)$ is the number

$$\varepsilon_{ij}^r = (t_i^{r+1}:t_j^r).$$

Knowing these matrices and using the first basic property of the incidence numbers, one can obviously obtain the incidence numbers of any two cells of consecutive dimensions and thus all the properties of the complex.

DEFINITION 4.21. The matrix

$$\mathfrak{E}^r = \left\| \begin{matrix} \varepsilon_{11}{}^r & \cdots & \varepsilon_{1j}{}^r & \cdots & \varepsilon_{1\rho(r)}{}^r \\ \cdot & \cdot & \cdot & \cdot & \cdot \\ \varepsilon_{i1}{}^r & \cdots & \varepsilon_{ij}{}^r & \cdots & \varepsilon_{i\rho(r)}{}^r \\ \cdot & \cdot & \cdot & \cdot & \cdot \\ \varepsilon_{\rho(r+1)1}{}^r & \cdots & \varepsilon_{\rho(r+1)j}{}^r & \cdots & \varepsilon_{\rho(r+1)\rho(r)}{}^r \end{matrix} \right\| \qquad \begin{matrix} \varepsilon_{ij}{}^r = (t_i{}^{r+1} : t_j{}^r), \\ r = 0, 1, \cdots, n-1, \end{matrix}$$

is called the rth *incidence matrix* of the cell complex $\mathfrak{K}$. (Whenever ρ occurs as a subscript, we shall write $\rho(r)$ instead of ρ^r.)

REMARK 1. If $\mathfrak{K}$ is a complex consisting of the oriented simplexes of a simplicial complex K, then the incidence matrices of the cell complex $\mathfrak{K}$ are known as the *incidence matrices of the simplicial complex* K.

We shall give several examples of cell complexes and their incidence matrices.

EXAMPLES. 1. The decomposition of a torus into four (curvilinear) rectangles is shown in Figs. 101 and 102. The orientations of these rectangles, their sides, and vertices are the elements of the cell complex $\mathfrak{K}$. Denoting by t_i^2, t_i^1, t_i^0 the oriented elements indicated in Fig. 102, we obtain the incidence matrices:

$\mathfrak{E}^1 =$

	t_1^1	t_2^1	t_3^1	t_4^1	t_5^1	t_6^1	t_7^1	t_8^1
t_1^2	1	−1	0	0	1	−1	0	0
t_2^2	−1	1	0	0	0	0	1	−1
t_3^2	0	0	1	−1	−1	1	0	0
t_4^2	0	0	−1	1	0	0	−1	1

$\mathfrak{E}^0 =$

	t_1^0	t_2^0	t_3^0	t_4^0
t_1^1	1	−1	0	0
t_2^1	0	0	1	−1
t_3^1	−1	1	0	0
t_4^1	0	0	−1	1
t_5^1	0	1	0	−1
t_6^1	1	0	−1	0
t_7^1	0	−1	0	1
t_8^1	−1	0	1	0

2. The cell complex $\mathfrak{K}$ consists of the elements $\pm t^2$, $\pm t_1^1$, $\pm t_2^1$, $\pm t^0$ with incidence numbers

$$(t^2 : t_1^1) = (t^2 : t_2^1) = (t_1^1 : t^0) = (t_2^1 : t^0) = 0.$$

3. The cell complex $\mathfrak{K}$ consists of the elements $\pm t^2$, $\pm t^1$, $\pm t^0$, with $(t^2 : t^1) = 2$, $(t^1 : t^0) = 0$.

4. The cell complex $\mathfrak{K}$ consists of the elements $\pm t^2$, $\pm t^1$, $\pm t^0$, with incidence numbers $(t^2 : t^1) = 6$, $(t^1 : t^0) = 0$.

REMARK 2. We arrive at a geometric interpretation of the last example (Example 4) by considering a hexagon with the identifications and orientations indicated in Fig. 103. In the same way, the geometric basis of Example 3 is the model of the projective plane obtained from a circle whose diametrically opposite points are identified; in Example 2 we have a torus represented as a square with identification of opposite sides. The precise meaning of these assertions will be considered in X, 2.4.

REMARK 3. The reader may easily verify that the cell complexes given in the preceding examples are a-complexes. An example of a cell complex which is not an a-complex was given in 4.1, Remark 1; it is left to the reader

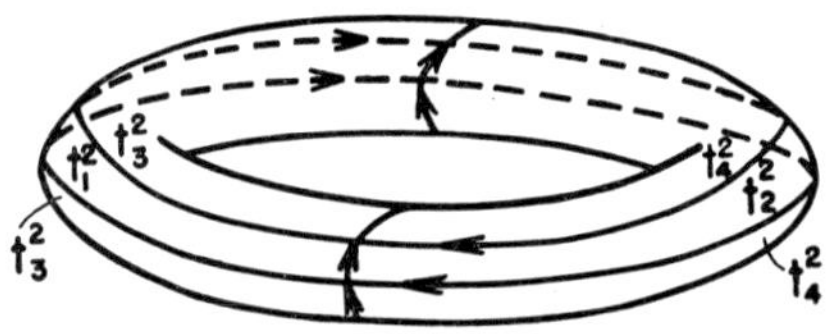

FIG. 101

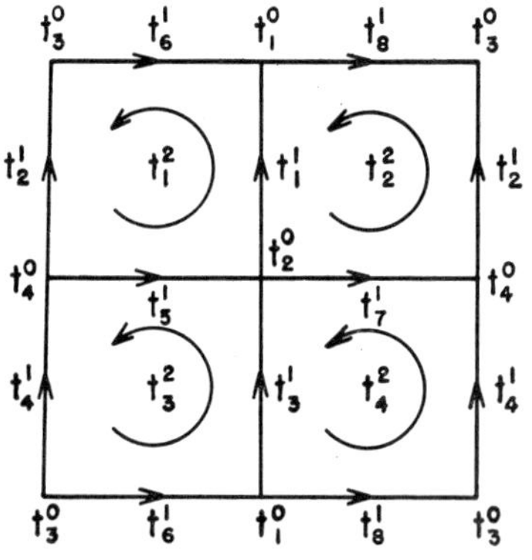

FIG. 102

to devise other similar examples as simple exercises. However, it should be kept in mind that the primary concept is that of an a-complex, while the concept of a cell complex has only secondary value; to what degree the properties of cell complexes which are not a-complexes are interesting and important is not well known.

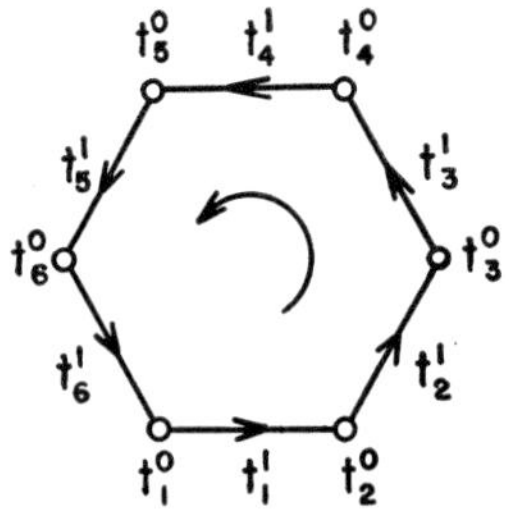

$t_1^1 = t_2^1 = \cdots = t_6^1 = t^1$

$t_1^0 = t_2^0 = \cdots = t_6^0 = t^0$

FIG. 103

§5. Chains

§5.1. Definition of a chain. Let us consider a polygonal line without self-intersections in the plane or in space.

The choice of a direction of motion along the polygonal line determines a definite orientation on each of its segments or links (1-simplexes). Hence a consideration of polygonal paths leads naturally to a consideration of sets whose elements are oriented 1-simplexes. This set of 1-simplexes could be naturally ordered, but we shall not do this since it would lead us to other concepts which we do not intend to investigate.

Let us now consider paths in which some links are traversed several times. A path of this sort is not simply a set of directed segments; it must be considered as a set of directed segments each of which is assigned an integer which indicates the number of times the segment is traversed; in other words, the multiplicity or coefficient with which the segment enters into the path must be indicated. Consider, for instance, the polygonal path $e_1e_2e_3e_1e_2e_4$ (Fig. 104). It can be considered as a collection of directed segments $|e_1e_2|$, $|e_2e_3|$, $|e_3e_1|$, $|e_2e_4|$, with multiplicities 2, 1, 1, 1, respectively. On the other hand, the directed segment $|e_1e_3|$ would have to be assigned the coefficient -1 in the given path.

Similarly, in the path $e_1e_2e_3e_1e_3e_2e_4$ the segments $|e_1e_2|$, $|e_2e_3|$, $|e_3e_1|$, $|e_4e_2|$ are assigned the coefficients 1, 0, 0, -1, respectively (the coefficients of $|e_2e_3|$ and $|e_3e_1|$ are zero since these segments are traversed twice and moreover in opposite directions).

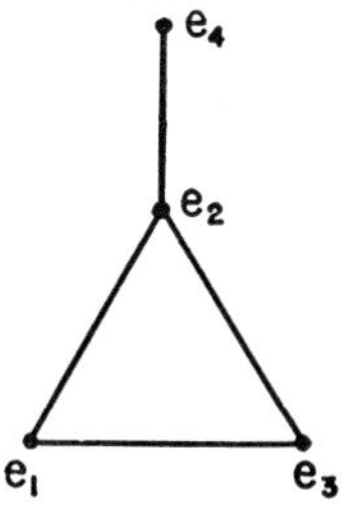

FIG. 104

The logical essence of the concept discussed above consists then in assigning to each oriented simplex (in the examples, a 1-simplex) of a given simplicial complex [in the examples, the complex consisting of the four segments (e_1e_2), (e_2e_3), (e_3e_1), (e_2e_4)] an integer, the "coefficient" with which the given oriented 1-simplex appears in the path; the integer-valued function so defined on a given set of oriented simplexes is odd since $x(t^1) = -x(-t^1)$. The latter fact expresses the equivalence of the two statements: the directed segment $|e_ie_j|$ appears in the path with coefficient a: the directed segment $|e_je_i|$ appears in the path with coefficient $-a$. Such integer-valued functions are called *integral chains*; in the given case, *one dimensional integral chains* (integral 1-chains).

The concept of integral 1-chain is susceptible of generalization in several directions. First, instead of an integral 1-chain, one can define an integral r-chain, and moreover on an arbitrary cell complex $\mathfrak{K}$, as an integer-valued odd function $x^r(t^r)$ whose value is defined for each cell t^r of $\mathfrak{K}$; the function is odd in the obvious sense that

$$x^r(-t^r) = -x^r(t^r),$$

and integer-valued in the sense that its values are integers. Secondly, one can relax the condition that the function be integer-valued and instead require that it take values in a given Abelian group $\mathfrak{A}$, which is then called the coefficient domain. We thus arrive at the following fundamental definition:

DEFINITION 5.1. *Let $\mathfrak{K}$ be a cell complex and let $\mathfrak{A} \neq 0$ be an Abelian group,* *referred to as a coefficient domain; suppose that each element of* $\mathfrak{K}$ *is assigned an element* $x^r(t^r)$ *of* $\mathfrak{A}$ *in such a way as to satisfy the condition*

$$x^r(-t^r) = -x^r(t^r) \qquad \text{(condition that the function be odd).}$$

The resulting function $x^r(t^r)$ *defined on* $\mathfrak{K}$ *is called an r-chain of* $\mathfrak{K}$ *over the coefficient domain* $\mathfrak{A}$.

REMARK 1. In this book we consider only finite chains, that is, chains which assume values different from zero only on a finite number of elements of $\mathfrak{K}$. Hence *chain* will always mean *finite chain*.

DEFINITION 5.11. The set $L^r(\mathfrak{K}, \mathfrak{A})$ of all r-chains of a complex $\mathfrak{K}$ over the coefficient domain $\mathfrak{A}$ forms a *group with respect to the operation of addition*: the sum of two chains $x_1^r \in L^r(\mathfrak{K}, \mathfrak{A})$ and $x_2^r \in L^r(\mathfrak{K}, \mathfrak{A})$ is the chain whose value on each element of $\mathfrak{K}$ is the sum of the values assumed on that element by x_1^r and x_2^r. The identity or zero of the group $L^r(\mathfrak{K}, \mathfrak{A})$ is the chain whose value is zero (the identity) on every $t^r \in \mathfrak{K}$. [On a first reading the reader may limit himself to a single coefficient domain, the additive group of integers, omitting everything pertaining to other coefficient domains. We shall usually omit the argument $\mathfrak{A}$ and write $L^r(\mathfrak{K})$ instead of $L^r(\mathfrak{K}, \mathfrak{A})$.]

REMARK 2. For convenience we shall not distinguish between the identities of the various Abelian groups which we shall consider as coefficient domains. This also makes it possible to regard as identical the zero elements of all the groups $L^r(\mathfrak{K}, \mathfrak{A})$ and to call them simply the zero or null r-chains of $\mathfrak{K}$ (without regard to the coefficient domains).

REMARK 3. If $\mathfrak{K}$ is an n-dimensional cell complex (cell n-complex), we introduce also for $r > n$ the group $L^r(\mathfrak{K}, \mathfrak{A})$ which, by definition, consists of the identity alone. In addition, for all cell complexes $\mathfrak{K}$ we define the group $L^{-1}(\mathfrak{K}, \mathfrak{A})$ consisting, by definition, also of the identity alone.

§5.2. Some remarks on chains. *The most important coefficient domains are the following additive groups:*

1) the group J of all integers;

2) the group J_m of integers (mod m) (m an integer);

3) the group $\mathfrak{R}$ of all rational numbers;

4) the group $\mathfrak{R}_1$ of rationals (mod 1), that is, the factor group of $\mathfrak{R}$ with respect to the subgroup J;

5) the group Π of all real numbers (mod 1); i.e., the factor group of the real numbers with respect to the subgroup J.

The group $L^r(\mathfrak{K}, J)$ is denoted simply by $L_0^r(\mathfrak{K})$; its elements are called *integral r-chains of* $\mathfrak{K}$.

The group $L^r(\mathfrak{K}, J_m)$ is denoted by $L_m^r(\mathfrak{K})$; its elements are referred to as *r-chains* (mod m). All the coefficient domains noted above, with the exception of $\mathfrak{R}_1$ and Π, are rings with unity. [In this book a ring is a group $\mathfrak{A}$ in which an associative and commutative multiplication, distributive with respect to addition, is defined. A ring with unity is a ring $\mathfrak{A}$ which contains an element, called a unit and denoted by 1, with the property that $1 \cdot a = a$ for every $a \in \mathfrak{A}$. It will be convenient for us to regard as identical the unit elements of all rings considered as coefficient domains.]

The groups J_m (m a prime) and $\mathfrak{R}$ are fields.

REMARK 1. The case $m = 2$ merits special attention. Since the group J_2 consists of the two elements 0 and 1, where $1 + 1 = 0$, we have for $x^r \in L_2^r(\mathfrak{K})$ and any cell $t^r \in \mathfrak{K}$,

$$x^r(-t^r) = -x^r(t^r) = x^r(t^r);$$

that is, an arbitrary chain (mod 2) has the same value on each of two opposite cells. Therefore, if a cell complex $\mathfrak{K}$ consists of the oriented simplexes of a simplicial complex K, the chain $x^r \in L_2^r(\mathfrak{K})$ assigns to each simplex $T^r \in K$ a definite value $x^r(T^r) = x^r(t^r) = x^r(-t^r)$. This fact enables us to regard chains (mod 2) of simplicial complexes as functions with values 0 and 1, defined on the set of simplexes T_1^r, T_2^r, $\cdots$ of the complex K.

There is a (1–1) correspondence between these functions and the subsets of the set of all r-dimensional elements of the complex; this correspondence may be realized by assigning to each chain $x^r \in L_2^r(\mathfrak{K})$ the set of r-simplexes $T_i^r \in K$ on which the chain x^r assumes the value 1.

REMARK 2. In the theory of sets the function defined on a set A which has the value 1 on all the elements of a subset B of A and the value 0 on each element of $A \setminus B$ is called the *characteristic function* of B. Hence the r-chains (mod 2) of a simplicial complex K may be defined as the *characteristic functions of the distinct subsets of the set of all r-simplexes of* K.

REMARK 3. (This remark is required for the later chapters, starting with Chapter XI.) Let K be an unrestricted skeleton complex or, in general, an unrestricted simplicial complex. Let x^r be a chain of K and denote by $| x^r |$ the combinatorial closure (see IV, Def. 1.84) of the subcomplex of K consisting of all simplexes on which x^r does not vanish. The elements (in particular, the vertices) of the complex $| x^r |$ are known as the *elements* (in particular, the *vertices*) *of the chain* x^r.

Finally, let the elements of the complex K be simplexes, perhaps degenerate, of some R^n. Let x^r be a chain of K. The union of the bodies of all the oriented simplexes (see 1.3) of the complex K on which the chain x^r does not vanish is called the body of the chain x^r and is denoted by $\bar{x}^r$. $\bar{x}^r$ is a compactum (and even a polyhedron).

§5.3. Monomial chains. Chains as linear forms. Let t^r be an element of a cell complex $\mathfrak{K}$ and let a be an element of a coefficient domain $\mathfrak{A}$. We shall denote by at^r the r-chain of $\mathfrak{K}$ which takes the value a on t^r and vanishes on all r-cells of $\mathfrak{K}$ distinct from $\pm t^r$.

It follows from 5.1 that the value of at^r on $-t^r$ is $-a$. Chains of the form

at^r are known as *monomial chains* (sometimes called cellular chains) or cells t^r with coefficient a. In particular, if $\mathfrak{A}$ is a ring with unity, the chain $1t^r$ (the chain which takes the value 1 on t^r and 0 on all cells distinct from $\pm t^r$) is denoted by t^r and is identified with the cell t^r.

5.31. *Every chain is a sum of monomial chains.*

Proof. Let us denote a definite cell of each pair of opposites of $\mathfrak{K}$ by t_i^r. If the value of the chain x^r on the cell t_i^r is a_i, we may obviously write the identity

$$x^r = \sum a_i t_i^r. \tag{5.31}$$

Indeed, on an arbitrary t_i^r the value of $a_i t_i^r$ is a_i, but the values of all the chains $a_j t_j^r$, $j \neq i$, are zero; consequently the value of the chain $\sum a_i t_i^r$ on t_i^r is a_i, that is, the value of the chain x^r on the cell t_i^r.

Both x^r and $\sum a_i t_i^r$ take the value $-a_i$ on $-t_i^r$. This proves 5.31, since t_i^r was arbitrary.

Hence every r-chain of $\mathfrak{K}$ may be written as a linear form (5.31), where t_i^r (for various values of i) is a representative of each pair of opposites of the cell complex $\mathfrak{K}$ and a_i is an element of the coefficient domain $\mathfrak{A}$. Distinct chains correspond to distinct linear forms, and conversely.

We see also that the addition of chains, defined as functions, corresponds to the usual addition of linear forms; hence the group $L^r(\mathfrak{K}, \mathfrak{A})$ may be thought of as the group of linear forms (5.31). It follows that *the group $L^r(\mathfrak{K}, \mathfrak{A})$ is a direct sum of ρ^r groups isomorphic to the group $\mathfrak{A}$*, where $2\rho^r$ is the number of r-elements of the cell complex $\mathfrak{K}$.

§5.4. Chains of a simplicial complex. If $\mathfrak{K}$ is a cell complex consisting of all the oriented simplexes of a simplicial complex K, the chains of K are called *simplicial chains* of K or chains of the simplicial complex K. Accordingly, the group $L^r(\mathfrak{K}, \mathfrak{A})$ is now written as $L^r(K, \mathfrak{A})$ and is known as the group of simplicial r-chains of K or the group of r-chains of the simplicial complex K. The number ρ^r is in this case simply the number of r-simplexes of K.

It is clear that in accordance with the abstract definition of a chain and the definition of an oriented simplex, the chains of a simplicial complex K may be directly defined as follows: *the r-chains of a simplicial complex K over a coefficient domain $\mathfrak{A}$* are functions x^r defined on the set of all ordered r-skeletons $| e_0, \cdots, e_r |$ of K satisfying the conditions:

1°. The functions x^r take values in the group $\mathfrak{A}$.

2°. If $\begin{pmatrix} e_0 & \cdots & e_r \\ e_{i(0)} & \cdots & e_{i(r)} \end{pmatrix}$ is an odd permutation of the vertices of an ordered skeleton $| e_0, \cdots, e_r |$, then

$$x^r | e_{i(0)}, \cdots, e_{i(r)} | = -x^r | e_0, \cdots, e_r |.$$

Remark. Since every even permutation is a product of an even number of odd permutations (transpositions), 2° implies

3°. If $\begin{pmatrix} e_0 & \cdots & e_r \\ e_{i(0)} & \cdots & e_{i(r)} \end{pmatrix}$ is an even permutation, $x^r \mid e_0, \cdots, e_r \mid = x^r \mid e_{i(0)}, \cdots, e_{i(r)} \mid$.

In the terminology of 5.3 we may say that a simplicial r-chain of K over $\mathfrak{A}$ is a linear form in the oriented simplexes t_i^r of K with coefficients $a_i \in \mathfrak{A}$.

§5.5. The scalar product of chains. We shall assume that the coefficient domain $\mathfrak{A}$ is a ring with unity.

Let

$$x^r = \sum a_i t_i^r, \qquad y^r = \sum b_i t_i^r$$

be two chains of a cell complex $\mathfrak{K}$ over $\mathfrak{A}$. We shall call

$$(x^r \cdot y^r) = \sum a_i b_i \in \mathfrak{A} \tag{5.51}$$

the *scalar product of the chains* x^r and y^r.

We have defined the scalar product of two chains over the same coefficient ring $\mathfrak{A}$. Now let $\mathfrak{A}$ be an arbitrary coefficient domain, that is, an arbitrary Abelian group. An arbitrary element a of $\mathfrak{A}$ can be multiplied by an arbitrary integer $\pm n$ $(n \geq 0)$ in accordance with the rule:

$$a \cdot n = na = a + a + \cdots + a \text{ (n times)}.$$

Consequently, if

$$x^r = \sum a_i t_i^r \in L^r(\mathfrak{K}, \mathfrak{A})$$

is a chain of $\mathfrak{K}$ over $\mathfrak{A}$ and

$$y^r = \sum b_i t_i^r$$

is an integral chain, then the scalar product

$$(x^r \cdot y^r) = \sum a_i b_i \in \mathfrak{A}$$

is defined; in particular, if $y^r = t_i^r$, then $(x^r \cdot y^r) = a_i$.

Hence, *the value of the chain x^r on the element t_i^r of a cell complex $\mathfrak{K}$ is the scalar product of x^r and t_i^r* (considered as an integral monomial chain). We can therefore denote the value of x^r on $t^r \in \mathfrak{K}$ by $(x^r \cdot t^r)$ [retaining, wherever convenient, also the notation $x^r(t^r)$]. The linear form (5.31) may now be rewritten as

$$x^r = \sum (x^r \cdot t_i^r) t_i^r.$$

§5.6. Extension of chains; restriction of chains to a subcomplex. The operators $\mathfrak{K}_0$ and $E_{\mathfrak{K}}$. [In what follows we recall that by a subcomplex of a cell complex we mean a *cell* subcomplex (see 4.1, Remark 5). The reader who has omitted §4 may think of a cell complex $\mathfrak{K}$ as the set of all oriented simplexes of a triangulation K and of a subcomplex $\mathfrak{K}_0$ of $\mathfrak{K}$ as the set of all oriented simplexes of a subcomplex $K_0 \subseteq K$.] Let $\mathfrak{K}_0$ be a subcomplex of a cell complex $\mathfrak{K}$ and let $x^r = \sum a_i t_i^r$ be a chain of $\mathfrak{K}$. If we consider the function x^r only on the cells of $\mathfrak{K}_0$, that is, if we retain in the linear form $x^r = \sum a_i t_i^r$ only those terms for which $t_i^r \in \mathfrak{K}_0$, we obtain a chain of the complex $\mathfrak{K}_0$ referred to as the *restriction of the chain* x^r *to* $\mathfrak{K}_0$ and denoted by $\mathfrak{K}_0 x^r$. In particular, if $x^r(t_i^r) = 0$ for every $t_i^r \in \mathfrak{K} \setminus \mathfrak{K}_0$, we shall say that the chain x^r is *on* $\mathfrak{K}_0$.

Now let x_0^r be a chain of $\mathfrak{K}_0$; defining the functions $x^r \in L^r(\mathfrak{K})$ by the equations

$$x^r(t^r) = x_0^r(t^r), \qquad \text{if } t^r \in \mathfrak{K}_0,$$

$$x^r(t^r) = 0, \qquad \text{if } t^r \notin \mathfrak{K}_0,$$

we obtain a chain x^r of $\mathfrak{K}$ called the *extension* of the chain $x_0^r \in L^r(\mathfrak{K}_0)$ over the complex $\mathfrak{K}$ and denoted by $E_{\mathfrak{K}} x_0^r$.

REMARK. If chains are thought of as linear forms, it is necessary to identify every chain of $\mathfrak{K}_0$ with its extension over $\mathfrak{K}$ (since two linear forms are identical if they differ only by terms which are equal to zero); in other words, it is necessary to regard $L^r(\mathfrak{K}_0)$ as a subgroup of $L^r(\mathfrak{K})$. From the point of view of functions, the chain $x_0^r \in L^r(\mathfrak{K}_0)$ is distinct from the chain $E_{\mathfrak{K}} x_0^r \in L^r(\mathfrak{K})$, since x_0^r is a function defined on $\mathfrak{K}_0$, while $E_{\mathfrak{K}} x_0^r$ is a function defined on $\mathfrak{K}$; two functions are identical if, and only if, they are defined on the same set and assume the same value on each element of this set. To emphasize this difference, chains defined as functions are sometimes called ∇-chains, while those defined as linear forms are referred to as Δ-chains.

If we assign to each chain $x^r \in L^r(\mathfrak{K})$ the chain $\mathfrak{K}_0 x^r \in L^r(\mathfrak{K}_0)$, we obtain a homomorphism of the group $L^r(\mathfrak{K})$ onto the group $L^r(\mathfrak{K}_0)$; the mapping is *onto* since, obviously,

$$x_0^r = \mathfrak{K}_0(E_{\mathfrak{K}} x_0^r)$$

for every $x_0^r \in L^r(\mathfrak{K}_0)$. Moreover, if we assign to each chain $x_0^r \in L^r(\mathfrak{K}_0)$ the chain $E_{\mathfrak{K}} x_0^r \in L^r(\mathfrak{K})$, we obtain a homomorphism (which is also an isomorphism) of $L^r(\mathfrak{K}_0)$ into $L^r(\mathfrak{K})$ (the homomorphism is not onto for $\mathfrak{K}_0 \neq \mathfrak{K}$). We shall call the first of these homomorphisms the *intersection-homomorphism* and the second the *extension-isomorphism* or refer to both, indifferently, as the *homomorphisms* (or *operators*) $\mathfrak{K}_0$ and $E_{\mathfrak{K}}$, respectively.

§6. The lower boundary operator (the operator Δ)

§6.1. Definition of the Δ-boundary.

DEFINITION 6.10. Let $\mathfrak{K}$ be a cell complex and let $t^r \in \mathfrak{K}$. We shall denote by $\Delta_{\mathfrak{K}} t^r$ the $(r - 1)$-chain which takes the value $(t^r : t^{r-1})$ on each cell t^{r-1}

of $\mathfrak{K}$. The chain $\Delta_{\mathfrak{K}} t^r$ is called the *boundary of the cell* t^r (or often: the lower boundary or the Δ-boundary) *in the complex* $\mathfrak{K}$.

If one of each pair of $(r-1)$-dimensional opposites of $\mathfrak{K}$ is denoted by t_j^{r-1} and the other by $-t_j^{r-1}$, then the chain $\Delta_{\mathfrak{K}} t^r$ may be written as the linear form

(6.10) $$\Delta_{\mathfrak{K}} t^r = \sum (t^r : t_j^{r-1}) t_j^{r-1}.$$

REMARK 1. If K is an unrestricted simplicial complex and $\mathfrak{K}$ is the cell complex of its oriented simplexes, then for

$$t^r = | e_0 \cdots e_r |, \qquad t_j^{r-1} = | e_0 \cdots \hat{e}_j \cdots e_r |$$

we have, by (3.11),

$$\Delta_{\mathfrak{K}} t^r = \sum (-1)^j t_j^{r-1}.$$

This expression is the same for all unrestricted simplicial complexes K containing the simplex t^r; consequently, we are justified in omitting the subscript $\mathfrak{K}$ and writing

(6.100) $$\Delta | e_0 \cdots e_r | = \sum (-1)^j | e_0 \cdots \hat{e}_j \cdots e_r |.$$

The chain (6.100) will be called simply the *boundary of the oriented simplex* $| e_0 \cdots e_r |$ (in an arbitrary unrestricted simplicial complex).

DEFINITION 6.11. Let

$$x^r = \sum a_i t_i^r \in L^r(\mathfrak{K})$$

be an arbitrary r-chain of a cell complex $\mathfrak{K}$. We define the $(r-1)$-chain $\Delta_{\mathfrak{K}} x^r$ by the equation

(6.11) $$\Delta_{\mathfrak{K}} x^r = \sum_i a_i \Delta_{\mathfrak{K}} t_i^r.$$

Substituting (6.10) into (6.11), we have

$$\Delta_{\mathfrak{K}} x^r = \sum_i a_i \sum_j (t_i^r : t_j^{r-1}) t_j^{r-1},$$

i.e.,

(6.111) $$\Delta_{\mathfrak{K}} x^r = \sum_{i,j} (t_i^r : t_j^{r-1}) a_i t_j^{r-1}.$$

For a fixed but arbitrary t_j^{r-1} (6.111) yields $\sum_i (t_i^r : t_j^{r-1}) a_i$ as the value of $\Delta_{\mathfrak{K}} x^r$ on t_j^{r-1}.

This result may be given another form. Let t^{r-1} be an arbitrary $(r-1)$-cell. Let us denote by t_i^r that cell of each pair of r-cells having a nonvanishing incidence number with respect to t^{r-1} for which the incidence number is positive. Then, since $a_i = (x^r \cdot t_i^r)$,

(6.12) $$(\Delta_{\mathfrak{K}} x^r \cdot t^{r-1}) = \sum (t_i^r : t^{r-1})(x^r \cdot t_i^r),$$

where the summation is extended over all t_i^r incident with t^{r-1}.

Equation (6.12) may be used to define the chain $\Delta_{\mathfrak{K}} x^r$.

Remark 2. If $\mathfrak{K}$ is the cell complex of all the oriented simplexes of a simplicial complex K, all the nonzero incidence numbers are equal to ± 1 and (6.12) becomes

$$(6.121) \qquad (\Delta_{\mathfrak{K}} x^r \cdot t^{r-1}) = \sum (x^r \cdot t_i^r),$$

where the sum is taken over all the cells t_i^r for which $(t_i^r : t^{r-1}) = +1$.

If $t^{r-1} = | e_1 \cdots e_r |$, then

$$(6.122) \qquad (\Delta_{\mathfrak{K}} x^r \cdot | e_1 \cdots e_r |) = \sum (x^r \cdot | e_i e_1 \cdots e_r |),$$

where the sum is taken over all the vertices $e_i \in | K |$, for which the simplex $| e_i e_1 \cdots e_r |$ is in K.

Remark 3. For 0-chains $x^0 \in L^0(\mathfrak{K})$ we set $\Delta_{\mathfrak{K}} x^0 = 0$ in agreement with the fact that, by definition, $L^{-1}(\mathfrak{K})$ consists of the identity alone.

6.13. An immediate consequence of the definition of the chain $\Delta_{\mathfrak{K}} x^r$ is that

$$\Delta_{\mathfrak{K}} (x_1^r + x_2^r) = \Delta_{\mathfrak{K}} x_1^r + \Delta_{\mathfrak{K}} x_2^r;$$

in other words, if every chain $x^r \in L^r(\mathfrak{K}, \mathfrak{A})$ is assigned its Δ-boundary $\Delta_{\mathfrak{K}} x^r$, the result is a homomorphism of the group $L^r(\mathfrak{K}, \mathfrak{A})$ into the group $L^{r-1}(\mathfrak{K}, \mathfrak{A})$. This homomorphism is referred to as the *homomoprhism* or *operator* Δ.

Remark 4. If $\mathfrak{K}$ is the cell complex consisting of the oriented simplexes of an unrestricted simplicial complex K, we shall write Δ_K instead of $\Delta_{\mathfrak{K}}$, and often we shall omit the subscript K entirely.

§6.2. Examples of chains and their boundaries.

1°. The complex $\mathfrak{K}$ is represented as the decomposition of the plane into congruent squares of side 1 with vertices at the lattice points (the points with integral coordinates). The oriented squares of this complex form an a-complex. The chain x^2 assumes the values indicated in Fig. 105 (the given values taken on by this chain on the squares oriented counterclockwise). The chain Δx^2 has the value 1 (2) on the sides marked with one or two arrows, respectively, and oriented by these arrows. On the remaining sides Δx^2 is zero.

2°. Fig. 106 shows a triangulation of a part of the plane. On the triangles (oriented counterclockwise) with values attached the chain x^2 has the indicated values; x^2 is zero on those triangles which have no values attached. The boundary Δx^2 is 1 on the segments marked with a single arrow, 2 on the segments with double arrows, and zero on the remaining segments.

3°. Fig. 107 shows a square divided into 24 triangles. These triangles, oriented counterclockwise, are denoted by $t_1^2, \cdots, t_{24}^2$. Let us denote by

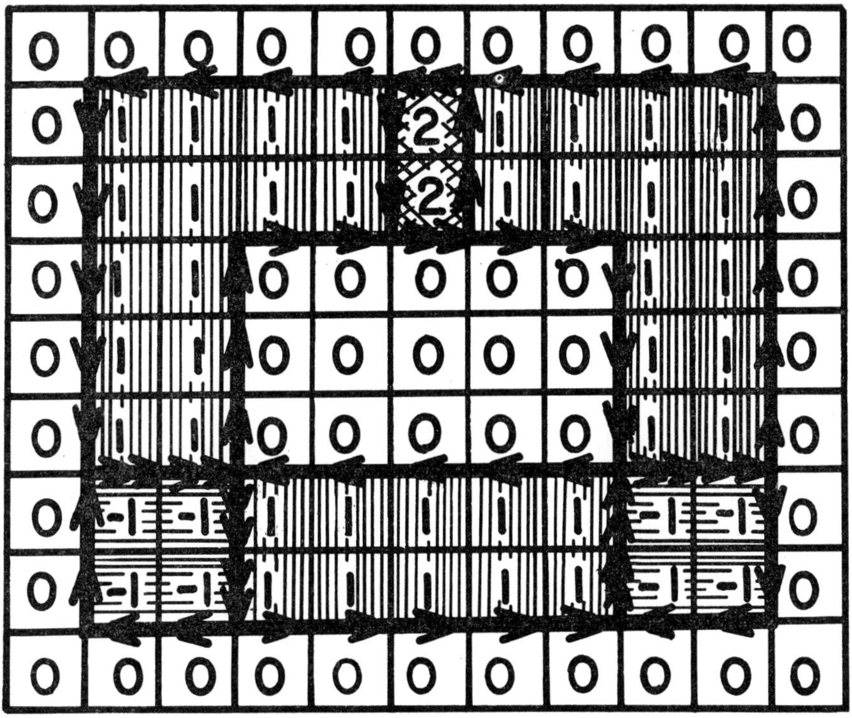

FIG. 105

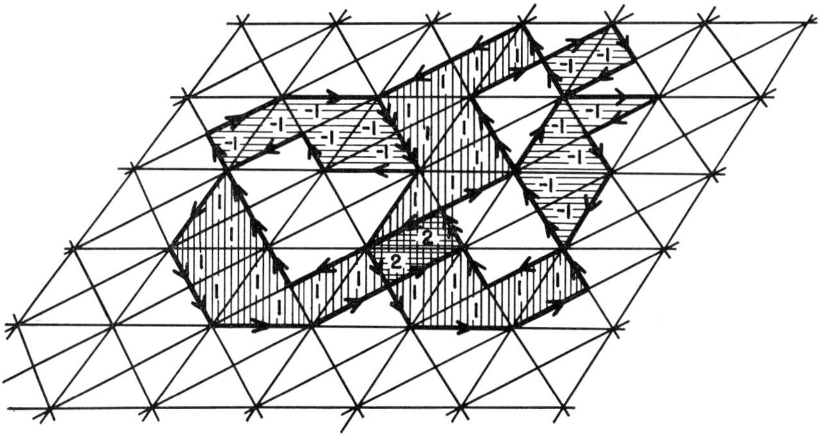

FIG. 106

$t_1^1, \cdots, t_{12}^1$ the segments on the boundary of the square, oriented by arrows in the indicated fashion.

We now identify t_1^1 with t_7^1, t_2^1 with t_8^1, t_3^1 with t_9^1, t_4^1 with t_{10}^1, t_5^1 with t_{11}^1, t_6^1 with t_{12}^1, so that instead of 12 segments $t_1^1, \cdots, t_{12}^1$, we now have 6, which we denote as before by $t_1^1, \cdots, t_6^1$ (with the same orientations which they had previously). This identification converts the complex consisting of 24 triangles, their sides, and vertices, into a triangulation K of the projective plane which has

$$\rho^2 = 24, \qquad \rho^1 = 36, \qquad \rho^0 = 13$$

2-elements, 1-elements, and 0-elements, respectively.

Let us set

$$x^2 = \textstyle\sum_{i=1}^{24} t_i^2, \qquad z^1 = \sum_{j=1}^{6} t_j^1.$$

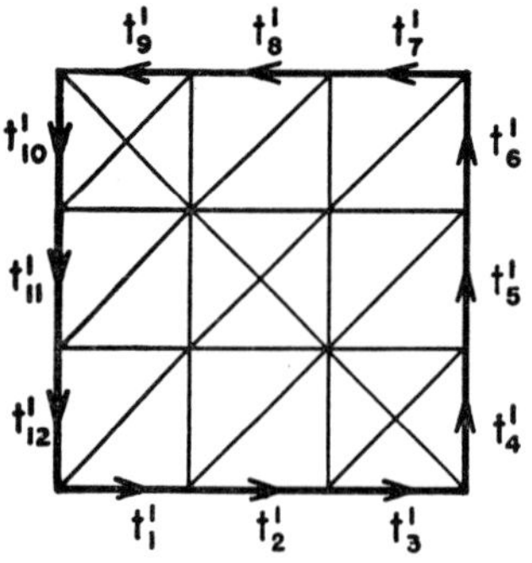

FIG. 107

Then

$$\Delta x^2 = 2z^1.$$

Hence the chain $2z^1$ is the boundary of the integral chain x^2 of the complex K. We shall prove that no integral chain of K has z^1 as its boundary. Indeed, let us suppose that K contained a chain y^2 such that $\Delta y^2 = z^1$. First, we shall show that then the chain y^2 assumes the same value on all t_i^2. For, in the contrary case there would be two triangles t_i^2 and t_j^2 on which the chain assumed different values and which had a common side in the interior of the square. But in that case the chain Δy^2 would necessarily have a nonzero value on the common side of these two triangles, which contradicts the assumption that $\Delta y^2 = z^1$.

Hence y^2 has the same value a on every t_i^2 and we may write

$$y^2 = ax^2.$$

Since $\Delta x^2 = 2z^1$, $\Delta y^2 = 2az^1$, which for a an integer contradicts the assumption that $\Delta y^2 = z^1$.

EXERCISE. Fig. 108 shows a triangulation K_{10} of the projective plane consisting of 10 triangles, their sides, and vertices. The chain $z^1 = t_1^1 + t_2^1 + t_3^1$ (the orientations of the segments t_i^1 are indicated in the figure by arrows) represents a cycle. Prove that $2z^1$ is the boundary of an integral chain and that the chain z^1 is not the boundary of any integral chain of the complex K_{10}.

4°. Fig. 109, after identification of the opposite sides of the square, represents a triangulation of the torus. The chain x^2 has the value 1 on the triangles marked with circular arrows and oriented by these arrows; x^2 is zero on the remaining triangles. The orientations of the segments t_i^1, $i = 1, \cdots, 12$, are indicated in the figure. Let us set

$$z^1 = \sum_{i=7}^{12} t_i^1, \qquad z_1^1 = \sum_{i=1}^{3} t_i^1, \qquad z_2^1 = \sum_{i=4}^{6} t_i^1.$$

Then $\Delta x^2 = z^1 - (2z_1^1 + z_2^1)$.

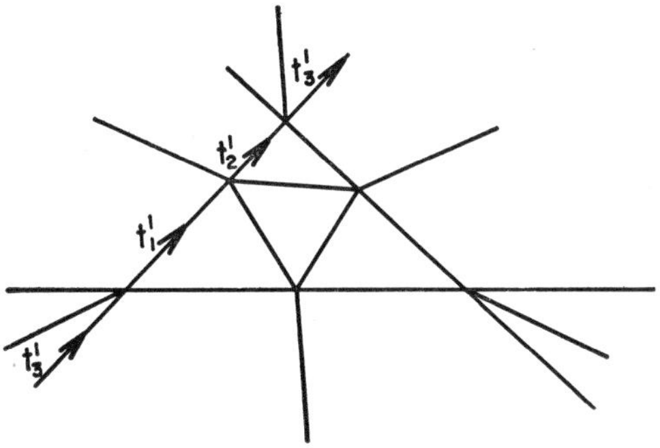

FIG. 108

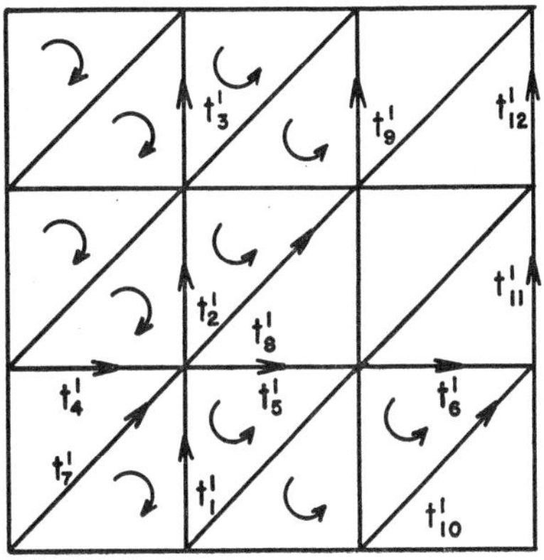

FIG. 109

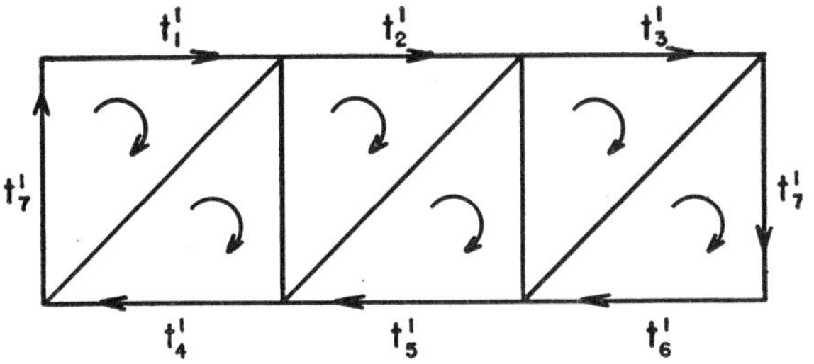

FIG. 110

5°. Fig. 110, after identification of the sides t_7^1 of the rectangle, gives a triangulation of the Möbius band. The orientations t_i^2 and t_j^1 of the triangles and segments, respectively, are indicated in the figure. Setting $x^2 = \sum t_i^2$, $z^1 = \sum_{j=1}^{6} t_j^1$, we have $\Delta x^2 = z^1 + 2t_7^1$.

6°. The complex K represents the space between two concentric spheres divided into prisms by radial cuts. The chain x^3 is $+1$ on all the prisms oriented coherently (i.e., all oriented in the same say). The chain Δx^3 is $+1$ on all the 2-faces of the prisms which lie on the inner sphere, -1 on all the faces which lie on the outer sphere, and zero on all other 2-faces.

§6.3. Cycles; chains homologous to zero; the groups $Z^r(\mathfrak{K})$ and $H^r(\mathfrak{K})$. Let us consider more closely the homomorphism Δ of the group $L^r(\mathfrak{K}, \mathfrak{A})$

into the group $L^{r-1}(\mathfrak{K}, \mathfrak{A})$. We shall denote the kernel of this homomorphism by $Z_\Delta{}^r(\mathfrak{K}, \mathfrak{A})$. The group $Z_\Delta{}^r(\mathfrak{K}, \mathfrak{A}) \subseteq L^r(\mathfrak{K}, \mathfrak{A})$ consists of all the chains x^r such that $\Delta_\mathfrak{K} x^r = 0$. These chains are called *r-cycles* (more precisely: r-dimensional Δ-cycles) of the cell complex $\mathfrak{K}$ over $\mathfrak{A}$.

Since $\Delta x^0 = 0$ for any 0-chain x^0, every 0-chain is a cycle, i.e.,

$$Z_\Delta{}^0(\mathfrak{K}, \mathfrak{A}) = L^0(\mathfrak{K}, \mathfrak{A}).$$

The image of the group $L^r(\mathfrak{K}, \mathfrak{A})$ under the homomorphism Δ is a subgroup of the group $L^{r-1}(\mathfrak{K}, \mathfrak{A})$. We shall denote this subgroup by $H_\Delta{}^{r-1}(\mathfrak{K}, \mathfrak{A})$. It is clear that in order for the chain x^r to be an element of the group $H_\Delta{}^r(\mathfrak{K}, \mathfrak{A})$ it is necessary and sufficient that there exist a chain $x^{r+1} \in L^{r+1}(\mathfrak{K}, \mathfrak{A})$ such that $\Delta_\mathfrak{K} x^{r+1} = x^r$.

THEOREM 6.3. *If $\mathfrak{K}$ is an n-dimensional cell complex, then $H_\Delta{}^n(\mathfrak{K}, \mathfrak{A})$ consists of the identity alone.*

For, in this case the group $L^{n+1}(\mathfrak{K})$ consists only of the identity. Hence its image under an arbitrary homomorphism, and consequently under Δ, is the identity. This proves the theorem.

REMARK. As in the case of the group $L^r(\mathfrak{K}, \mathfrak{A})$, we shall often omit the argument $\mathfrak{A}$ and also, whenever possible, the subscript Δ. Hence, instead of $Z_\Delta{}^r(\mathfrak{K}, \mathfrak{A})$ and $H_\Delta{}^r(\mathfrak{K}, \mathfrak{A})$ we shall usually write simply $Z^r(\mathfrak{K})$ and $H^r(\mathfrak{K})$. The especially important special cases of $\mathfrak{A} = J$ and $\mathfrak{A} = J_m$ will be emphasized by using the abbreviated notation $Z_0{}^r(\mathfrak{K})$, $H_0{}^r(\mathfrak{K})$ instead of $Z_\Delta{}^r(\mathfrak{K}, J)$, $H_\Delta{}^r(\mathfrak{K}, J)$ and $Z_m{}^r(\mathfrak{K})$, $H_m{}^r(\mathfrak{K})$ instead of $Z_\Delta{}^r(\mathfrak{K}, J_m)$, $H_\Delta{}^r(\mathfrak{K}, J_m)$.

If the cell complex $\mathfrak{K}$ consists of all the oriented simplexes of a simplicial complex K, we shall write $Z^r(K)$, $H^r(K)$, etc. instead of $Z^r(\mathfrak{K})$, $H^r(\mathfrak{K})$, etc.

§6.4. Homologies. The symbol $\sim$. Linear independence of chains with respect to homology. The elements of $H^r(\mathfrak{K}, \mathfrak{A})$ are known as r-chains *homologous to zero in $\mathfrak{K}$ over $\mathfrak{A}$*. To denote the fact that a chain x^r is homologous to zero, we shall write

(6.41) $$x^r \sim 0 \quad \text{in } \mathfrak{K} \text{ (over } \mathfrak{A}).$$

Since $H^r(\mathfrak{K})$ is a group, it follows immediately that the symbol $\sim$ has the following properties:

1°) $0 \sim 0$;

2°) $x^r \sim 0$ implies that $-x^r \sim 0$;

3°) $x_1{}^r \sim 0$ and $x_2{}^r \sim 0$ imply that $x_1{}^r \pm x_2{}^r \sim 0$.

We shall say that the chains $x_1{}^r$ and $x_2{}^r$ are *homologous* in $\mathfrak{K}$ (over $\mathfrak{A}$) and write

$$x_1{}^r \sim x_2{}^r \quad \text{in } \mathfrak{K} \text{ (over } \mathfrak{A})$$

if

$$x_1{}^r - x_2{}^r \sim 0 \quad \text{in } \mathfrak{K} \text{ (over } \mathfrak{A}).$$

From the above properties of the symbol ~ 0 it follows that

1°) $x \sim x$;

2°) if $x_1 \sim x_2$, then $x_2 \sim x_1$;

3°) if $x_1 \sim x_2$ and $x_2 \sim x_3$, then $x_1 \sim x_3$.

Moreover, if $x_1 \sim y_1$ and $x_2 \sim y_2$, then $x_1 \pm x_2 \sim y_1 \pm y_2$.

Let

$$x_1^r, \cdots, x_s^r \tag{6.42}$$

be chains of a cell complex $\mathfrak{K}$ over $\mathfrak{R}$ (the most important special case occurs when the chains in (6.42) are integral chains). Let $c_1, \cdots, c_s$ be integers. A linear combination $\sum c_i x_i^r$ of the chains (6.42) is said to be *trivial* if all the coefficients c_i are equal to zero; in the contrary case it is said to be *nontrivial.*

If all the chains x_i^r are chains over J_m, m a prime, the linear combination $\sum c_i x_i^r$ (c_i an integer) is said to be trivial if every c_i is divisible by m, i.e., if every $c_i \equiv 0 \pmod{m}$.

DEFINITION 6.42. Let $\mathfrak{A}$ be one of the coefficient domains J, $\mathfrak{R}$, or J_m. The chains (6.42) over $\mathfrak{A}$ are said to be *linearly independent with respect to homology* (*lirh*) in $\mathfrak{K}$ if no nontrivial linear combination of these chains is homologous to zero in $\mathfrak{K}$ over $\mathfrak{A}$.

REMARK 1. Especially important is the case of systems of integral chains *lirh.*

REMARK 2. Since all the coefficient domains $\mathfrak{A}$ mentioned in Def. 6.42 are rings, and the chains $x^r \in L^r(\mathfrak{K}, \mathfrak{A})$ are linear forms with coefficients in $\mathfrak{A}$, we can speak of linear combinations $\sum a_i x_i^r$ of the chains (6.42) with coefficients $a_i \in \mathfrak{A}$. Hence we may give Def. 6.42 the following form:

6.420. The chains (6.42) of a cell complex $\mathfrak{K}$ over $\mathfrak{A}$ (where $\mathfrak{A}$ is one of the rings J, $\mathfrak{R}$, or J_m) are said to be *lirh* in $\mathfrak{K}$ if a chain of the form $\sum a_i x_i^r$, where $a_i \in \mathfrak{A}$, is homologous to zero in $\mathfrak{K}$ over $\mathfrak{A}$ only if each of the elements a_i is the identity of the group $\mathfrak{A}$.

The proof of the equivalence of Defs. 6.42 and 6.420 is left to the reader.

§6.5. Restricted chains and cycles. Let $\mathfrak{K}$ be a cell complex and let $\mathfrak{M}$ be a cell subcomplex of $\mathfrak{K}$. Let $x^r \in L^r(\mathfrak{K})$. In general, $\Delta_{\mathfrak{M}} \mathfrak{M} x^r \neq \mathfrak{M} \Delta_{\mathfrak{K}} x^r$, as the following elementary example shows. Let the cell complex $\mathfrak{K}$ consist of all the oriented sides and vertices of the triangle $(e_0e_1e_2)$; let $\mathfrak{M}$ be the closed subcomplex of $\mathfrak{K}$ consisting of $\pm \mid e_0e_1 \mid$, $\pm \mid e_0 \mid$, $\pm \mid e_1 \mid$.

Let us define $x^1 \in L^1(\mathfrak{K})$ as the linear form

$$x^1 = \mid e_0e_1 \mid + \mid e_1e_2 \mid + \mid e_2e_0 \mid.$$

It is clear that $\mathfrak{M}x^1 = \mid e_0e_1 \mid$, $\Delta_{\mathfrak{K}} x^1 = 0$, $\mathfrak{M}\Delta_{\mathfrak{K}} x^1 = 0$, and $\Delta_{\mathfrak{M}} \mathfrak{M} x^1 = e_1 - e_0 \neq 0$. This example indicates the importance of the following theorem:

THEOREM 6.51. *If $\mathfrak{M}$ is an open subcomplex of a complex $\mathfrak{K}$ and $x^r \in L^r(\mathfrak{K})$, then*

$$\Delta_{\mathfrak{M}}\mathfrak{M}x^r = \mathfrak{M}\Delta_{\mathfrak{K}}x^r.$$

Proof. Let t^{r-1} be an arbitrary $(r - 1)$-cell of $\mathfrak{M}$. Since $\mathfrak{M}$ is an open subcomplex of $\mathfrak{K}$, every cell t_i^r for which $(t_i^r : t^{r-1}) \neq 0$ belongs to $O_{\mathfrak{K}}t^{r-1}$ and consequently to $\mathfrak{M}$. Hence the value of $(t_i^r : t^{r-1})$ does not depend on whether the cells t_i^r and t^{r-1} are considered as elements of $\mathfrak{K}$ or as elements of $\mathfrak{M}$.

In particular, the set of cells t_i^r incident (see Def. 4.12) with the cell t^{r-1} in $\mathfrak{K}$ is identical with the set of cells incident with t^{r-1} in $\mathfrak{M}$ and the corresponding incidence numbers $(t_i^r : t^{r-1})$ are equal in both complexes.

Therefore, $\sum (t_i^r : t^{r-1})$ has the same value whether the sum is extended over all $t_i^r \in \mathfrak{K}$ incident with t^{r-1} or whether it is extended over all $t_i^r \in \mathfrak{M}$ incident with t^{r-1}. But the first sum is equal to the value of $\mathfrak{M}\Delta_{\mathfrak{K}}x^r$ on t^{r-1}, while the second is the value of $\Delta_{\mathfrak{M}}\mathfrak{M}x^r$ on t^{r-1}.

COROLLARY. If $z^r \in Z_{\Delta}^r(\mathfrak{K})$ and $\mathfrak{M}$ is an open cell subcomplex of $\mathfrak{K}$, then $\mathfrak{M}z^r \in Z_{\Delta}^r(\mathfrak{M})$.

Indeed,

$$\Delta_{\mathfrak{M}}\mathfrak{M}z^r = \mathfrak{M}\Delta_{\mathfrak{K}}z^r.$$

§6.6. Extension of chains and cycles. Let $\mathfrak{M}$ be a subcomplex (see the parenthetical remark at the beginning of 5.6) of a cell complex $\mathfrak{K}$ and let $x_0^r \in L^r(\mathfrak{M})$. Let us consider the extension $E_{\mathfrak{K}}x_0^r$ of the chain x_0^r over the complex $\mathfrak{K}$, i.e., the chain $E_{\mathfrak{K}}x_0^r \in L^r(\mathfrak{K})$ defined as

$$E_{\mathfrak{K}}x_0^r(t^r) = x_0^r(t^r), \qquad \text{if } t^r \in \mathfrak{M},$$

$$E_{\mathfrak{K}}x_0^r(t^r) = 0, \qquad \text{if } t^r \notin \mathfrak{M}.$$

In general,

$$\Delta_{\mathfrak{K}}E_{\mathfrak{K}}x_0^r \neq E_{\mathfrak{K}}\Delta_{\mathfrak{M}}x_0^r.$$

To see this it is enough to define $\mathfrak{K}$ as the cell complex consisting of the two oriented 1-simplexes $| e_0e_1 |$ and $| e_1e_0 |$ and their oriented vertices $\pm | e_0 |$, $\pm | e_1 |$. The open subcomplex $\mathfrak{M}$ of $\mathfrak{K}$ consists, by definition, only of the two oriented 1-simplexes $| e_0e_1 |$ and $| e_1e_0 |$. Let us set $x_0^1 = | e_0e_1 |$. Clearly, $\Delta_{\mathfrak{M}}x_0^1 = 0$ and hence $E_{\mathfrak{K}}\Delta_{\mathfrak{M}}x_0^1 = 0$.

But

$$\Delta_{\mathfrak{K}}E_{\mathfrak{K}}x_0^1 = e_1 - e_0 \neq 0.$$

This shows the importance of the following theorem, the dual of Theorem 6.51:

THEOREM 6.61. *If $\mathfrak{K}$ is a cell complex and $\mathfrak{M}$ is a closed subcomplex of $\mathfrak{K}$, then*

$$\Delta_{\mathfrak{K}} E_{\mathfrak{K}} x_0^r = E_{\mathfrak{K}} \Delta_{\mathfrak{M}} x_0^r$$

for every chain $x_0^r \in L^r(\mathfrak{M})$.

To prove this consider an arbitrary $(r - 1)$-cell $t^{r-1} \in \mathfrak{K}$. Then

$$\Delta_{\mathfrak{K}} E_{\mathfrak{K}} x_0^r(t^{r-1}) = \textstyle\sum_i (t_i^r : t^{r-1}) E_{\mathfrak{K}} x_0^r(t_i^r)$$

(summed over all $t_i^r \in \mathfrak{K}$ incident with t^{r-1}). The terms of this sum corresponding to cells t_i^r not in $\mathfrak{M}$ are equal to zero, since $E_{\mathfrak{K}} x_0^r(t_i^r) = 0$ for such t_i^r. The terms corresponding to cells $t_i^r \in \mathfrak{M}$ are $(t_i^r : t^{r-1}) E_{\mathfrak{K}} x_0^r(t_i^r) = (t_i^r : t^{r-1}) x_0^r(t_i^r)$.

In particular, if t^{r-1} is not in $\mathfrak{M}$, then, since $\mathfrak{M}$ is a closed subcomplex of $\mathfrak{K}$, none of the cells t_i^r incident with t^{r-1} can belong to $\mathfrak{M}$. Hence

$$(1) \quad \begin{cases} \text{if } t^{r-1} \notin \mathfrak{M}, & \text{then} \quad \Delta_{\mathfrak{K}} E_{\mathfrak{K}} x_0^r(t^{r-1}) = 0; \\ \text{if } t^{r-1} \in \mathfrak{M}, & \text{then} \quad \Delta_{\mathfrak{K}} E_{\mathfrak{K}} x_0^r(t^{r-1}) = \sum (t_i^r : t^{r-1}) x_0^r(t_i^r), \end{cases}$$

where the summation is extended over all $t_i^r \in \mathfrak{M}$ incident with t^{r-1}.

On the other hand, in accordance with the definitions of the operators $E_{\mathfrak{K}}$ and $\Delta_{\mathfrak{K}}$,

$$(2) \quad \begin{cases} \text{if } t^{r-1} \notin \mathfrak{M}, & \text{then} \quad E_{\mathfrak{K}} \Delta_{\mathfrak{M}} x_0^r(t^{r-1}) = 0, \\ \text{if } t^{r-1} \in \mathfrak{M}, & \text{then} \quad E_{\mathfrak{K}} \Delta_{\mathfrak{M}} x_0^r(t^{r-1}) = \Delta_{\mathfrak{M}} x_0^r(t^{r-1}) \\ & \qquad = \sum (t_i^r : t^{r-1}) x_0^r(t_i^r), \end{cases}$$

where the summation is extended over all $t_i^r \in \mathfrak{M}$ incident with t^{r-1}. Comparing (1) with (2), we see that

$$\Delta_{\mathfrak{K}} E_{\mathfrak{K}} x_0^r(t^{r-1}) = E_{\mathfrak{K}} \Delta_{\mathfrak{M}} x_0^r(t^{r-1}).$$

This completes the proof.

Considering chains as linear forms (see 5.6, Remark), that is, regarding $L^r(\mathfrak{M})$ as a subset of $L^r(\mathfrak{K})$, $L^r(\mathfrak{M}) \subseteq L^r(\mathfrak{K})$, and consequently thinking of $E_{\mathfrak{K}}$ as the identity mapping of $L^r(\mathfrak{M})$ $[L^{r-1}(\mathfrak{M})]$ onto itself, we may express Theorem 6.61 as

6.610. If $\mathfrak{M}$ is a closed subcomplex of a cell complex $\mathfrak{K}$ and $x_0^r \in L^r(\mathfrak{M})$, then

$$\Delta_{\mathfrak{K}} x_0^r = \Delta_{\mathfrak{M}} x_0^r.$$

In particular, if $\mathfrak{M}$ is an unrestricted simplicial complex M, it is closed in every simplicial complex K containing it. Therefore

6.6100. If K is any simplicial complex containing the unrestricted simplicial complex M as a subcomplex, then

$$\Delta_K x_0^r = \Delta_M x_0^r$$

for every chain $x_0^r \in L^r(M)$.

In particular, if $\Delta_M x_0^r = 0$, then $\Delta_K x_0^r = 0$, i.e.,

6.62. Every Δ-cycle of a closed subcomplex $M \subset K$ is also a Δ-cycle of K.

6.63. A Δ-cycle of an unrestricted simplicial complex M is a Δ-cycle of every simplicial complex K containing M as a subcomplex.

§7. The fundamental case: $\mathfrak{K}$ is an a-complex

§7.1. The fundamental formula $\Delta\Delta x^r = 0$. Let t^r be an element of a cell complex $\mathfrak{K}$. Instead of $\Delta_{\mathfrak{K}}$ we shall write simply Δ. Let us calculate the value of the chain $\Delta\Delta t^r$ on any element $t^{r-2} \in \mathfrak{K}$. By (6.12) we have

$$(\Delta\Delta t^r \cdot t^{r-2}) = \textstyle\sum_j (t_j^{r-1} : t^{r-2})\,(\Delta t^r \cdot t_j^{r-1}),$$

where the sum is taken over all t_j^{r-1} incident with t^{r-2}. But according to the definition of Δt^r,

$$(\Delta t^r \cdot t_j^{r-1}) = (t^r : t_j^{r-1});$$

hence,

$$(\Delta\Delta t^r \cdot t^{r-2}) = \textstyle\sum_j (t^r : t_j^{r-1})\,(t_j^{r-1} : t^{r-2}). \tag{7.110}$$

[The reader who has omitted §4 may finish this subsection as follows: (7.110) and (3.25) imply that $\Delta\Delta t^r = 0$, which means that

$$\Delta\Delta x^r = 0 \tag{7.1}$$

for every chain $x^r \in L^r(K)$. In other words,

7.1′. *The boundary of every chain is a cycle*, i.e.,

$$H^r(K) \subseteq Z^r(K).] \tag{7.1″}$$

From (7.110) and the definition of an a-complex (see 4.1) it immediately follows that:

7.11. *In order that a cell complex $\mathfrak{K}$ be an a-complex it is necessary and sufficient that*

$$\Delta\Delta t^r = 0 \tag{7.11}$$

for every $t^r \in \mathfrak{K}$.

(7.11) implies that $\Delta\Delta x^r = \sum a_i \Delta\Delta t_i^r = 0$ for every chain $x^r = \sum a_i t_i^r$, i.e.,

7.1. *If $\mathfrak{K}$ is an a-complex, then*

$$\Delta\Delta x^r = 0 \tag{7.1}$$

for every chain x^r *of* $\mathfrak{K}$.

In other words,

7.1′. *In an a-complex the boundary of every chain is a cycle.*

Or, finally,

7.1″. *If* $\mathfrak{K}$ *is an a-complex, then*

$$H_\Delta{}^r(\mathfrak{K}, \mathfrak{A}) \subseteq Z_\Delta{}^r(\mathfrak{K}, \mathfrak{A}). \tag{7.1″}$$

REMARK. Condition 7.1 = 7.1′ = 7.1″ is obviously necessary and sufficient for a cell complex $\mathfrak{K}$ to be an a-complex.

§7.2. Closed and open subcomplexes of an a-complex. [The reader who has omitted §4 can omit this subsection also.]

THEOREM 7.2. *Every closed and every open subcomplex of an a-complex is an a-complex.*

Proof. Let $\mathfrak{M}$ be an open subcomplex of an a-complex $\mathfrak{K}$; let t^r and t^{r-2} be any elements of dimensions r and $r - 2$, respectively, of $\mathfrak{M}$. All the elements t_j^{r-1} of $\mathfrak{K}$ which are incident with t^{r-2} in $\mathfrak{K}$ belong to $\mathfrak{M}$ and the incidence numbers $(t^r:t_j^{r-1})$ and $(t_j^{r-1}:t^{r-2})$ are the same in $\mathfrak{K}$ as they are in $\mathfrak{M}$. Consequently, the sum

$$\sum (t^r:t_j^{r-1})\,(t_j^{r-1}:t^{r-2}) \tag{7.21}$$

does not depend on whether it is taken in $\mathfrak{K}$ or in $\mathfrak{M}$. Since $\mathfrak{K}$ is an a-complex, the sum (7.21) is equal to zero in $\mathfrak{K}$ (or in $\mathfrak{M}$). Hence $\mathfrak{M}$ is an a-complex.

Now let $\mathfrak{M}$ be a closed subcomplex of the a-complex $\mathfrak{K}$ and let $t^r \in \mathfrak{M}$. By Theorem 6.610,

$$\Delta_{\mathfrak{M}}\Delta_{\mathfrak{M}}t^r = \Delta_{\mathfrak{K}}\Delta_{\mathfrak{K}}t^r = 0,$$

so that $\mathfrak{M}$ is an a-complex.

Since the complex consisting of the oriented elements of an unrestricted simplicial (or polyhedral) complex is an a-complex, 7.2 implies

7.20. *The complex consisting of the oriented elements of an arbitrary closed or open subcomplex of an unrestricted simplicial (or polyhedral) complex is an a-complex.*

§7.3. Weak homology of integral cycles; the dual coefficient domain. Let us return to 6.2, Example 3°. There we considered an integral cycle z^1 such that the cycle $2z^1$ was homologous to zero in $\mathfrak{K}$, although the cycle z^1 was itself not homologous to zero in $\mathfrak{K}$. In other words, the cycle z^1, while not an element of the group $H_\Delta{}^1(\mathfrak{K}, J)$, is an element of the division closure (see Appendix 2, 1.2) of $H_\Delta{}^1(\mathfrak{K}, J)$ in the group $Z_\Delta{}^1(\mathfrak{K}, J)$.

DEFINITION 7.31. Let $\mathfrak{K}$ be any a-complex. The division closure of $H_0{}^r(\mathfrak{K})$ in $Z_0{}^r(\mathfrak{K})$, which is obviously identical with the division closure of $H_0{}^r(\mathfrak{K})$ in $L_0{}^r(\mathfrak{K})$, is denoted by $\hat{H}_0{}^r(\mathfrak{K})$; the cycles of $\hat{H}_0{}^r(\mathfrak{K})$ are called *cycles weakly homologous to zero in* $\mathfrak{K}$.

If $z^r \in \hat{H}_0^r(\mathfrak{K})$, there exists a chain $x^{r+1} \in L_0^{r+1}(\mathfrak{K})$ such that $\Delta x^{r+1} = az^r$, where a is a natural number. Then $(1/a)x^{r+1} \in L^r(\mathfrak{K}, \mathfrak{R})$ and, obviously,

$$\Delta[(1/a)x^{r+1}] = z^r.$$

Hence, cycles weakly homologous to zero in $\mathfrak{K}$ may be defined as integral cycles which are boundaries of rational chains (i.e., chains with rational coefficients) $x^{r+1} \in L^r(\mathfrak{K}, \mathfrak{R})$.

We may say then

$$\hat{H}_0^r(\mathfrak{K}) = Z_0^r(\mathfrak{K}) \cap H_\Delta^r(\mathfrak{K}, \mathfrak{R}). \tag{7.31}$$

§8. Simplicial images of chains

§8.1. Simplicial images of oriented simplexes. Let t_β^r be an oriented simplex of a complex K_β,

$$t_\beta^r = | e_{\beta 0} \cdots e_{\beta r} |,$$

and let S_α^β be a simplicial mapping of K_β into a complex K_α. We shall define the integral chain

$$S_\alpha^\beta t_\beta^r \in L^r(K_\alpha)$$

by giving its values on all the oriented simplexes t_α^r of K_α as follows:

1°. If $S_\alpha^\beta | t_\beta^r | \neq | t_\alpha^r |$, put

$$(S_\alpha^\beta t_\beta^r \cdot t_\alpha^r) = 0.$$

2°. Let $t_\alpha^r = | e_{\alpha 0} \cdots e_{\alpha r} |$ and $S_\alpha^\beta | t_\beta^r | = | t_\alpha^r |$. Then

$$S_\alpha^\beta e_{\beta 0} = e_{\alpha i(0)}, \cdots, S_\alpha^\beta e_{\beta r} = e_{\alpha i(r)},$$

where $e_{\alpha i(0)}, \cdots, e_{\alpha i(r)}$ are the same as $e_{\alpha 0}, \cdots, e_{\alpha r}$ but, perhaps, in a different order; then, obviously,

$$| S_\alpha^\beta e_{\beta 0} \cdots S_\alpha^\beta e_{\beta r} | = \varepsilon t_\alpha^r,$$

where $\varepsilon = \pm 1$ is the sign of the permutation

$$\begin{pmatrix} 0 & 1 & \cdots & r \\ i(0) & i(1) & \cdots & i(r) \end{pmatrix}.$$

In this case we put

$$(S_\alpha^\beta t_\beta^r \cdot t_\alpha^r) = \varepsilon.$$

In accordance with this definition the chain $S_\alpha^\beta t_\beta^r = 0$ if there is no simplex t_α^r in K_α which satisfies the condition

$$S_\alpha^\beta | t_\beta^r | = | t_\alpha^r |,$$

that is, if the image of the simplex $| t_\beta^r |$ under the mapping S_α^β is a simplex of dimension less than r.

We may therefore say that $S_\alpha^\beta t_\beta^r = 0$ if, and only if, at least two of the vertices

$$S_\alpha^\beta e_{\beta 0}, \cdots, S_\alpha^\beta e_{\beta r}$$

coincide.

If, however, the vertices $S_\alpha^\beta e_{\beta 0}, \cdots, S_\alpha^\beta e_{\beta r}$ are all distinct, then

$$(S_\alpha^\beta e_{\beta 0} \cdots S_\alpha^\beta e_{\beta r})$$

is an r-simplex of K_α and $S_\alpha^\beta t_\beta^r$ is the oriented simplex

$$S_\alpha^\beta t_\beta^r = | S_\alpha^\beta e_{\beta 0} \cdots S_\alpha^\beta e_{\beta r} |.$$

§8.2. The homomorphism S_α^β of the group $L^r(K_\beta)$ into the group $L^r(K_\alpha)$ induced by a simplicial mapping S_α^β of a complex K_β into a complex K_α. Let

$$x_\beta^r = \sum_j a_j t_{\beta j}^r$$

be a chain of K_β over $\mathfrak{A}$. Let us set

$$S_\alpha^\beta x_\beta^r = \sum_j a_j S_\alpha^\beta t_{\beta j}^r. \tag{8.21}$$

We shall call the chain $S_\alpha^\beta x_\beta^r \in L^r(K_\alpha)$ the image of the chain $x_\beta^r \in L^r(K_\beta)$ under the simplicial mapping S_α^β.

It is clear that for two chains $x_{\beta 1}^r \in L^r(K_\beta)$, $x_{\beta 2}^r \in L^r(K_\beta)$,

$$S_\alpha^\beta(x_{\beta 1}^r \pm x_{\beta 2}^r) = S_\alpha^\beta x_{\beta 1}^r \pm S_\alpha^\beta x_{\beta 2}^r;$$

hence the mapping of $L^r(K_\beta)$ into $L^r(K_\alpha)$ defined by (8.21) is a homomorphism.

We shall calculate the value of the chain

$$S_\alpha^\beta x_\beta^r = \sum_j a_j S_\alpha^\beta t_{\beta j}^r$$

on any oriented simplex $t_\alpha^r = t_{\alpha h}^r$ of K_α. To this end, denoting an arbitrary but definite orientation of each simplex $T_{\alpha i}^r$ of K_α by $t_{\alpha i}^r$, we obtain

$$(S_\alpha^\beta x_\beta^r \cdot t_{\alpha i}^r) = \sum_j a_j (S_\alpha^\beta t_{\beta j}^r \cdot t_{\alpha i}^r)$$

or, since $a_j = (x_\beta^r \cdot t_{\beta j}^r)$,

$$(S_\alpha^\beta x_\beta^r \cdot t_{\alpha i}^r) = \sum_j (S_\alpha^\beta t_{\beta j}^r \cdot t_{\alpha i}^r)(x_\beta^r \cdot t_{\beta j}^r). \tag{8.22}$$

This formula can be somewhat simplified by choosing the orientations $t_{\beta j}^r$ of the simplexes $T_{\beta j}^r \in K_\beta$ so that if

$$S_\alpha^\beta T_{\beta j}^r = | t_{\alpha i}^r |,$$

then

$$S_\alpha^\beta t_{\beta j}^r = t_{\alpha i}^r.$$

For this choice of the orientations $t_{\beta j}{}^r$ (8.22) becomes

$$(8.23) \qquad (S_\alpha{}^\beta x_\beta{}^r \cdot t_{\alpha i}{}^r) = \sum_j (x_\beta{}^r \cdot t_{\beta j}{}^r),$$

where the summation is extended over all $t_{\beta j}{}^r$ such that

$$S_\alpha{}^\beta \mid t_{\beta j}{}^r \mid = \mid t_{\alpha i}{}^r \mid.$$

Giving the orientations of the simplexes to the corresponding ordered skeletons we may, finally, rewrite (8.23) as

$$(8.24) \qquad S_\alpha{}^\beta x_\beta{}^r \mid e_{\alpha 0} \cdots e_{\alpha r} \mid = \sum x_\beta{}^r \mid e_{\beta j(0)} \cdots e_{\beta j(r)} \mid,$$

where $\mid e_{\alpha 0}, \cdots, e_{\alpha r} \mid$ is any ordered skeleton of K_α and the sum on the right is taken over all ordered skeletons $\mid e_{\beta j(0)}, \cdots, e_{\beta j(r)} \mid$ of K_β for which

$$S_\alpha{}^\beta e_{\beta j(0)} = e_{\alpha 0}, \cdots, S_\alpha{}^\beta e_{\beta j(r)} = e_{\alpha r}.$$

§8.3. Commutativity of the operators Δ and $S_\alpha{}^\beta$. We shall prove that the identity

$$(8.3) \qquad \Delta S_\alpha{}^\beta x_\beta{}^r = S_\alpha{}^\beta \Delta x_\beta{}^r$$

holds for every chain $x_\beta{}^r \in L^r(K_\beta)$.

In view of the linearity of the operators $S_\alpha{}^\beta$ and Δ it is sufficient to prove (8.3) for the case $x_\beta{}^r = t_{\beta j}{}^r$, $\mid t_{\beta j}{}^r \mid \in K_\beta$.

Let $t_\beta{}^r = \mid e_{\beta 0} \cdots e_{\beta r} \mid$. We shall consider two cases:

1°. $S_\alpha{}^\beta t_\beta{}^r = 0$;

2°. $S_\alpha{}^\beta t_\beta{}^r \neq 0$.

Case 1°. $S_\alpha{}^\beta t_\beta{}^r = 0$. Then

$$(8.30) \qquad S_\alpha{}^\beta e_{\beta i} = S_\alpha{}^\beta e_{\beta k}$$

for at least one pair of vertices $e_{\beta i}$, $e_{\beta k}$ of the simplex $\mid t_\beta{}^r \mid$. In this case $\Delta S_\alpha{}^\beta t_\beta{}^r = 0$ and we must show that $S_\alpha{}^\beta \Delta t_\beta{}^r = 0$ also. We subdivide the proof of this into two subcases:

1a) There is precisely one pair of vertices $e_{\beta i}$, $e_{\beta k}$ of the simplex $t_\beta{}^r$ satisfying (8.30).

1b) There are at least two such pairs of vertices.

In Case 1a) we may assume without loss of generality that the unique pair of vertices in question is the pair $e_{\beta 0}$, $e_{\beta 1}$, so that $S_\alpha{}^\beta e_{\beta 0} = S_\alpha{}^\beta e_{\beta 1}$; while $S_\alpha{}^\beta e_{\beta i} \neq S_\alpha{}^\beta e_{\beta k}$ for every pair of vertices $e_{\beta i}$, $e_{\beta k}$ distinct from $e_{\beta 0}$, $e_{\beta 1}$. Then

$$(8.301) \qquad \Delta t_\beta{}^r = \sum_k (-1)^k \mid e_{\beta 0} \cdots \hat{e}_{\beta k} \cdots e_{\beta r} \mid;$$

where

$$S_\alpha{}^\beta \mid e_{\beta 0} \cdots \hat{e}_{\beta k} \cdots e_{\beta r} \mid = 0$$

for $k \neq 0$, $k \neq 1$, since $e_{\beta 0}$, $e_{\beta 1}$ are among the vertices

$$e_{\beta 0}, \cdots, \hat{e}_{\beta k}, \cdots, e_{\beta r} \quad \text{and} \quad S_\alpha{}^\beta e_{\beta 0} = S_\alpha{}^\beta e_{\beta 1}.$$

Consequently, (8.301) yields

$$S_\alpha^\beta \Delta t_\beta^r = S_\alpha^\beta \mid e_{\beta 1} e_{\beta 2} \cdots e_{\beta r} \mid - S_\alpha^\beta \mid e_{\beta 0} e_{\beta 2} \cdots e_{\beta r} \mid$$
$$= \mid S_\alpha^\beta e_{\beta 1} S_\alpha^\beta e_{\beta 2} \cdots S_\alpha^\beta e_{\beta r} \mid - \mid S_\alpha^\beta e_{\beta 0} S_\alpha^\beta e_{\beta 2} \cdots S_\alpha^\beta e_{\beta r} \mid.$$

But $S_\alpha^\beta e_{\beta 0} = S_\alpha^\beta e_{\beta 1}$ and so

$$\mid S_\alpha^\beta e_{\beta 1} S_\alpha^\beta e_{\beta 2} \cdots S_\alpha^\beta e_{\beta r} \mid = \mid S_\alpha^\beta e_{\beta 0} S_\alpha^\beta e_{\beta 2} \cdots S_\alpha^\beta e_{\beta r} \mid,$$

whence

$$S_\alpha^\beta \Delta t_\beta^r = 0.$$

In Case 1b) the vertices $S_\alpha^\beta e_{\beta 0}, \cdots, S_\alpha^\beta e_{\beta(k-1)}, S_\alpha^\beta e_{\beta(k+1)}, \cdots, S_\alpha^\beta e_{\beta r}$ contain at least one identical pair for arbitrary k; consequently,

$$S_\alpha^\beta \mid e_{\beta 0} \cdots \hat{e}_{\beta k} \cdots e_{\beta r} \mid = 0$$

and

$$S_\alpha^\beta \Delta t_\beta^r = \sum (-1)^k S_\alpha^\beta \mid e_{\beta 0} \cdots \hat{e}_{\beta k} \cdots e_{\beta r} \mid = 0.$$

This proves (8.3) in Case 1°.

Case 2°. If $S_\alpha^\beta t_\beta^r \neq 0$, then

$$S_\alpha^\beta t_\beta^r = \mid S_\alpha^\beta e_{\beta 0} \cdots S_\alpha^\beta e_{\beta r} \mid,$$
$$\Delta S_\alpha^\beta t_\beta^r = \sum (-1)^k \mid S_\alpha^\beta e_{\beta 0} \cdots \widehat{S_\alpha^\beta e_{\beta k}} \cdots S_\alpha^\beta e_{\beta r} \mid$$

and

$$\Delta t_\beta^r = \sum (-1)^k \mid e_{\beta 0} \cdots \hat{e}_{\beta k} \cdots e_{\beta r} \mid,$$
$$S_\alpha^\beta \Delta t_\beta^r = \sum (-1)^k S_\alpha^\beta \mid e_{\beta 0} \cdots \hat{e}_{\beta k} \cdots e_{\beta r} \mid$$
$$= \sum (-1)^k \mid S_\alpha^\beta e_{\beta 0} \cdots \widehat{S_\alpha^\beta e_{\beta k}} \cdots S_\alpha^\beta e_{\beta r} \mid,$$

i.e.,

$$\Delta S_\alpha^\beta t_\beta^r = S_\alpha^\beta \Delta t_\beta^r.$$

This completes the proof of (8.3).

It follows from (8.3) that if x_β^r is a cycle, then $S_\alpha^\beta x_\beta^r$ is also a cycle; if $x_\beta^r \sim 0$ in K_β, then $S_\alpha^\beta x_\beta^r \sim 0$ in K_α; that is,

8.31. *The homomorphism* S_α^β *of* $L^r(K_\beta)$ *into* $L^r(K_\alpha)$ *maps* $Z^r(K_\beta)$ *into* $Z^r(K_\alpha)$ *and* $H^r(K_\beta)$ *into* $H^r(K_\alpha)$.

§8.4. The case of open subcomplexes. Let K_α and K_β be two unrestricted simplicial complexes and let $K_{\alpha 0} \subset K_\alpha$, $K_{\beta 0} \subset K_\beta$ be closed subcomplexes of K_α and K_β. For brevity we set

$$G_\alpha = K_\alpha \setminus K_{\alpha 0}, \qquad G_\beta = K_\beta \setminus K_{\beta 0}.$$

Let S_α^β be a simplicial mapping of K_β into K_α such that

(8.40) $$S_\alpha^\beta(K_{\beta 0}) \subseteq K_{\alpha 0}\,.$$

We shall associate with every chain $x_\beta^r \in L^r(G_\beta)$ the chain $G_\alpha S_\alpha^\beta x_\beta^r \in L^r(G_\alpha)$. This mapping induces a homomorphism $G_\alpha S_\alpha^\beta$ of $L^r(G_\beta)$ into $L^r(G_\alpha)$. We shall prove that

8.4. *The homomorphism* $G_\alpha S_\alpha^\beta$ *commutes with the boundary* operator Δ:

(8.4) $$G_\alpha S_\alpha^\beta \Delta_{G_\beta} x_\beta^r = \Delta_{G_\alpha} G_\alpha S_\alpha^\beta x_\beta^r.$$

Indeed, since G_β is an open subcomplex of K_β, 6.51 implies that

$$\Delta_{G_\beta} x_\beta^r = G_\beta \Delta x_\beta^r = (K_\beta \setminus K_{\beta 0})\Delta x_\beta^r = \Delta x_\beta^r - K_{\beta 0}\Delta x_\beta^r,$$

so that

(8.41) $$S_\alpha^\beta \Delta_{G_\beta} x_\beta^r = S_\alpha^\beta \Delta x_\beta^r - S_\alpha^\beta K_{\beta 0} \Delta x_\beta^r.$$

(8.40) implies that $S_\alpha^\beta K_{\beta 0}\Delta x_\beta^r \in L^{r-1}(K_{\alpha 0})$, so that $G_\alpha S_\alpha^\beta K_{\beta 0} \Delta x_\beta^r = 0$ and

(8.42) $$G_\alpha S_\alpha^\beta \Delta_{G_\beta} x_\beta^r = G_\alpha S_\alpha^\beta \Delta x_\beta^r.$$

On the other hand

(8.43) $$\Delta_{G_\alpha} G_\alpha S_\alpha^\beta x_\beta^r = G_\alpha \Delta S_\alpha^\beta x_\beta^r = G_\alpha S_\alpha^\beta \Delta x_\beta^r.$$

(8.4) follows from (8.42) and (8.43).

An immediate consequence of Theorem 8.4 is

8.40. *The homomorphism* $G_\alpha S_\alpha^\beta$ *of* $L^r(G_\beta)$ *into* $L^r(G_\alpha)$ *maps* $Z^r(G_\beta)$ *into* $Z^r(G_\alpha)$ *and* $H^r(G_\beta)$ *into* $H^r(G_\alpha)$.

§9. Auxiliary constructions

§9.1. Cone over a chain. Let K be a skeleton complex and let $<oK>$ be a cone over K, i.e., a cone with vertex o and base K (see IV, 2.3). *The oriented simplex* (see 3.1) (ot^r) *of the complex* $<oK>$ is called *the cone over the oriented simplex* t^r.

If

$$x^r = \sum a_i t_i^r$$

is a chain of K, the chain

(9.11) $$(ox^r) = \sum a_i (ot_i^r)$$

of the complex $<oK>$ is referred to as *the cone over the chain* x^r.

For $r > 0$ it is easily seen that

(9.12) $$\Delta_{<oK>}(ot^r) = t^r - (o\Delta_K t^r),$$

whence

(9.13) $$\Delta_{<oK>}(ox^r) = x^r - (o\Delta_K x^r).$$

In particular, if z^r is a cycle, then

$$\Delta_{<oK>}(oz^r) = z^r. \tag{9.14}$$

For $r = 0$,

$$\Delta_{<oK>}(ot^0) = t^0 - (o), \tag{9.120}$$

and if $x^0 = \sum a_i t_i^0$,

$$\Delta_{<oK>}(ox^0) = x^0 - (\textstyle\sum_i a_i)\,(o). \tag{9.130}$$

REMARK. For an open cone oK (IV, 2.3) we have the analogous formulas:

$$\Delta_{oK}(ot^r) = -o\Delta_K t^r, \tag{9.12'}$$

$$\Delta_{oK}(ox^r) = -o\Delta_K x^r, \tag{9.13'}$$

$$\Delta_{oK}(oz^r) = 0, \tag{9.14'}$$

$$\Delta_{oK}(ot^0) = -(o), \tag{9.120'}$$

$$\Delta_{oK}(ox^0) = -(\textstyle\sum_i a_i)\,(o). \tag{9.130'}$$

§9.2. Application of the constructions of 9.1. Let r be a natural number and let K be an unrestricted skeleton complex with the property

9.210. Every $r + 2$ vertices of K form a skeleton of K.

Under these conditions we shall prove

$$Z_\Delta^r(K) = H_\Delta^r(K), \tag{9.21}$$

that is, every r-cycle of K is homologous to zero in K.

In the proof of (9.21) we shall denote by K^r the closed subcomplex of K consisting of all the r-skeletons of K and all their faces.

Let us construct the cone $<oK^r>$ and associate with each vertex e of the complex $<oK^r>$ a vertex Se of K in accordance with the following rule:

1. If $e \in K^r$, set $Se = e$.
2. Let So be an arbitrarily chosen fixed vertex e_0 of K.

By 9.210, S is a simplicial mapping of $<oK^r>$ into K.

Now let z^r be an arbitrary r-cycle of K. z^r is at the same time a cycle of the complexes K^r and $<oK^r>$ and is homologous to zero in $<oK^r>$ by (9.14); consequently, by 8.31, the cycle $Sz^r = z^r$ is homologous to zero in K and (9.21) follows.

SPECIAL CASES. 1) If $K = |T^n|$, then (9.21) holds for every r.

Since the complex $|T^n|$ satisfies 9.210 for $r \leq n - 1$, (9.21) holds; for $r \geq n$ (9.21) follows from the fact that $Z_\Delta^r(|T^n|)$ is the null group for $r \geq n$.

2) Let $K = K(\Gamma^n)$, where Γ^n is a convex open set of R^n [for the definition of the complex $K(\Gamma^n)$ see IV, 1.5, Example 2]. Condition 9.210 is obviously satisfied for every r and hence (9.21) is also true for arbitrary r.

3) Let $K = \dot{T}^n = | T^n | \setminus T^n$ be the complex consisting of all the proper faces of an n-simplex T^n; 9.210 is satisfied for $r < n - 1$; consequently,

$$Z_\Delta{}^r(\dot{T}^n) = H_\Delta{}^r(\dot{T}^n)$$

for $0 < r < n - 1$.

4) *Application to the elementary subdivisions of simplexes* (see IV, 4.3 for terminology and notation). We shall apply the notion of a cone over a chain to prove the following proposition which is required in Chapter X.

9.22. *Let V^n be an elementary subdivision of an n-simplex $T^n = (e_0 \cdots e_n)$ relative to a face $T^p = (e_0 \cdots e_p)$. Then every r-cycle $(0 \leq r \leq n - 1)$ of V^n is homologous to zero in V^n, i.e.,*

$$Z^r(V^n) = H^r(V^n).$$

Proof. Let z^r, $0 < r \leq n - 1$, be any r-cycle of V^n (there are no 0-elements in V^n). Let Q^r denote the combinatorial closure of the complex consisting of all the simplexes on which the chain z^r does not vanish.

The cone

$$K_\beta = <oQ^r>$$

contains the cone $K_{\beta 0}$ over $Q^r \cap | V^n | \setminus V^n$ as a closed subcomplex. The complementary subcomplex $K_\beta \setminus K_{\beta 0}$ will be denoted by G_β.

Denoting the boundary operator in K_β by Δ, we have

$$\Delta(oz^r) = z^r - o\Delta z^r,$$

where $o\Delta z^r$ is in $K_{\beta 0}$, so that

$$\Delta_\beta(oz^r) = z^r, \tag{9.221}$$

where Δ_β denotes the boundary operator in G_β. Let us now set $K_\alpha = | V^n |$, $S_\alpha{}^\beta e = e$ (where e is the center of T^p; see IV, 4.3, Remark 1), $S_\alpha{}^\beta e_i = e_i$, $S_\alpha{}^\beta o = e_0$. In virtue of IV, 4.3, Remark 3, none of the simplexes of Q^r have among their vertices more than $p - 1$ vertices of T^p. Hence $S_\alpha{}^\beta$ defines a simplicial mapping, also denoted by $S_\alpha{}^\beta$, of K_β into K_α, which maps $K_{\beta 0}$ into $K_{\alpha 0} = | V^n | \setminus V^n$ (this assertion follows immediately from the definition of $K_{\beta 0}$ and IV, 4.3, Remarks 1 and 2). Consequently, replacing $x_\beta{}^r$ by oz^r and putting $G_\alpha = V^n$ in (8.4), we get

$$V^n S_\alpha{}^\beta \Delta_\beta(oz^r) = \Delta_n V^n S_\alpha{}^\beta(oz^r),$$

where we have written the boundary operator in V^n as Δ_n. Hence, in virtue of (9.221),

$$V^n S_\alpha{}^\beta z^r = \Delta_n V^n S_\alpha{}^\beta(oz^r);$$

or

$$z^r = \Delta_n V^n S_\alpha^{\ \beta}(oz^r),$$

since it is clear that

$$S_\alpha^{\ \beta} z^r = z^r, \qquad V^n z^r = z^r.$$

This completes the proof of 9.22.

§9.3. Prism over a chain. Let K_0 be a finite unrestricted skeleton complex; the vertices of K_0 will be numbered as:

(9.30) $$a_1, \cdots, a_s.$$

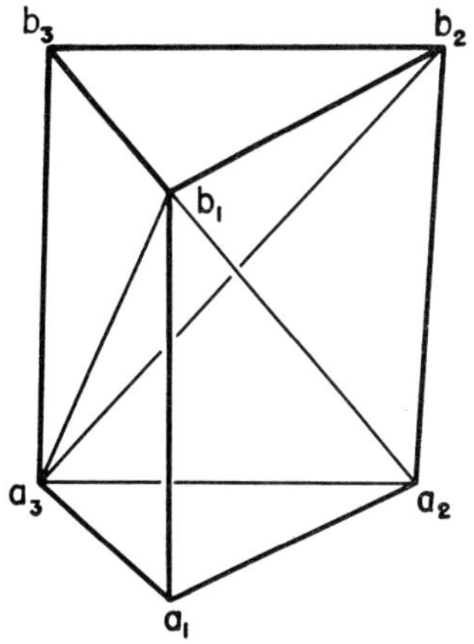

FIG. 111

Using this enumeration of the vertices we construct the prism $K_{[01]}$ over K_0 (see IV, 2.4).

Let the vertices of the upper base of the prism $K_{[01]}$, corresponding to the vertices (9.30), be

(9.31) $$b_1, \cdots, b_s.$$

If $T_{0h}^{\ r} = (a_{h(0)} \cdots a_{h(r)}) \in K_0$, $h(0) < \cdots < h(r)$, set $t_{0h}^{\ r} = | a_{h(0)} \cdots a_{h(r)} |$, $t_{1h}^{\ r} = | b_{h(0)} \cdots b_{h(r)} |$. The oriented simplex $t_{1h}^{\ r}$ is referred to as the *projection* of the oriented simplex $t_{0h}^{\ r}$ of K_0 on the upper base of $K_{[01]}$. If $x_0^{\ r} = \sum c_h t_{0h}^{\ r} \in L^r(K_0)$, set $x_1^{\ r} = \sum c_h t_{1h}^{\ r}$ and call the chain $x_1^{\ r}$ the *projection* of $x_0^{\ r}$ on the upper base of $K_{[01]}$.

Let us now put, for $i = 0, \cdots, r$ (Fig. 111),

(9.32) $$\Pi_i t_{0h}^{\ r} = | b_{h(0)} \cdots b_{h(i)} a_{h(i)} \cdots a_{h(r)} |;$$

and further

$$\Pi t_{0h}^{\ r} = \sum_{i=0}^{r} (-1)^i \Pi_i t_{0h}^{\ r}.$$

Finally, if $x_0^{\ r} = \sum c_h t_{0h}^{\ r}$, let

(9.33) $$\Pi x_0^{\ r} = \sum c_h \Pi t_{0h}^{\ r} \in L^{r+1}(K_{[01]}).$$

The chain $\Pi x_0^{\ r}$ is known as *the prism over the chain* $x_0^{\ r} \in L^r(K_0)$ (in the prism $K_{[01]}$). We shall prove the *fundamental identity*

(9.34) $$\Delta \Pi x_0^{\ r} = x_0^{\ r} - x_1^{\ r} - \Pi \Delta x_0^{\ r}.$$

In view of the linearity of both operators Δ and Π, it is enough to prove (9.34) for the special case $x_0^{\ r} = t_0^{\ r}$.

For brevity let

$$t_0^r = | a_0 \cdots a_r |;$$

then

$$t_1^r = | b_0 \cdots b_r |.$$

Let us set

$$t_{0i}^{r-1} = | a_0 \cdots \hat{a}_i \cdots a_r |, \qquad {}_jt^{r+1} = | b_0 \cdots b_j a_j \cdots a_r |,$$

$$(9.35) \qquad {}_jt_i^r = | b_0 \cdots b_j a_j \cdots \hat{a}_i \cdots a_r | \; (i \geq j),$$

$${}_jt^{r,i} = | b_0 \cdots \hat{b}_i \cdots b_j a_j \cdots a_r | \; (i \leq j).$$

Since the vertex b_j is in the $(j+1)$st and a_j is in the $(j+2)$nd position in ${}_jt^{r+1}$,

$$(9.36) \qquad \begin{aligned} \Delta_j t^{r+1} &= \textstyle\sum_{i=1}^{j} (-1)^i {}_jt^{r,i} - \sum_{i=j}^{r} (-1)^i {}_jt_i^r \\ &= (-1)^j({}_jt^{r,j} - {}_jt_j^r) + \textstyle\sum_{i<j} (-1)^i {}_jt^{r,i} - \sum_{i>j} (-1)^i {}_jt_i^r. \end{aligned}$$

Since ($\sum_j$ denotes $\sum_{j=0}^{r}$)

$$\Pi t_0^r = \textstyle\sum_j (-1)^j {}_jt^{r+1},$$

it follows from (9.36) that

$$\begin{aligned} \Delta\Pi t_0^r = \textstyle\sum_j (-1)^j \Delta_j t^{r+1} &= \textstyle\sum_j {}_jt^{r,j} - \sum_j {}_jt_j^r \\ &+ \textstyle\sum_j (-1)^j[\sum_{i<j} (-1)^i {}_jt^{r,i} - \sum_{i>j} (-1)^i {}_jt_i^r]. \end{aligned}$$

But

$${}_jt_j^r = {}_{j+1}t^{r,j+1}, \qquad {}_0t^{r,0} = t_0^r, \qquad {}_rt_r^r = t_1^r.$$

Hence

$$\textstyle\sum_j {}_jt^{r,j} - \sum_j {}_jt_j^r = t_0^r - t_1^r$$

and so

$$\begin{aligned} \Delta\Pi t_0^r &= t_0^r - t_1^r + \textstyle\sum_j (-1)^j[\sum_{i<j} (-1)^i {}_jt^{r,i} - \sum_{i>j} (-1)^i {}_jt_i^r] \\ &= t_0^r - t_1^r - \textstyle\sum_j [\sum_{i>j} (-1)^{i+j} {}_jt_i^r - \sum_{i<j} (-1)^{i+j} {}_jt^{r,i}]. \end{aligned}$$

It is required to prove that

$$\Delta\Pi t_0^r = t_0^r - t_1^r - \Pi\Delta t_0^r.$$

Hence, it remains to be proved that

$$\Pi\Delta t_0^r = \textstyle\sum_j [\sum_{i>j} (-1)^{i+j} {}_jt_i^r - \sum_{i<j} (-1)^{i+j} {}_jt^{r,i}].$$

But since $\Delta t_0^r = \sum_i (-1)^i t_{0i}^{r-1}$ and the operator Π is linear, it is sufficient to prove that

$$\Pi t_{0i}^{r-1} = \sum_{j=1}^{i-1} (-1)^j {}_j t_i^r - \sum_{j=i+1}^{r} (-1)^j {}_j t^{r,i}.$$

In virtue of (9.32) and (9.35),

$$\Pi_j\, t_{0i}^{r-1} = {}_j t_i^r, \qquad j = 0, \cdots, i-1,$$

$$\Pi_j\, t_{0i}^{r-1} = {}_{j+1} t^{r,i}, \qquad j = i, \cdots, r-1.$$

But, by definition,

$$\Pi t_{0i}^{r-1} = \sum_j (-1)^j \Pi_j\, t_{0i}^{r-1},$$

so that

$$\Pi t_{0i}^{r-1} = \sum_{j=0}^{i-1} (-1)^j {}_j t_i^r + \sum_{j=i}^{r-1} (-1)^j {}_{j+1} t^{r,i}$$
$$= \sum_{j=0}^{i-1} (-1)^j {}_j t_i^r - \sum_{j=i+1}^{r} (-1)^j {}_j t^{r,i}.$$

This completes the proof of (9.34).

If $z_0^r \in Z^r(K_0)$ is a cycle, then $\Pi\Delta z_0^r = 0$, (9.34) becomes

(9.37) $$\Delta\Pi z_0^r = z_0^r - z_1^r,$$

and we have

9.37. *Every cycle of K_0 is homologous in $K_{[01]}$ to its projection on the upper base of $K_{[01]}$.*

§9.4. Application to simplicial mappings. As another application, we shall prove the following theorem, required in Chapter XIV.

THEOREM 9.4. *Let S_α^β and $\bar{S}_\alpha^\beta$ be two simplicial mappings of a complex K_β into a complex K_α with the following property: for every simplex $T_\beta \in K_\beta$ there exists a simplex $T_\alpha \in K_\alpha$ such that $S_\alpha^\beta T_\beta$ and $\bar{S}_\alpha^\beta T_\beta$ are both faces of T_α. Then, if $z_\beta^r \in Z_\Delta^r(K_\beta)$ is any cycle of K_β, the cycles $S_\alpha^\beta z_\beta^r$ and $\bar{S}_\alpha^\beta z_\beta^r$ are homologous in K_α.*

Proof. Let us index the vertices of K_β in a definite order $e_{\beta 1}, \cdots, e_{\beta s(\beta)}$ and construct, on the basis of this enumeration of the vertices, the prism $K_{\beta[01]}$ over K_β (see IV, 2.4).

Let us associate with each vertex $e_{\beta 0i} = e_{\beta i}$ of the lower base $K_{\beta 0} = K_\beta$ of the prism $K_{\beta[01]}$ the vertex $S_\alpha^\beta e_{\beta i}$ of K_α and with each vertex $e_{\beta 1i}$ (the projection of the vertex $e_{\beta i}$ on the upper base) of the upper base $K_{\beta 1}$ the vertex $\bar{S}_\alpha^\beta e_{\beta i}$ of K_α.

We shall show that the mapping thus defined is a simplicial mapping S of the prism $K_{\beta[01]}$ into K_α. For, if

$$T = \{e_{\beta 00}, \cdots, e_{\beta 0k} e_{\beta 1k}, \cdots, e_{\beta 1n}\}$$

is an arbitrary skeleton of the prism $K_{\beta[01]}$, then, according to the definition of prisms, $\{e_{\beta 0}, \cdots, e_{\beta n}\}$ is a skeleton of K_β. The mapping S transforms the skeleton

$$T = \{e_{\beta 00}, \cdots, e_{\beta 0k}, e_{\beta 1k}, \cdots, e_{\beta 1n}\}$$

into the set of vertices

$$S(T) = (S_\alpha^\beta e_{\beta 0}, \cdots, S_\alpha^\beta e_{\beta k}, \bar{S}_\alpha^\beta e_{\beta k}, \cdots, \bar{S}_\alpha^\beta e_{\beta n})$$

of K_α. But this set is a subset of the set

$$(S_\alpha^\beta e_{\beta 0}, \bar{S}_\alpha^\beta e_{\beta 0}, \cdots, S_\alpha^\beta e_{\beta n}, \bar{S}_\alpha^\beta e_{\beta n}),$$

which is a skeleton of K_α by the hypothesis of the theorem. Consequently, $S(T)$ is also a skeleton of K_α.

For the simplicial mapping S we obviously have

$$S z_{\beta 0}^r = S_\alpha^\beta z_\beta^r, \qquad S z_{\beta 1}^r = \bar{S}_\alpha^\beta z_\beta^r.$$

Since (by 9.37)

$$z_{\beta 0}^r \sim z_{\beta 1}^r \quad \text{in} \quad K_{\beta[01]},$$

it follows that

$$S z_{\beta 0}^r \sim S z_{\beta 1}^r \quad \text{in} \quad K_\alpha,$$

i.e., $S_\alpha^\beta z_\beta^r \sim \bar{S}_\alpha^\beta z_\beta^r$ in K_α. This proves the theorem.

Addendum to Chapter VII

THE a-COMPLEX OF THE ORIENTED ELEMENTS OF A POLYHEDRAL COMPLEX

§1. The incidence numbers of a half-space and a plane, of a convex polyhedral domain and one of its faces. Let $|E^n|$ be a half-space of the Euclidean n-space $|R^n|$ determined by a plane $|E^{n-1}|$, or a convex polyhedral domain of which $|E^{n-1}|$ is a face. Let $E^n(E^{n-1})$ be any orientation of $|E^n|(|E^{n-1}|)$. We shall define the incidence number $(E^n:E^{n-1})$. To this end, let T^n be a simplex in $|E^n|$ with a face T^{n-1} in $|E^{n-1}|$. Let $t^n(t^{n-1})$ be the orientation of $T^n(T^{n-1})$ coherent with the orientation $E^n(E^{n-1})$. We set

$$(1.11) \qquad (E^n:E^{n-1}) = (t^n:t^{n-1})$$

by definition. It is easily seen that (1.11) depends only on the orientations E^n and E^{n-1} and is independent of the choice of the simplex T^n.

Let $|t^{n-1}|$ be any simplex in $|E^{n-1}|$ oriented coherently with E^{n-1}. Let e be a point in the interior of $|E^n|$. Then $(E^n:E^{n-1}) = +1$ (-1) if the orientations (et^{n-1}) and E^n are coherent (noncoherent), that is,

$$(1.12) \qquad (E^n:E^{n-1}) = (et^{n-1})E^n.$$

But if the orientation t^{n-1} is opposite to E^{n-1}, then

$$(1.13) \qquad (E^n:E^{n-1}) = -(et^{n-1})E^n;$$

consequently,

1.1. *If t^{n-1} is any orientation of a simplex $T^{n-1} \subset |E^{n-1}|$ and e is a point in the interior of $|E^n|$, then*

$$(1.1) \qquad (E^n:E^{n-1}) = (et^{n-1})E^n \cdot t^{n-1}E^{n-1}.$$

If E^n is an oriented convex polyhedral domain, and E^{n-1} is an arbitrarily oriented face of E^n, then, as we shall prove, *the incidence numbers $(E^n:E^{n-1})$ have the same properties as the incidence numbers of oriented simplexes.*

Property 1.

$$(-E^n:E^{n-1}) = (E^n:-E^{n-1}) = -(E^n:E^{n-1}).$$

This property needs no proof.

Property 2.

$$(1.14) \qquad \sum_i (E^n:E_i^{n-1})\,(E_i^{n-1}:E^{n-2}) = 0,$$

where E^n and E^{n-2} are arbitrary orientations of the polyhedral domain

$|E^n|$ and one of its $(n-2)$-faces $|E^{n-2}|$, and the sum is taken over all the $(n-1)$-faces $|E_i^{n-1}|$, with E_i^{n-1} any one of the orientations of the face $|E_i^{n-1}|$.

We shall prove this property.

Since only two of the $(n-1)$-faces of the polyhedral domain $|E^n|$, say $|E_0^{n-1}|$ and $|E_1^{n-1}|$, have $|E^{n-2}|$ on their boundary, (1.14) may be rewritten as

$$(1.15) \qquad (E^n:E_0^{n-1})\,(E_0^{n-1}:E^{n-2}) + (E^n:E_1^{n-1})\,(E_1^{n-1}:E^{n-2}) = 0.$$

Since the left side of (1.15) remains unchanged if E_0^{n-1} is replaced by $-E_0^{n-1}$ or E_1^{n-1} by $-E_1^{n-1}$, it is enough to prove (1.15) for a single choice of the orientations E_0^{n-1}, E_1^{n-1}. Let us choose the orientations E_0^{n-1} and E_1^{n-1} so that

$$(E_0^{n-1}:E^{n-2}) = +1, \qquad (E_1^{n-1}:E^{n-2}) = +1;$$

then (1.15) becomes simply

$$(1.150) \qquad (E^n:E_0^{n-1}) + (E^n:E_1^{n-1}) = 0.$$

To prove (1.150), let us take in $|E^{n-2}|$ a simplex T^{n-2} with orientation $t^{n-2} = |e_2e_3 \cdots e_n|$ coherent with the orientation E^{n-2}.

Let $e_0(e_1)$ be a point inside $|E_0^{n-1}|$ ($|E_1^{n-1}|$) and denote by $t_0^{n-1}(t_1^{n-1})$ the orientation of the simplex $T_0^{n-1} = (e_0e_2e_3 \cdots e_n)$ $[T_1^{n-1} = (e_1e_2e_3 \cdots e_n)]$ coherent with the orientation $E_0^{n-1}(E_1^{n-1})$. Since

$$(t_0^{n-1}:t^{n-2}) = (E_0^{n-1}:E^{n-2}) = 1$$

and

$$(t_1^{n-1}:t^{n-2}) = (E_1^{n-1}:E^{n-2}) = 1$$

and

$$t^{n-2} = |e_2e_3 \cdots e_n|,$$

it follows that

$$t_0^{n-1} = |e_0e_2e_3 \cdots e_n|,$$

$$t_1^{n-1} = |e_1e_2e_3 \cdots e_n|.$$

Finally, let e be a point inside $|E^n|$ and denote by t_0^n, t_1^n the orientations of the simplexes $T_0^n = (ee_0e_2 \cdots e_n)$ and $T_1^n = (ee_1e_2 \cdots e_n)$ coherent with the orientation E^n. Then

$$(E^n:E_0^{n-1}) = (t_0^n:t_0^{n-1}), \qquad (E^n:E_1^{n-1}) = (t_1^n:t_1^{n-1}).$$

Since the points e_0 and e_1, and consequently the simplexes T_0^n and T_1^n, are on different sides of the plane $| eT^{n-2} |$, the orientations $| e_0ee_2e_3 \cdots e_n |$ and $| e_1ee_2e_3 \cdots e_n |$ of these simplexes are opposite; but the orientations t_0^n and t_1^n are coherent, so that if

$$t_0^n = \varepsilon \, | e_0ee_2e_3 \cdots e_n |,$$

then

$$t_1^n = -\varepsilon \, | e_1ee_2e_3 \cdots e_n |$$

and

$$\begin{aligned}(E^n : E_0^{n-1}) &= (t_0^n : t_0^{n-1}) = \varepsilon(| e_0ee_2e_3 \cdots e_n | : | e_0e_2e_3 \cdots e_n |) \\ &= \varepsilon(-1) = -\varepsilon, \\ (E^n : E_1^{n-1}) &= (t_1^n : t_1^{n-1}) = -\varepsilon(| e_1ee_2e_3 \cdots e_n | : | e_1e_2e_3 \cdots e_n |) \\ &= (-\varepsilon)(-1) = \varepsilon,\end{aligned}$$

i.e.,

$$(E^n : E_0^{n-1}) = -(E^n : E_1^{n-1}).$$

This completes the proof.

§2. Since incidence numbers have been defined for convex polyhedral domains and both properties of the incidence numbers of simplexes remain true in this case, *the oriented elements of a polyhedral complex K form an a-complex*; the chains, cycles, etc. of this a-complex are referred to as the chains, cycles, etc. of the polyhedral complex K, or simply as *polyhedral chains, polyhedral cycles*, etc.

Chapter VIII

Δ-GROUPS OF COMPLEXES (LOWER BETTI OR HOMOLOGY GROUPS)

§1. Definitions. Examples. Simplest general properties

§1.1. Definition of the group $\Delta^r(\mathfrak{K}, \mathfrak{A})$. Let $\mathfrak{K}$ be an a-complex. [See VII, 4.1; to read Chapter VIII the reader may think of an a-complex $\mathfrak{K}$ as simply the set of all oriented simplexes of a triangulation K and of $\mathfrak{A}$ as the group J. He may then replace $\mathfrak{K}$ by K and omit $\mathfrak{A}$ in all expressions of the form $L^r(\mathfrak{K}, \mathfrak{A})$, $Z^r(\mathfrak{K}, \mathfrak{A})$, $H^r(\mathfrak{K}, \mathfrak{A})$, $\Delta^r(\mathfrak{K}, \mathfrak{A})$, etc.]

DEFINITION 1.1. The group

$$\Delta^r(\mathfrak{K}, \mathfrak{A}) = Z_\Delta{}^r(\mathfrak{K}, \mathfrak{A})/H_\Delta{}^r(\mathfrak{K}, \mathfrak{A}) \tag{1.1}$$

is called the *r-dimensional* (*lower*) *Betti group* or the r-dimensional Δ-group or simply the Δ^r-group of the complex $\mathfrak{K}$ over the coefficient domain $\mathfrak{A}$ (this group is also referred to as the rth Betti or homology group over $\mathfrak{A}$). The elements of the group $\Delta^r(\mathfrak{K}, \mathfrak{A})$ are known as the *r-dimensional homology classes of the complex $\mathfrak{K}$ over the coefficient domain $\mathfrak{A}$* (the rth homology classes of $\mathfrak{K}$ over $\mathfrak{A}$). In particular, we shall denote the group $\Delta^r(\mathfrak{K}, J)$ simply by $\Delta_0{}^r(\mathfrak{K})$; its rank is written as $\pi^r(\mathfrak{K})$ and is called the *r-dimensional* (or rth) *Betti number of* $\mathfrak{K}$; its elements are known as integral homology classes. In general, instead of $\Delta^r(\mathfrak{K}, \mathfrak{A})$ we shall usually write simply $\Delta^r(\mathfrak{K})$.

Let K be a simplicial complex. We shall assume that the complex consisting of the oriented simplexes of K is an a-complex $\mathfrak{K}$. Then, instead of $\Delta^r(\mathfrak{K}, \mathfrak{A})$, $\pi^r(\mathfrak{K})$, etc., we shall write $\Delta^r(K, \mathfrak{A})$, $\pi^r(K)$, etc. and speak of the Δ^r-groups and Betti numbers of K. This definition enables us, in particular, to speak of the Δ-groups of unrestricted simplicial complexes and of open and closed subcomplexes of unrestricted simplicial complexes.

REMARK. The groups $\Delta^r(\mathfrak{K}, J_m)$ are called the *Betti groups* (mod m) and for brevity are denoted simply by $\Delta_m{}^r(\mathfrak{K})$. If m is a prime, the rank (mod m) (see Appendix 2, 3.3) of the group $\Delta_m{}^r(\mathfrak{K})$ is called the rth *Betti number* (mod m) *of* $\mathfrak{K}$ and is denoted by $\pi_m{}^r(\mathfrak{K})$.

§1.2. The group $\Delta^n(\mathfrak{K}^n, \mathfrak{A})$. If $\mathfrak{K}^n$ is an n-complex, the group $L^{n+1}(\mathfrak{K}^n, \mathfrak{A})$ is the null group; consequently the group $H^n(\mathfrak{K}^n, \mathfrak{A})$ is also the null group and

$$\Delta^n(\mathfrak{K}^n, \mathfrak{A}) = Z^n(\mathfrak{K}^n, \mathfrak{A}). \tag{1.2}$$

§1.3. The groups $\Delta^0(K, \mathfrak{A})$. In this article K will denote an unrestricted simplicial complex.

DEFINITION 1.31. We shall say that a 0-cycle $z^0 = \sum_{i=1}^{\rho(0)} a_i e_i$ of K is *normal* if $\sum_i a_i = 0$.

It is clear that the sum of two normal cycles is a normal cycle; if z^0 is a normal cycle, $-z^0$ is a normal cycle; the 0-cycle identically equal to zero is a normal cycle. Hence the normal 0-cycles form a subgroup $Z^{00}(K, \mathfrak{A})$ of the group $Z^0(K, \mathfrak{A})$.

Again, if x^1 is a monomial 1-chain

$$x^1 = at_i^1, \qquad t_i^1 = | e_{i0} e_{i1} |,$$

then

$$\Delta x^1 = a \Delta t_i^1 = a(e_{i1} - e_{i0}) = ae_{i1} - ae_{i0},$$

that is, the boundary of a monomial 1-chain is a normal cycle. Since every chain is a sum of monomial chains, we have proved the following proposition:

1.32. *The boundary of every* 1-*chain is a normal* 0-*cycle.*

COROLLARY. A 0-cycle z_1^0, homologous in K to a normal cycle z^0, is normal.

For, 1.32 implies that $z^0 - z_1^0$ is a normal cycle. Hence the sum of the values of the cycle z^0 is the same as that of z_1^0. Since this sum is zero for z^0, the same is true for z_1^0.

We shall now prove the converse of Theorem 1.32 for a connected K.

1.33. *Every normal* 0-*cycle of a connected complex is homologous to zero.*

Let

$$z^0 = \sum_{i=1}^{\rho(0)} a_i e_i$$

be a normal cycle. Since $\sum_{i=1}^{\rho(0)} a_i = 0$, i.e., $a_1 = -a_2 - a_3 - \cdots - a_{\rho(0)}$, it follows that

$$z^0 = \sum_{i=2}^{\rho(0)} a_i(e_i - e_1),$$

and it is enough to prove that the 0-cycle $e_i - e_1 \sim 0$.

Since K is a connected complex, there is a broken line connecting the vertices e_1 and e_i in K.

Let $[e_1 e_{i(1)} \cdots e_{i(k)} e_i]$ be such a broken line. Let us set

$$x^1 = | e_1 e_{i(1)} | + | e_{i(1)} e_{i(2)} | + | e_{i(k-1)} e_{i(k)} | + | e_{i(k)} e_i |.$$

Clearly,

$$\begin{aligned} \Delta x^1 &= \Delta | e_i e_{i(1)} | + \Delta | e_{i(1)} e_{i(2)} | + \cdots + \Delta | e_{i(k-1)} e_{i(k)} | + \Delta | e_{i(k)} e_i | \\ &= (e_{i(1)} - e_1) + (e_{i(2)} - e_{i(1)}) + \cdots + (e_{i(k)} - e_{i(k-1)}) \\ &\qquad + (e_i - e_{i(k)}) \\ &= e_i - e_1 . \end{aligned}$$

This proves Theorem 1.33.

1.34. Let e_1 be any vertex of a connected complex K; every 0-cycle z^0 of K is homologous to a monomial cycle ae_1, where $a \in \mathfrak{A}$.

To prove this, let $z^0 = \sum a_i e_i$ and set $\sum a_i = a$. Then $z^0 - ae_1$ is a normal cycle and, therefore,

$$z^0 - ae_1 \sim 0,$$

which was to be proved.

Hence, there is a (1−1) correspondence between the 0-dimensional homology classes of K and the cycles ae_1, that is, the elements of the group $\mathfrak{A}$. We have proved the proposition:

1.35. *If K is a connected complex, then the group $\Delta^0(K, \mathfrak{A})$ is isomorphic to the group $\mathfrak{A}$.*

Corollary. The 0th Betti number and the 0th Betti number (mod m), m a prime, of a connected complex are both equal to 1.

The following theorem is a generalization of this proposition:

1.36. *The 0th Betti number and the 0th Betti number* (mod m), *m a prime, of an arbitrary unrestricted simplicial complex are equal to the number of components of the complex.*

Proof. Let $\{K_\nu\}$ be the components of the complex K and choose a vertex e_ν in each K_ν. Then, for any 0-cycle

$$z^0 \in Z^0(K, \mathfrak{A})$$

we have

$$K_\nu z^0 \sim a_\nu e_\nu \text{ in } K_\nu, \qquad a_\nu \in \mathfrak{A},$$

$$z^0 = \textstyle\sum_\nu K_\nu z^0 \sim \sum_\nu a_\nu e_\nu \text{ in } K,$$

where only a finite number of terms can be different from zero. It remains to be proved that all the 0-cycles e_ν are *lirh*. Suppose a linear combination $\sum_\nu a_\nu e_\nu$ is the boundary of a 1-chain x^1:

$$\Delta x^1 = \textstyle\sum_\nu a_\nu e_\nu .$$

Since every component K_ν is both a closed and open subcomplex of K, we have, by VII, 6.51,

$$\Delta K_\nu x^1 = a_\nu e_\nu .$$

Hence, in virtue of 1.32, $a_\nu = 0$. This proves 1.36.

The following proposition is easily proved from 1.36:

1.37. *If $\mathfrak{A} = J$ and $cz^0 \sim 0$ in K, c an integer $\neq 0$, then $z^0 \sim 0$ in K.*

We may therefore say (Appendix 2, 1.2): *$H_0^0(K)$ is a division closed subgroup of $Z_0^0(K)$, $H_0^0(K) = \hat{H}^0(K)$*, or: *the group $\Delta_0^0(K)$ does not contain elements of finite order different from zero.*

To prove this, let us assume that $cz^0 \sim 0$ in K,

$$cz^0 = \Delta x^1.$$

Then, if the components of K are denoted by K_i,

$$K_i cz^0 = \Delta K_i x^1;$$

whence it follows, by 1.32, that $K_i cz^0 = cK_i z^0$ is a normal cycle. But then $K_i z^0$ is also a normal cycle, and consequently by 1.33

$$K_i z^0 \sim 0 \text{ in } K_i;$$

hence

$$z^0 = \sum_i K_i z^0 \sim 0 \text{ in } K.$$

This proves 1.37.

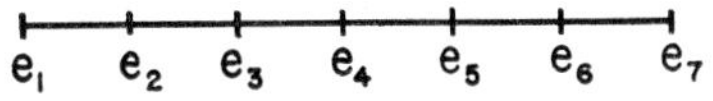

FIG. 112

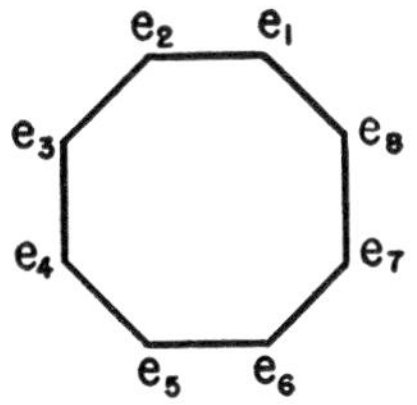

FIG. 113

§1.4. Simplest examples of the groups Δ^r.

1. Let K be the complex consisting of all the segments (e_1e_2), (e_2e_3), $\cdots$, $(e_{s-1}e_s)$ and vertices $e_1, e_2, \cdots, e_s$ of a simple nonclosed broken line $[e_1 \cdots e_s]$. In Fig. 112, $s = 7$.

We shall prove that the group $\Delta^1(K) = Z^1(K)$ is the null group. We set

$$t_i^1 = | e_i e_{i+1} |, \qquad i = 1, 2, \cdots, s - 1.$$

Let $z^1 = \sum_{i=1}^{s-1} a_i t_i^1$ be a cycle. Then all the a_i are equal. For, in the contrary case, there would be two adjacent segments t_{i-1}^1 and t_i^1 with $a_{i-1} \neq a_i$. But in that case the cycle Δz^1 would have the value $a_{i-1} - a_i \neq 0$ on the common vertex e_i of these two segments and z^1 would not be a cycle.

Hence, all the a_i are equal to a single $a \in \mathfrak{A}$:

$$z^1 = a \sum_{i=1}^{s-1} t_i^1, \qquad a \in \mathfrak{A}.$$

But $\Delta \sum_{i=1}^{s-1} t_i^1 = e_s - e_1$, so that $\Delta z^1 = a(e_s - e_1)$. Since z^1 is a cycle, $a = 0$, and the assertion is proved.

2. The complex K consists of all the segments and vertices of a simple closed broken line (Fig. 113). We shall orient all its segments in the same

direction, for instance (regarding K as a plane broken line) counterclockwise. We shall denote the resulting oriented 1-simplexes by t_i^1. Let

$$z^1 = \sum a_i t_i^1$$

be a cycle. Exactly as in the preceding example we prove that all the a_i must be equal, so that every 1-cycle of K must have the form

$$z^1 = a \sum t_i^1, \qquad a \in \mathfrak{A}.$$

On the other hand, it is easy to see that $\sum t_i^1$, and hence also $z^1 = a \sum t_i^1$, is a cycle for arbitrary a. Therefore the group $Z^1(K)$ consists of all the chains of the form $a \sum t_i^1$, where a is an arbitrary element of $\mathfrak{A}$. Consequently, $\Delta^1(K) = Z^1(K) \approx \mathfrak{A}$ (the symbol $\approx$ denotes the isomorphism of the two groups). In particular, if $\mathfrak{A} = J$, the group $\Delta^1(K) = Z^1(K)$ is an infinite cyclic group.

3. The complex K consists of all the segments and vertices of two closed broken lines having no elements in common except for one vertex (see Fig. 114).

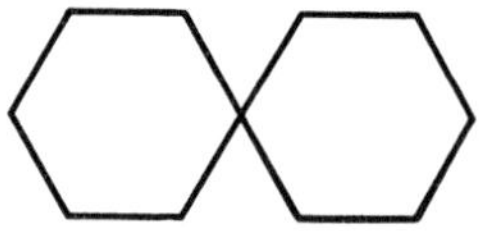

FIG. 114

Let us orient each broken line in a definite direction (for instance, regarding both of them as situated in the same plane, counterclockwise). The oriented segments of the first (second) broken line will be designated by $t_{1i}^1 (t_{2i}^1)$

Then every cycle $z^1 \in Z^1(K)$ assumes the same value on all the segments of any one of the two broken lines; for instance, the value a_1 on all t_{1i}^1 and the value a_2 on all t_{2i}^1. Hence

$$z^1 = a_1 \sum t_{1i}^1 + a_2 \sum t_{2i}^1,$$

or, setting

$$z_1^1 = \sum t_{1i}^1, \qquad z_2^1 = \sum t_{2i}^1,$$

$$z^1 = a_1 z_1^1 + a_2 z_2^2.$$

On the other hand, since z_1^1 and z_2^1 are cycles, every chain of the form $a_1 z_1^1 + a_2 z_2^1$ is a cycle; furthermore, it is easy to see that

$$a_1 z_1^1 + a_2 z_2^1 = 0$$

if, and only if, $a_1 = a_2 = 0$. In other words, the group

$$\Delta^1(K) = Z^1(K)$$

is isomorphic to the direct sum $\mathfrak{A} + \mathfrak{A}$; if $\mathfrak{A} = J$, the group $\Delta^1(K) = Z^1(K)$ is the free Abelian group of rank 2.

4. The following remark is required for the later examples (starting with Example 5) and by its nature is directly relevant to what we have been discussing.

Let K be the 1-complex consisting of all the edges and vertices shown in Fig. 115.

Every 1-cycle z^1 of this complex is identically equal to zero if it is equal to zero on all the horizontal edges of K which do not lie on the side AB of the square $ABCD$.

Indeed, considering separately each of the verticals AD, EF, PQ, and BC, we may show by repeating the reasoning in Example 1 that z^1 vanishes on all the edges of K which lie on these verticals. After this, the argument of Example 1 shows that z^1 is zero also on all the edges of K lying on AB.

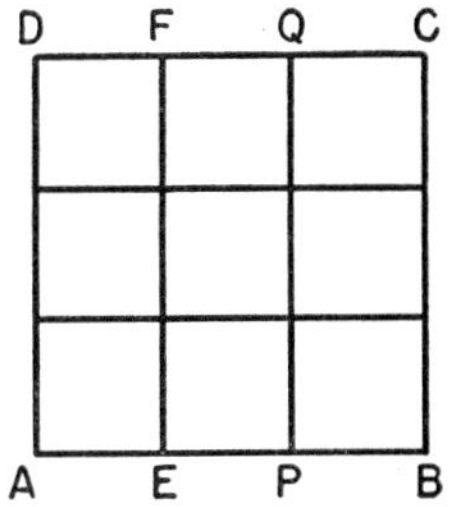

FIG. 115

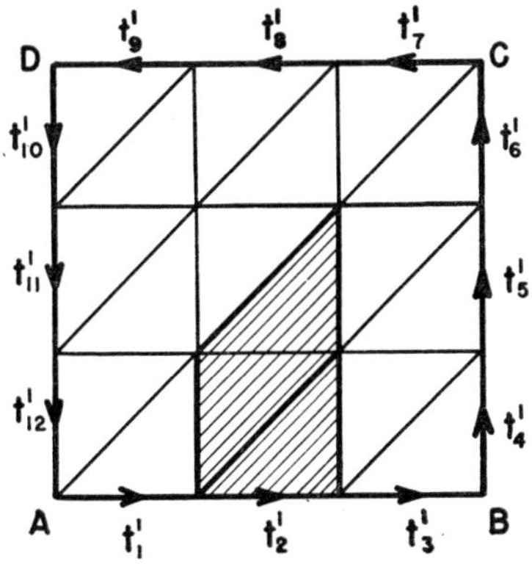

FIG. 116

EXERCISE. Let z^1 be a 1-cycle of K (Fig. 115). Why is it not possible to prove that $z^1 = 0$ by applying the reasoning in Example 1 and considering z^1 on each of the verticals and horizontals of Fig. 115? Find several non-vanishing 1-cycles of K and explain why the reasoning mentioned above is not applicable to them.

5. Let K_0 be the 2-complex consisting of the 18 triangles, their sides, and vertices shown in Fig. 116. We shall denote by $\dot{K}_0$ the subcomplex of K_0 consisting of all the elements of K_0 on the boundary of the square $ABCD$. Let us orient the boundary of the square $ABCD$ counterclockwise and denote the correspondingly oriented 12 segments of $\dot{K}_0$ by $t_1^1, \cdots, t_{12}^1$. Denote the cycle $\sum_{i=1}^{12} t_i^1$ by z_0^1. Let t_i^2, $i = 1, 2, \cdots, 18$, stand for the 18 triangles of K_0 oriented counterclockwise. We shall first prove a lemma.

LEMMA. *Every 1-cycle is homologous to zero in K_0.*

Proof. We shall define for each oriented segment t of K_0 which, in Fig.

116, is directed horizontally or diagonally the 1-cycle $z(t)$ as the chain whose value is 1 on the following oriented segments:

1) the segment t;

2) the vertical segments (directed downward) from the terminal point of t to AB;

3) The projection of the segment $-t$ on AB;

4) the vertical segments (directed upwards) from AB up to the initial point of t.

The cycle $z(t)$ is zero by definition on the remaining segments of K_0 (here, contrary to our usual practice, $z(t)$ denotes a cycle and not the value of the cycle z on t).

Now let $z^1 = \sum_{i=1}^{33} a_i t_i^1$ be a 1-cycle of K_0 with the orientations t_i^1, $i = 1, 2, \cdots, 12$ as chosen above; the orientations of the remaining 21 segments are arbitrary, for instance, from left to right and upward.

Let z' be the cycle

$$z' = z^1 - \sum{}' a_i z(t_i^1),$$

where the sum

$$\sum{}' a_i z(t_i^1)$$

is taken over all horizontal and diagonal segments. It is clear that every cycle $z(t_i^1)$ is the boundary of a quadrilateral consisting of triangles of K_0 [that is, the boundary of a chain which assumes the same value, 1 or -1, on the triangles (oriented counterclockwise) making up the quadrilateral and is zero on all the other triangles (one such quadrilateral is hatched in Fig. 116)]. Hence it is easily shown that $z^1 \sim z'$. Since the elements on which z' is different from zero are either vertical segments or horizontal segments on AB, it follows from Example 4 that $z' = 0$. Consequently $z^1 \sim 0$.

Now we identify AB with DC and AD with BC to obtain a torus. This identification transforms the complex K_0 (Fig. 116) into a triangulation K of a torus, which consists of 18 triangles, 27 segments, and 9 vertices. The complex $\dot{K}_0$ is converted into a complex $\dot{K}$ consisting of 6 segments and 5 vertices, altogether making up two closed broken lines with a single common vertex. As a result $AB^{\rightarrow}$ and $DC^{\rightarrow}$ become the cycle $z_1^1 = t_1^1 + t_2^1 + t_3^1 = -t_7^1 - t_8^1 - t_9^1$, $AD^{\rightarrow}$ and $BC^{\rightarrow}$ are transformed into the cycle $z_2^1 = t_4^1 + t_5^1 + t_6^1 = -t_{10}^1 - t_{11}^1 - t_{12}^1$.

Both cycles z_1^1 and z_2^1 are obviously cycles of $\dot{K}$ and moreover (see Example 3) every 1-cycle of $\dot{K}$ is of the form $a_1 z_1^1 + a_2 z_2^1$.

We shall calculate the group $\Delta_0^1(K)$. Let z^1 be any 1-cycle of K. Set $Q = K_0 \setminus \dot{K}_0$ and

$$x_0^1 = Qz^1 \in L^1(K_0).$$

The same chain x_0^1 regarded as a chain of K will be denoted by x^1.

Since $\Delta z^1 = 0$, it follows easily that Δx_0^1 is a 0-cycle of $\dot{K}_0$. Since Δx_0^1 is a normal cycle and $\dot{K}_0$ is a connected complex, $\Delta x_0^1 \sim 0$ on $\dot{K}_0$. Hence there exists a 1-chain y_0^1 of $\dot{K}_0$ whose boundary is Δx_0^1, so that $x_0^1 - y_0^1$ is a cycle. It follows from the Lemma that there is a chain x_0^2 of K_0 whose boundary is the cycle $x_0^1 - y_0^1$:

$$\Delta x_0^2 = x_0^1 - y_0^1,$$

where

$$x_0^1 - \Delta x_0^2 = y_0^1 \in L^1(\dot{K}_0).$$

In Fig. 117, z^1 is 1 on the heavily drawn segments oriented as indicated by arrows, and is zero on the remaining 1-simplexes; $x_0^1 = Qz^1$ is 1 on the segments mentioned above which are inside the square; Δx_0^1 consists of the

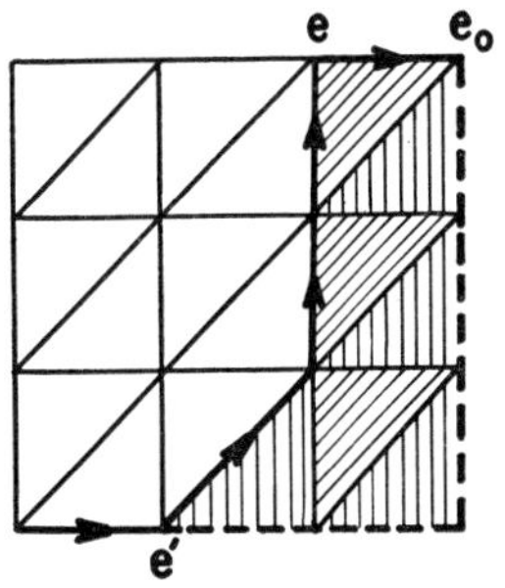

FIG. 117

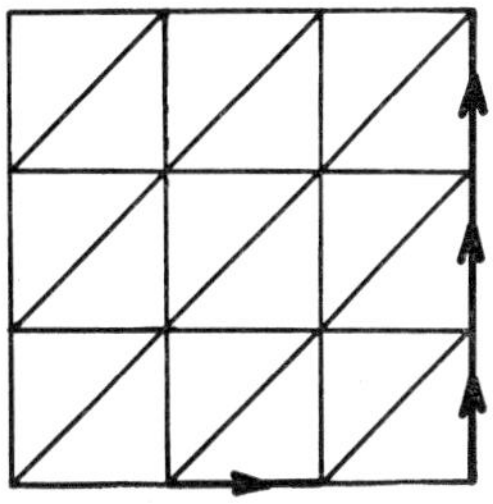

FIG. 118

two vertices e and e' with coefficients 1 and -1, respectively. The chain y_0^1 is 1 on the heavily drawn dotted segments of $\dot{K}_0$ and on $|e_0e|$, and is zero on the remaining 1-simplexes. The chain x_0^2 is -1 on the hatched triangles oriented counterclockwise, and is zero on all the other triangles.

Let us denote by x^2 the chain x_0^2 regarded as a chain of K. The chains $x^1 - \Delta x^2$ and $z' = z^1 - \Delta x^2$ are on $\dot{K}$.

For the case shown in Fig. 117, the chain $x^1 - \Delta x^2$ (on K) has the form shown in Fig. 118; the edges on which this chain does not vanish, but assumes the value 1, are heavily drawn; their orientations are indicated by arrows.

The cycle $z' = z^1 - \Delta x^2$ on $\dot{K}$ is shown in Figs. 119 and 120 (by heavily drawn segments oriented by arrows).

Since $z' = z^1 - \Delta x^2$ (homologous by definition to the cycle z^1) is on $\dot{K}$, z' has the form $a_1z_1^1 + a_2z_2^1$.

In the case shown in the figures it is clear that $a_1 = a_2 = 1$. It is left to the reader to construct several cycles z^1 homologous to the cycle $z_1^1 + 2z_2^1$ on K.

Hence we have proved that *every 1-cycle z^1 of K satisfies*

$$z^1 \sim a_1 z_1^1 + a_2 z_2^1. \tag{1.40}$$

We shall now prove that every cycle z^1 of K satisfies *precisely one* such homology, that is, that the coefficients a_1 and a_2 in (1.40) are uniquely determined by the cycle z^1. This will prove that the group $\Delta^1(K)$ is isomorphic to the group of all integral linear forms in two variables, that is, the free Abelian group of rank 2.

It is enough to prove that

$$a_1 z_1^1 + a_2 z_2^1 \sim 0 \text{ in } K \tag{1.41}$$

implies

$$a_1 = a_2 = 0. \tag{1.42}$$

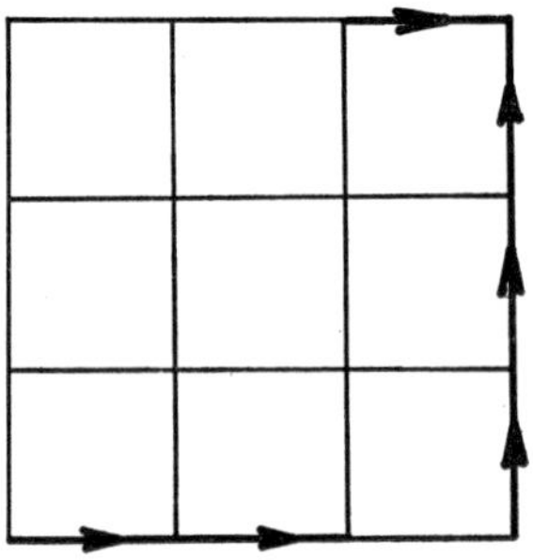

FIG. 119

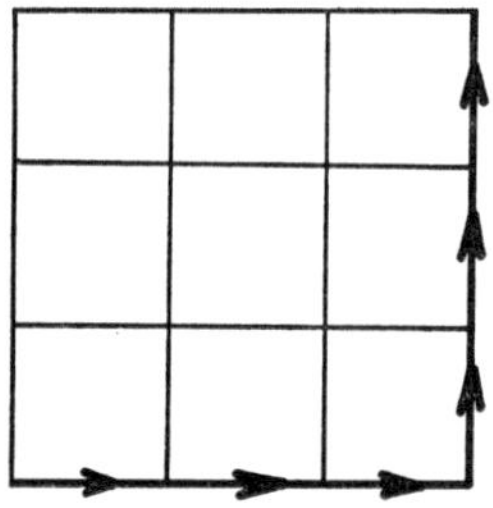

FIG. 120

By (1.41), there is a 2-chain x^2 of K bounded by the cycle $a_1 z_1^1 + a_2 z_2^1$. It is easy to see that the values of x^2 on all the t_i^2 are the same number a; but then, if x_0^2 is the chain of K_0 corresponding to x^2, $\Delta x_0^2 = a z_0^1$.

The transformation of the complex K_0 into the complex K effected by identification is a simplicial mapping of K_0 onto K which maps $\dot{K}_0$ into $\dot{K}$ and the cycle z_0^1 into zero. On the other hand, since x_0^2 is mapped into x^2 and the boundary of the image is always the image of the boundary (commutativity of the boundary operator with a simplicial mapping),

$$\Delta x^2 = a_1 z_1^1 + a_2 z_2^1 = 0.$$

According to the definition of the cycles z_1^1 and z_2^1 this is possible only if $a_1 = a_2 = 0$. This proves the assertion.

6. Let us identify the side $\overrightarrow{AB}$ of the square of Fig. 116 with the side $\overrightarrow{DC}$ and the side $\overrightarrow{AD}$ with the side $\overrightarrow{CB}$ (watch the directions!). The result is a triangulation K' of a Klein bottle (III, 3.1). We shall calculate the group $\Delta^1(K')$. First, exactly as in the case of the torus, we shall prove that

every 1-cycle z^1 satisfies (1.40); in this case, however, the numbers a_1 and a_2 are not uniquely determined by (1.40). Instead, we have the following result:

In order that

$$a_1 z_1^{\ 1} + a_2 z_2^{\ 2} \sim 0 \text{ in } K',$$

it is necessary and sufficient that $a_1 = 0$ *and* $a_2 \equiv 0 \pmod 2$.

This is proved in the same way as in the case of the torus except that the new result is obtained as a consequence of the fact that in this case the simplicial mapping transforming K_0 into K' maps $z_0^{\ 1}$ not into zero but into the cycle $2z_2^{\ 1}$. Hence in place of the identity $a_1 z_1^{\ 1} + a_2 z_2^{\ 1} = 0$ we now obtain the identity

$$a_1 z_1^{\ 1} + a_2 z_2^{\ 1} = 2a z_2^{\ 1},$$

whence

$$a_1 = 0, \qquad a_2 = 2a.$$

Since a is an integer, our assertion is proved. We then have:

There is a (1–1) correspondence between the homology classes of K' and the linear forms $a_1 z_1^{\ 1} + a_2 z_2^{\ 1}$, where a_1 is an integer and $a_2 \equiv 0 \pmod 2$.

This implies that the group $\Delta^1(K')$ is the direct sum of an infinite cyclic group and a group of order 2; the 1st Betti number of the triangulation K' of the Klein bottle is 1.

7. Let us return to Fig. 116 and subdivide the triangulation indicated in the figure by drawing the second diagonal of the square. This triangulation (consisting now of 24 triangles, their sides, and vertices) we again denote by K_0. Let us now identify the directed segments $t_1^{\ 1}$ and $t_7^{\ 1}$, $t_2^{\ 1}$ and $t_8^{\ 1}$, $t_3^{\ 1}$ and $t_9^{\ 1}$, $t_4^{\ 1}$ and $t_{10}^{\ 1}$, $t_5^{\ 1}$ and $t_{11}^{\ 1}$, and $t_6^{\ 1}$ and $t_{12}^{\ 1}$. This identification converts K_0 into a triangulation K'' of the projective plane and the complex $\dot{K}_0$ into a closed polygon $\dot{K}''$; the counterclockwise orientation of $ABCD$ is transformed into a definite orientation of the polygon $\dot{K}''$. [If it were not for the subdivision of the original triangulation K_0, the identification would not yield a triangulation (the resulting complex would contain two triangles with the same vertices).] The closed polygon $\dot{K}''$ with the orientation induced by the identification is a 1-cycle $z_1^{\ 1}$ of $\dot{K}''$. The simplicial mapping of K_0 onto K'' induced by the identification transforms the cycle $z_0^{\ 1}$ of $\dot{K}_0$ into the cycle $2z_1^{\ 1}$; on the other hand, reasoning as above, we can prove that every 1-cycle z^1 of K'' is homologous to a cycle of the form $a_1 z_1^{\ 1}$. Now however, $a_1 z_1^{\ 1} \sim 0$ in K'' if, and only if, $a_1 \equiv 0 \pmod 2$. Hence

The group $\Delta^1(K'')$ *is the group of order* 2; *the* 1*st Betti number of the triangulation* K'' *of the projective plane is zero.*

Remark 1. It is easily proved that every integral 2-cycle of each of the complexes K, K', K'' has the same value on all the oriented triangles t_i^2, $i = 1, 2, \cdots, 18$ (for K'' the number of triangles is 24), that is, every 2-cycle is of the form

$$z^2 = a \sum_{i=1}^{18} t_i^2,$$

where a is an integer. It is also easily seen that chains of this form on K are indeed cycles, while on K'

$$\Delta z^2 = 2az_2^1$$

and on K''

$$\Delta z^2 = 2az_1^1.$$

It follows that the group $\Delta^2(K) = Z_\Delta^2(K)$ is infinite cyclic while the groups $\Delta^2(K') = Z_\Delta^2(K')$ and $\Delta^2(K'') = Z_\Delta^2(K'')$ are null groups. The 2nd Betti number of the triangulation K of the torus is 1, while the 2nd Betti number of the triangulation K' (K'') of the Klein bottle (projective plane) is zero.

REMARK 2. In Chapter X we shall see that all the triangulations of the same polyhedron or of homeomorphic polyhedra have isomorphic groups Δ^r and consequently the same Betti numbers. It is therefore justifiable to speak of the Betti groups and numbers of the torus, the Klein bottle, and the projective plane.

EXERCISE 1. If the directed sides $\overrightarrow{AD}$ and $\overrightarrow{BC}$ of the square in Fig. 116 are identified, but the sides AB and DC are left free, the result is a triangulation of a plane circular ring. If, however, $\overrightarrow{AD}$ and $\overrightarrow{CB}$ are identified, and AB and DC are again left free, the result is a triangulation of the Möbius band. Prove that in both cases the group Δ^1 is infinite cyclic.

EXERCISE 2. Prove that the groups $\Delta_2^2(K)$, $\Delta_2^2(K')$, and $\Delta_2^2(K'')$ (where K, K', K'' are the triangulations of the torus, Klein bottle, and projective plane considered above) are of order 2 and that the 2nd Betti number (mod 2) is in each case equal to 1.

EXERCISE 3. Starting with a triangulation K of a cube analogous to the triangulation of the square $ABCD$ of Fig. 116, prove that the 2nd Betti number of the 3-dimensional torus M_1^3 (I, 5.2, Example 6) is 3. The generators of the group $\Delta^2(M_1^3)$ are the homology classes of the three 2-cycles resulting from the identification of opposite faces of the cube.

The group $\Delta^2(M_1^3)$ is the free Abelian group of rank 3. The group $\Delta^1(M_1^3)$ is also the free Abelian group of rank 3. As its generators we may take the homology classes of the three 1-cycles obtained by identifying the edges of the cube.

It can be shown in an analogous way that the groups $\Delta^1(M^3)$ and $\Delta^2(M^3)$ of the manifold M^3 defined as the topological product of a 1-sphere and a 2-sphere (I, 5.2, Example 7) are infinite cyclic, so that the 1st and 2nd Betti numbers of this manifold are both equal to 1.

If we use the first model of this manifold (the region bounded by two concentric 2-spheres S^2 and s^2 with identification of the two spheres), we may take as the generator of $\Delta^2(M^3)$ the homology class of the cycle z_0^2 obtained as follows: take a triangulation of any 2-sphere S_0^2 concentric with the spheres S^2 and s^2, orient all the triangles of this triangulation (e.g., counterclockwise) and assign the chain z_0^2 the value 1 on all these triangles.

As the generator of the group $\Delta^1(M^3)$ we may take the homology class of the 1-cycle z_0^1 obtained after the identification of S^2 and s^2 from the directed segment joining two corresponding points of the two spheres (I, 5.2, Fig. 6).

Find cycles z_0^2 and z_0^1 on the second model of M^3 (doubled torus).

EXERCISE 4. Prove that the a-complexes defined in VII, 4.2, Examples 1 and 2, have the same Betti groups as the triangulation of the torus considered in Example 5 above. Prove that the a-complex defined in VII, 4.2, Example 3 has the same Betti groups as the triangulation of the projective plane considered in Example 7 above. Finally, show that the a-complex $\mathfrak{K}$ of VII, 4.2, Example 4 has the following Betti groups:

$\Delta_0^0(\mathfrak{K})$ is infinite cyclic,

$\Delta_0^1(\mathfrak{K})$ is the cyclic group of order 6,

$\Delta_0^2(\mathfrak{K})$ is the null group,

$\Delta_6^2(\mathfrak{K})$ is the cyclic group of order 6.

§1.5. Some elementary n-complexes and their Betti groups.

1.51. Let the complex K consist of a single element, an n-simplex. The group $\Delta^n(K, \mathfrak{A}) = Z^n(K, \mathfrak{A})$ is isomorphic to the group $\mathfrak{A}$. The group $\Delta^r(K, \mathfrak{A})$, $r \neq n$, is the null group.

1.52. Let $K = | T^n |$. The group $\Delta^n(K, \mathfrak{A}) = Z^n(K, \mathfrak{A})$ is the null group. $\Delta^r(K, \mathfrak{A})$ $(0 < r < n)$ is the null group (by VII, 9.2, Case 1).

1.53. Let $K = \dot{T}^n = | T^n | \setminus T^n$. By VII, 9.2, Case 3, $\Delta^r(\dot{T}^n, \mathfrak{A})$, $0 < r < n - 1$, is the null group. We shall see in 4.1 that $\Delta_0^{n-1}(K)$ is infinite cyclic (this could also be easily proved directly at this point). It is also easily shown that $\Delta^{n-1}(K, \mathfrak{A})$ is isomorphic to $\mathfrak{A}$ for arbitrary $\mathfrak{A}$ (see 4.1).

1.54. Let O_α be the star of a p-simplex T^p in any n-dimensional triangulation K_α. We shall assume that O_α contains at least one element different from T^p; hence, at least one $(p + 1)$-simplex T^{p+1} of K_α.

1.54 $(r \leq p)$. If $r \leq p$, $\Delta^r(O_\alpha, \mathfrak{A})$ is the null group.

Indeed, the complex O_α contains no r-simplexes for $r < p$; hence $L^r(O_\alpha, \mathfrak{A})$, $Z^r(O_\alpha, \mathfrak{A})$, $\Delta^r(O_\alpha, \mathfrak{A})$ are null groups.

For $r = p$, T^p is the only r-simplex of O_α. If T^p is given an arbitrary orientation t^p and the orientation t^{p+1} is chosen so that $(t^{p+1}:t^p) = 1$, then $\Delta_\alpha t^{p+1} = t^p$, where Δ_α is the boundary operator in O_α. Hence $Z^p(O_\alpha, \mathfrak{A}) = H^p(O_\alpha, \mathfrak{A})$ and $\Delta^p(O_\alpha, \mathfrak{A})$ is the null group. We therefore have

1.54$_0$. The group $\Delta^0(O_\alpha, \mathfrak{A})$ of the star O_α of an arbitrary element of a triangulation K_α is different from the null group only if the star O_α consists of a single vertex of K_α.

1.54 $(n, n - 1)$. Let K_α be an n-complex and let the number of n-simplexes of K_α having a given $(n - 1)$-simplex $T^{n-1} \in K_\alpha$ as a face be k. Then $\Delta_0{}^n(O_\alpha)$ is the free Abelian group of rank $k - 1$.

Proof. Let $T_1{}^n, \cdots, T_k{}^n$ be all the n-simplexes of K_α which have T^{n-1} on their boundaries. Let T^{n-1} have an arbitrary orientation t^{n-1} and orient the simplexes $T_i{}^n$, $i = 1, 2, \cdots, k$, so that $(t_i{}^n : t^{n-1}) = 1$. The chains

$$z_i{}^n = t_i{}^n - t_1{}^n, \qquad i = 2, 3, \cdots, k,$$

are integral cycles of the n-complex O_α. Since the simplexes $T_2{}^n, \cdots, T_k{}^n$ are distinct, the chains $z_2{}^n, \cdots, z_k{}^n$ are linearly independent elements of the group $Z_0{}^n(O_\alpha)$ and therefore belong to linearly independent cosets of the group $\Delta_0{}^n(O_\alpha)$. It remains to be proved that they form a system of generators of $Z_0{}^n(O_\alpha)$. We shall consider the case $k = 1$ first. Then there are no cycles $z_i{}^n$ and it is required to prove that $Z_0{}^n(O_\alpha)$ is the null group. This follows from the fact that on our assumptions every n-chain of O_α is of the form $at_1{}^n$, where a is an integer, and that for $a \neq 0$ none of these chains is a cycle, since $\Delta_\alpha a t_1{}^n = at^{n-1}$ (Δ_α is the boundary operator in O_α). We shall now assume that the assertion is true for $k = m$ and prove it for $k = m + 1$. By our assumptions, every cycle $z^n \in Z_0{}^n(O_\alpha)$ in which the simplex $t_{m+1}{}^n$ has coefficient zero is a linear combination of the cycles $z_2{}^n, \cdots, z_m{}^n$.

Now let

$$z^n = \textstyle\sum_{i=1}^{m+1} a_i t_i{}^n \in Z^n(O_\alpha), \qquad a_{m+1} \neq 0;$$

then

$$\Delta z^n = (\textstyle\sum_{i=1}^{m+1} a_i) t^{n-1} = 0$$

and

$$\textstyle\sum_{i=1}^{m+1} a_i = 0.$$

Setting

$$z'^n = (z^n - a_{m+1} t_{m+1}{}^n) + a_{m+1} t_1{}^n,$$

we get

$$\Delta z'^n = (\textstyle\sum_{i=1}^{m+1} a_i) t^{n-1} = 0,$$

i.e., z'^n is a cycle of O_α and moreover obviously such that the coefficient of $t_{m+1}{}^n$ in z'^n is zero. Hence z'^n is a linear combination with integral coefficients of the cycles $z_2{}^n, \cdots, z_m{}^n$ and the cycle

$$z^n = z'^n + a_{m+1}(t_{m+1}{}^n - t_1{}^n) = z'^n + a_{m+1} z_{m+1}$$

is a linear combination with integral coefficients of the cycles $z_2{}^n, \cdots, z_{m+1}{}^n$. This is what we wished to show.

1.55. *Let K be a simplicial complex, $T^p \in K$, and $r > p$. Let $Q = O_K(T^p)$ and $B = B_K(T^p)$* (see IV, Def. 1.86).

Suppose that K has the following property: if $T^m \in Q$, the face of T^m opposite T^p is contained in K. Then the groups $\Delta^{r+1}(Q, \mathfrak{A})$ and $\Delta^{r-p}(B, \mathfrak{A})$ are isomorphic.

Proof. With each simplex $T^m = (T^pT^{m-p-1}) \in Q$ of dimension $m \geq p + 1$ associate the simplex $T^{m-p-1} \in B$ which is the face of T^m opposite T^p; this establishes a $(1-1)$ correspondence between the complex $Q \setminus T^p$ and B in which each simplex T^h, $h \geq 0$, of the complex B is made to correspond to a unique simplex $T^m = T^{p+h+1}) \in Q$. Let us choose a definite orientation t^p of T^p; then to each orientation t^m of T^m there corresponds a unique orientation $t^{m-p-1} = ft^m$ of T^{m-p-1} such that

$$t^m = | t^pt^{m-p-1} |;$$

if

$$t^p = | e_0 \cdots e_p |, \qquad t^m = \varepsilon | e_0 \cdots e_pe_{p+1} \cdots e_m |, \qquad \varepsilon = \pm 1,$$

then

$$ft^m = t^{m-p-1} = \varepsilon | e_{p+1} \cdots e_m |.$$

Hence f is a $(1-1)$ mapping of the set of all oriented simplexes of the complex $Q \setminus T^p$ onto the set of all oriented simplexes of the complex B. We shall prove that

$$(1.551) \qquad (ft^m : ft^{m-1}) = (-1)^{p+1}(t^m : t^{m-1}).$$

If

$$t^m = \varepsilon | e_0 \cdots e_pe_{p+1} \cdots e_{m-1}e_m |$$

and

$$t^{m-1} = \varepsilon' | e_0 \cdots e_pe_{p+1} \cdots e_{i-1}e_{i+1} \cdots e_m |, \qquad i \geq p + 1,$$

then

$$ft^m = \varepsilon | e_{p+1} \cdots e_{m-1}e_m |$$

and

$$ft^{m-1} = \varepsilon' | e_{p+1} \cdots e_{i-1}e_{i+1} \cdots e_m |.$$

Applying VII, (3.11), we get

$$(t^m : t^{m-1}) = (-1)^i\varepsilon\varepsilon'; \qquad (ft^m : ft^{m-1}) = (-1)^{i-p-1}\varepsilon\varepsilon'.$$

Hence (1.551) follows.

REMARK. We shall write f^{-1} for the inverse of f, i.e., the mapping of B onto $Q \setminus T^p$.

Now let $r \geq p + 1$. To each chain

$$x^{r+1} \in L^{r+1}(Q, \mathfrak{A})$$

there corresponds the chain

$$fx^{r+1} \in L^{r-p}(B, \mathfrak{A})$$

which, on each oriented simplex t^{r-p} of B, has the value of x^{r+1} on $f^{-1}t^{r-p}$.

In other words, if $x^{r+1} = \sum a_i t_i^{r+1}$, then

$$fx^{r+1} = \sum a_i f t_i^{r+1}.$$

Hence f is an isomorphism of the group $L^{r+1}(Q, \mathfrak{A})$ onto $L^{r-p}(B, \mathfrak{A})$.

We shall prove that

$$\Delta f x^{r+1} = (-1)^{p+1} f \Delta x^{r+1} \tag{1.552}$$

for every chain $x^{r+1} \in L^{r+1}(Q, \mathfrak{A})$. It is sufficient to show that

$$\Delta f t^{r+1} = (-1)^{p+1} f \Delta t^{r+1} \tag{1.5520}$$

for an arbitrary oriented simplex t^{r+1} of the complex Q.

To prove (1.5520) let us consider any oriented simplex t^{r-p-1} of B and the simplex $t^r = f^{-1}(t^{r-p-1})$. We shall calculate the values of both chains $\Delta f t^{r+1}$ and $f\Delta t^{r+1}$ on t^{r-p-1}.

If

$$ft^{r+1} = t^{r-p},$$

then

$$(\Delta f t^{r+1} \cdot t^{r-p-1}) = (t^{r-p} : t^{r-p-1}) = (ft^{r+1} : ft^r) = (-1)^{p+1}(t^{r+1} : t^r),$$

$$(f\Delta t^{r+1} \cdot t^{r-p-1}) = (\Delta t^{r+1} \cdot t^r) = (t^{r+1} : t^r).$$

This proves (1.5520), and hence (1.552).

It follows from (1.552) that the isomorphism f of the group $L^{r+1}(Q, \mathfrak{A})$ onto $L^{r-p}(B, \mathfrak{A})$ maps $Z^{r+1}(Q, \mathfrak{A})$ $[H^{r+1}(Q, \mathfrak{A})]$ onto $Z^{r-p}(B, \mathfrak{A})$ $[H^{r-p}(B, \mathfrak{A})]$. Hence f induces an isomorphism, denoted by the same letter, of the group $\Delta^{r+1}(Q, \mathfrak{A})$ onto the group $\Delta^{r-p}(B, \mathfrak{A})$. This is what we wished to show.

COROLLARY. Let K be a triangulation and let e be a vertex of K. The group $\Delta^{r+1}(O_K e, \mathfrak{A})$ is isomorphic to $\Delta^r(B_K e, \mathfrak{A})$ for every $r \geq 1$.

In Chapter XIII we shall need the following proposition which is a special case of the preceding theorem:

1.550. *Let $K' = eK$ be an open cone over a simplicial complex K* (see IV, 2.3). *The groups $\Delta^r(K)$ and $\Delta^{r+1}(eK)$ are isomorphic for $r > 0$.*

Indeed, in the cone $<eK>$ the open cone eK is the star and K is the outer boundary of the star of the vertex e and the face of an arbitrary simplex

$(eT) \in eK$ opposite e is the simplex $T \in K$. Hence all the hypotheses of Theorem 1.55 are satisfied (for $p = 0$, $T^p = e$), whence 1.550 follows.

§1.6. The group $\Delta^{00}(K, \mathfrak{A})$. (See Alexandroff-Hopf [A-H; V, §1, 5, p. 209]; Bibliography, Vol. 1. §1.6 is required for applications in Chapters XIV and XV, and its reading may be deferred till then.) In the sequel K is an unrestricted simplicial complex.

The group $\Delta^{00}(K, \mathfrak{A})$ is defined as:

$$\Delta^{00}(K, \mathfrak{A}) = Z^{00}(K, \mathfrak{A})/H^0(K, \mathfrak{A}), \tag{1.61}$$

where $Z^{00}(K, \mathfrak{A})$ is the group of all normal 0-cycles of K over $\mathfrak{A}$. In virtue of 1.32,

$$H^0(K, \mathfrak{A}) \subseteq Z^{00}(K, \mathfrak{A}).$$

Hence the factor group (1.61) has meaning.

1.62. *The group $\Delta^{00}(K, \mathfrak{A})$ is a direct sum of groups isomorphic to $\mathfrak{A}$ whose number is* 1 *less than the number of components of* K.

Proof (for the case of a finite number of components). The group $\Delta^0(K, \mathfrak{A})$ is the direct sum of s groups isomorphic to $\mathfrak{A}$, where s is the number of components of K; that is, the group $\Delta^0(K, \mathfrak{A})$ can be thought of as the group of linear forms $z = \sum_{i=1}^{s} a_i x_i$ in s variables x_i with coefficients in $\mathfrak{A}$.

According to its definition, the group $\Delta^{00}(K, \mathfrak{A})$ is isomorphic to the group of those linear forms $\sum_{i=1}^{s} a_i x_i$ for which $\sum_{i=1}^{s} a_i = 0$. Hence Theorem 1.62 is a special case of the following general theorem of group theory:

LEMMA 1.63. *Let $\mathfrak{A}^n$ be the group of all linear forms*

$$z = \sum_{i=1}^{n} a_i x_i \tag{1.63}$$

in n variables x_i with coefficients in an Abelian group $\mathfrak{A}$. Let $\mathfrak{B}$ be the subgroup of $\mathfrak{A}$ consisting of all the linear forms (1.63) *the sum of whose coefficients vanishes; then $\mathfrak{B}$ is isomorphic to the group $\mathfrak{A}^{n-1}$ of linear forms $y = \sum_{i=2}^{n} a_i x_i$ in the $n - 1$ variables $x_2, \cdots, x_n$ with coefficients in $\mathfrak{A}$.*

For the proof we consider the mapping f of $\mathfrak{B}$ into $\mathfrak{A}^{n-1}$ which associates with each element

$$z = \sum_{i=1}^{n} a_i x_i$$

of $\mathfrak{B}$ the element

$$y = f(z) = \sum_{i=2}^{n} a_i x_i$$

of $\mathfrak{A}^{n-1}$. The mapping f is *onto* $\mathfrak{A}^{n-1}$.

Indeed, if

$$y = \sum_{i=2}^{n} b_i x_i$$

is an arbitrary element of $\mathfrak{A}^{n-1}$, then f maps

$$z = (-\sum_{i=2}^{n} b_i)x_1 + y \in \mathfrak{B}$$

into y.

To prove that f is an isomorphism it is sufficient to show that the kernel of f is the identity of $\mathfrak{B}$. In fact, if

$$y = \sum_{i=2}^{n} b_i x_i = 0,$$

i.e., $b_2 = b_3 = \cdots = b_n = 0$, and $f(z) = y$, then $x_2, x_3, \cdots, x_n$ have in the representation of z the same coefficients as in that of y (that is, zero). Since the sum of all the coefficients of z is zero, the coefficient of x_1 in z is also zero and consequently z is zero. This completes the proof.

1.64. *The rank* $\pi^{00}(K)$ *of the group* $\Delta^{00}(K)$ *and the rank* $\pi_m^{00}(K)$ (mod m) (m a prime) *of the group* $\Delta_m^{00}(K)$ *are both equal to the number of components of* K *less* 1.

Proof. The theorem follows from 1.62 if 1.62 is proved without the assumption that the number of components of K is finite. Since we have not proved Theorem 1.62 in such generality, we shall prove 1.64 directly.

Let $\{K_\nu\}$ be the components of K, and denote (a definite) one of these by K_0. In each K_ν choose a vertex e_ν and consider the 0-cycles.

$$z_\nu^0 = e_\nu - e_0, \qquad \nu \neq 0.$$

To prove 1.64 it is enough to show that:

1°. Every normal 0-cycle z^0 is homologous in K to a linear combination of cycles z_ν^0.

2°. The cycles z_ν^0 are lirh (see VII, Def. 6.42).

Proof of 1°. To begin with, $z^0 = \sum_\nu K_\nu z^0$. Furthermore, in virtue of (1.34), $K_\nu z^0 \sim a_\nu e_\nu$ in K_ν, so that

$$(1.64) \qquad z^0 = \sum_\nu K_\nu z^0 \sim \sum_\nu a_\nu e_\nu = \sum_\nu a_\nu(e_\nu - e_0) + (\sum_\nu a_\nu) e_0 .$$

Since z^0 is a normal cycle and a cycle homologous to a normal cycle is also normal (in consequence of the Corollary to 1.32), the cycle $\sum_\nu a_\nu e_\nu$ is normal. Hence $\sum_\nu a_\nu = 0$. Taking account of this in (1.64), we obtain the required homology

$$z^0 \sim \sum_\nu a_\nu(e_\nu - e_0).$$

Proof of 2°. Let us assume that $\sum_\nu a_\nu z_\nu^0 \sim 0$, i.e.,

$$\sum_\nu a_\nu(e_\nu - e_0) \sim 0;$$

then

$$K_\mu \sum_\nu a_\nu(e_\nu - e_0) = a_\mu e_\mu \sim 0$$

for an arbitrary component K_μ, $\mu \neq 0$. Hence $a_\mu = 0$, which was to be proved.

§1.7. Decomposition of the group $\Delta^r(\mathfrak{K}, \mathfrak{A})$ into a direct sum over the components of the complex $\mathfrak{K}$. Let $\mathfrak{K}$ be an arbitrary a-complex. We shall prove the following theorem for a complex $\mathfrak{K}$ with a finite number of components (the theorem is also true for a complex with an infinite number of components).

1.71. *The group* $\Delta^r(\mathfrak{K}, \mathfrak{A})$ *is the direct sum of the groups* $\Delta^r(\mathfrak{K}_\nu, \mathfrak{A})$, *where the* $\mathfrak{K}_\nu$ *are the components of* $\mathfrak{K}$.

This theorem follows easily from:

1.72. *If an a-complex* $\mathfrak{K}$ *is the union of two disjoint closed subcomplexes* $\mathfrak{K}'$ *and* $\mathfrak{K}''$, *then*

$$L^r(\mathfrak{K}, \mathfrak{A}) = L^r(\mathfrak{K}', \mathfrak{A}) + L^r(\mathfrak{K}'', \mathfrak{A}), \tag{1.721}$$

$$Z^r(\mathfrak{K}, \mathfrak{A}) = Z^r(\mathfrak{K}', \mathfrak{A}) + Z^r(\mathfrak{K}'', \mathfrak{A}), \tag{1.722}$$

$$H^r(\mathfrak{K}, \mathfrak{A}) = H^r(\mathfrak{K}', \mathfrak{A}) + H^r(\mathfrak{K}'' \mathfrak{A}), \tag{1.723}$$

$$\Delta^r(\mathfrak{K}, \mathfrak{A}) = \Delta^r(\mathfrak{K}', \mathfrak{A}) + \Delta^r(\mathfrak{K}'', \mathfrak{A}), \tag{1.724}$$

where the sums are direct sums.

To prove (1.721) it suffices to note that

$$x^r = \mathfrak{K}'x^r + \mathfrak{K}''x^r$$

for an arbitrary $x^r \in L^r(\mathfrak{K}, \mathfrak{A})$. Hence $L^r(\mathfrak{K}, \mathfrak{A})$ is the sum of its subgroups $L^r(\mathfrak{K}', \mathfrak{A})$ and $L^r(\mathfrak{K}'', \mathfrak{A})$. Since $L^r(\mathfrak{K}', \mathfrak{A})$ and $L^r(\mathfrak{K}'', \mathfrak{A})$, as subgroups of $L^r(\mathfrak{K}, \mathfrak{A})$, have, in consequence of $\mathfrak{K}' \cap \mathfrak{K}'' = 0$, only the element 0 in common, $L^r(\mathfrak{K}, \mathfrak{A})$ is the direct sum of the subgroups $L^r(\mathfrak{K}', \mathfrak{A})$ and $L^r(\mathfrak{K}'', \mathfrak{A})$.

If $z^r \in Z^r(\mathfrak{K})$ [$z^r \in H^r(\mathfrak{K})$, $z^r = \Delta x^{r+1}$], then in virtue of the fact that $\mathfrak{K}'$ and $\mathfrak{K}''$ are open subcomplexes, we have (VII, Theorem 6.51)

$$\mathfrak{K}'z^r \in Z^r(\mathfrak{K}') \; [\Delta\mathfrak{K}'x^{r+1} = \mathfrak{K}'z^r, \; \mathfrak{K}'z^r \in H^r(\mathfrak{K}')]$$

and

$$\mathfrak{K}''z^r \in Z^r(\mathfrak{K}'') \; [\Delta\mathfrak{K}''x^{r+1} = \mathfrak{K}''z^r, \; \mathfrak{K}''z^r \in H^r(\mathfrak{K}'')],$$

so that z^r may be represented uniquely in the form

$$z^r = z'^r + z''^r,$$

where $z'^r \in Z^r(\mathfrak{K}')$, $z''^r \in Z^r(\mathfrak{K}'')$ [$z'^r \in H^r(\mathfrak{K}')$, $z''^r \in H^r(\mathfrak{K}'')$]. Since $\mathfrak{K}'$ and $\mathfrak{K}''$ are closed subcomplexes of $\mathfrak{K}$,

$$Z^r(\mathfrak{K}') \subseteq Z^r(\mathfrak{K}), \qquad Z^r(\mathfrak{K}'') \subseteq Z^r(\mathfrak{K}),$$
$$H^r(\mathfrak{K}') \subseteq H^r(\mathfrak{K}), \qquad H^r(\mathfrak{K}'') \subseteq H^r(\mathfrak{K}).$$

This proves (1.722) and (1.723), and (1.724) follows from these.

§1.8. The homomorphism of the group $\Delta^r(K_\beta, \mathfrak{A})$ into $\Delta^r(K_\alpha, \mathfrak{A})$ induced by a simplicial mapping S_α^β of a simplicial complex K_β into a simplicial complex K_α. Let S_α^β be a simplicial mapping of a simplicial complex

K_β into a simplicial complex K_α. We recall (VII, 8.2) that S_α^β induces a homomorphism (denoted by the same letter) of $L^r(K_\beta, \mathfrak{A})$ into $L^r(K_\alpha, \mathfrak{A})$ which maps $Z_\Delta^r(K_\beta, \mathfrak{A})$ into $Z_\Delta^r(K_\alpha, \mathfrak{A})$ and $H_\Delta^r(K_\beta, \mathfrak{A})$ into $H_\Delta^r(K_\alpha, \mathfrak{A})$. Consequently (Appendix 2, 1.1), the homomorphism S_α^β of $L^r(K_\beta, \mathfrak{A})$ into $L^r(K_\alpha, \mathfrak{A})$ induces a homomorphism of $\Delta^r(K_\beta, \mathfrak{A})$ into $\Delta^r(K_\alpha, \mathfrak{A})$, likewise denoted by S_α^β and referred to as the *induced homomorphism of* $\Delta^r(K_\beta, \mathfrak{A})$ *into* $\Delta^r(K_\alpha, \mathfrak{A})$ *of the simplicial mapping* S_α^β of K_β into K_α.

REMARK 1. Two simplicial mappings S_α^β and $\bar{S}_\alpha^\beta$ of K_β into K_α are said to be *$(r, \mathfrak{A})$-homologous* if they induce identical homomorphisms of $\Delta^r(K_\beta, \mathfrak{A})$ into $\Delta^r(K_\alpha, \mathfrak{A})$.

Two mappings, $(r, \mathfrak{A})$-homologous for arbitrary r (for arbitrary $\mathfrak{A}$), are said to be *$\mathfrak{A}$-homologous* (*r-homologous*).

Finally, two simplicial mappings of K_β into K_α, $(r, \mathfrak{A})$-homologous for every r and $\mathfrak{A}$, are said to be *completely homologous*.

REMARK 2. Theorem 8.40 of Chapter VII implies the following proposition:

If K_α and K_β are unrestricted simplicial complexes, $K_{\alpha 0} \subset K_\alpha$ and $K_{\beta 0} \subset K_\beta$ closed subcomplexes, S_α^β a simplicial mapping of K_β into K_α such that $S_\alpha^\beta(K_{\beta 0}) \subseteq K_{\alpha 0}$, and $G_\alpha = K_\alpha \setminus K_{\alpha 0}$, $G_\beta = K_\beta \setminus K_{\beta 0}$, then the homomorphism $G_\alpha S_\alpha^\beta$ of $L^r(G_\beta, \mathfrak{A})$ into $L^r(G_\alpha, \mathfrak{A})$ induces a homomorphism (denoted by the same symbol) of $\Delta^r(G_\beta, \mathfrak{A})$ into $\Delta^r(G_\alpha, \mathfrak{A})$.

§2. The groups $\Delta_0^r(\mathfrak{K})$

[See Appendix 2, 4.]

§2.1. The torsion groups. Let $\mathfrak{K}$ be an arbitrary a-complex. The elements of finite order of $\Delta_0^r(\mathfrak{K})$ form a subgroup of $\Delta_0^r(\mathfrak{K})$; this subgroup, denoted by $\Theta^r(\mathfrak{K})$, is called the *r-dimensional* (rth) *torsion group of* $\mathfrak{K}$; if the group $\Theta^r(\mathfrak{K})$ is not the null group, we shall say that $\mathfrak{K}$ has r-dimensional (r-) torsion; if $\Theta^r(\mathfrak{K})$ is the null group, we shall say that $\mathfrak{K}$ is r-torsion free. It follows from 1.37 that for an arbitrary unrestricted simplicial complex K, $\Theta^0(K)$ is the null group.

If $\mathfrak{K}$ is an n-complex, $\Delta_0^n(\mathfrak{K})$ coincides with $Z_0^n(\mathfrak{K})$ and is therefore also a free group. Hence

2.11. *Every unrestricted simplicial complex is* 0-*torsion free.* [This theorem is also true for simplicial complexes which are not unrestricted. It is left to the reader to prove this. On the other hand, the a-complex consisting of the cells $\pm t^1$, $\pm t^0$ with incidence number $(t^1 : t^0) = 2$ has 0-torsion.]

2.12. *Every n-complex is n-torsion free.*

COROLLARY 2.13. 1-complexes are 1-torsion free.

EXAMPLES (see 1.4, Example 7). The projective plane and the Klein bottle have 1-torsion; for both surfaces the 1st torsion group is the group of order 2.

Let us consider any triangulation of the Möbius band (for instance, the triangulation of 1.4, Exercise 1, or the simpler triangulation of III, 3.1). Let us delete from the triangulation all boundary elements of the surface (in the case of the triangulation of 1.4, all elements on the sides AB and DC). The 1st torsion group of the resulting open subcomplex of the triangulation K is of order 2.

DEFINITION 2.14. If z^r is an integral cycle of $\mathfrak{K}$, the *order* of z^r is the order of the homology class, as an element of $\Delta_0^r(\mathfrak{K})$, containing z^r.

Clearly, only cycles weakly homologous to zero have finite order; among these, the cycles homologous to zero have order 1. In other words, the homology classes of cycles weakly homologous to zero are elements of $\Theta^r(\mathfrak{K})$ (in particular, the homology class of cycles homologous to zero is the identity of $\Theta^r(\mathfrak{K})$). By assigning to each cycle weakly homologous to zero, i.e., to each element u^r of the group $\hat{H}^r(\mathfrak{K})$, the homology class which contains it, we obtain a homomorphism of $\hat{H}^r(\mathfrak{K})$ onto $\Theta^r(\mathfrak{K})$; the kernel of this homomorphism is $H^r(\mathfrak{K})$.

Consequently,

2.1. *The groups $\Theta^r(\mathfrak{K})$ and $\hat{H}^r(\mathfrak{K})/H^r(\mathfrak{K})$ are isomorphic for every a-complex $\mathfrak{K}$.*

§2.2. The groups $\Delta_{00}^r(\mathfrak{K})$. The factor group

$$\Delta_0^r(\mathfrak{K})/\Theta^r(\mathfrak{K})$$

contains no nonzero element of finite order; this follows from the fact that $\Theta^r(\mathfrak{K})$ is, by definition, a division closed subgroup of $\Delta_0^r(\mathfrak{K})$.

2.21. *The groups $\Delta_0^r(\mathfrak{K})/\Theta^r(\mathfrak{K})$ and $\Delta_{00}^r(\mathfrak{K}) = Z_0^r(\mathfrak{K})/\hat{H}^r(\mathfrak{K})$ are isomorphic.*

Proof. Let us consider the homomorphism of $Z_0^r(\mathfrak{K})$ onto $\Delta_0^r(\mathfrak{K})$ which assigns to every cycle $z^r \in Z_0^r(\mathfrak{K})$ its homology class. Since $\hat{H}^r(\mathfrak{K})$ is the inverse image of $\Theta^r(\mathfrak{K})$ under this homomorphism, we have, by Appendix 2, Theorem 1.1, an isomorphism between

$$Z_0^r(\mathfrak{K})/\hat{H}^r(\mathfrak{K}) \quad \text{and} \quad \Delta_0^r(\mathfrak{K})/\Theta^r(\mathfrak{K}).$$

This completes the proof.

Since $\Theta^r(\mathfrak{K})$ consists of elements of finite order, its rank is zero. Hence, by Appendix 2, 3.2, $\Delta_{00}^r(\mathfrak{K}) = \Delta_0^r(\mathfrak{K})/\Theta^r(\mathfrak{K})$ and $\Delta_0^r(\mathfrak{K})$ have the same rank $\pi^r(\mathfrak{K})$.

Summing up, we have

2.2. *Let $\mathfrak{K}$ be an arbitrary a-complex. The groups*

$$\Delta_0^r(\mathfrak{K})/\Theta^r(\mathfrak{K}) \quad \text{and} \quad \Delta_0^r(\mathfrak{K}) = Z_0^r(\mathfrak{K})/\hat{H}^r(\mathfrak{K})$$

are isomorphic; they contain no nonzero elements of finite order; their rank is the rth *Betti number of $\mathfrak{K}$.*

§2.3. **Finite a-complexes. Homology bases.** We shall now assume until the end of this section that $\mathfrak{K}$ is a finite a-complex and that $2\rho^r$ is the number of r-elements in $\mathfrak{K}$; in particular, if $\mathfrak{K}$ is the complex of all the oriented elements of a simplicial complex (or a complex of convex polyhedral domains) K, then ρ^r is the number of r-elements of K; the groups $L_0^r(\mathfrak{K})$, $Z_0^r(\mathfrak{K})$, $H_0^r(\mathfrak{K})$, $\Delta_0^r(\mathfrak{K})$, etc. will be written simply as L_0^r, Z_0^r, H_0^r, Δ_0^r.

The rank of L_0^r is obviously ρ^r; consequently the subgroups Z_0^r and H_0^r of L_0^r and the factor group Δ_0^r of Z_0^r also have finite rank (see Appendix 2, 3.2). Hence

2.31. *The Betti numbers of a finite complex are finite.*

The group L_0^r and hence also Z_0^r, H_0^r, and Δ_0^r have a finite number of generators: the chains t_i^r, $i = 1, 2, \cdots, \rho^r$, are the generators of L_0^r. Since Δ_{00}^r, by Theorem 2.2, contains no nonzero elements of finite order and has rank π^r, we have

2.32. *If $\mathfrak{K}$ is a finite complex, Δ_{00}^r is the free Abelian group with π^r independent generators.*

Since all the elements of Θ^r are of finite order and Θ^r has a finite number of generators, it follows that:

2.33. *The torsion groups of a finite a-complex are finite.*

Moreover, it follows from Appendix 2, Theorem 4.35, that

2.34. *The group Δ_0^r of a finite a-complex is the direct sum of the finite torsion group Θ^r and a free group of rank π^r isomorphic to Δ_{00}^r.*

If $\mathfrak{K}$ is a finite a-complex, the group Δ_0^r, as is the case with all Abelian groups with a finite number of generators, is uniquely determined by its rank π^r and its torsion coefficients (Appendix 2, Theorem 4.351). The torsion coefficients of $\Delta_0^r(\mathfrak{K})$ are referred to as the *torsion coefficients of the complex* $\mathfrak{K}$.

Finally, we shall introduce the following definition, which is frequently applied:

2.35. A system of lirh (VII, Def. 6.42) integral r-cycles

$$z_1^r, \cdots, z_s^r \tag{2.35}$$

of a complex $\mathfrak{K}$ is called an *r-dimensional homology basis*, or a *$(J, \mathfrak{R})$-basis*, of $\mathfrak{K}$ if the homology classes of the cycles z_i^r form a basis of the free group $\Delta_{00}^r(\mathfrak{K}) = Z^r(\mathfrak{K})/\hat{H}^r(\mathfrak{K})$.

The same definition may obviously also be expressed as follows:

2.350. The lirh integral cycles (2.35) form a $(J, \mathfrak{R})$-basis for $\mathfrak{K}$ if every integral r-cycle $z^r \in Z_0^r(\mathfrak{K})$ can be written in the form

$$z^r = v^r + \sum c_i z_i^r,$$

where $v^r \in \hat{H}^r(\mathfrak{K})$ and the coefficients c_i are integers (uniquely determined, as is easily seen).

REMARK. Let $\mathfrak{A}$ be either of the algebraic fields $\mathfrak{R}$ or J_m, m a prime. *Every maximal set of lirh r-cycles* (2.35) *of* $\mathfrak{K}$ *over the coefficient field* $\mathfrak{A}$ *is called an r-dimensional homology basis of* $\mathfrak{K}$ *over the coefficient field* $\mathfrak{A}$ (if $\mathfrak{A} = J_m$, we say *homology basis* (mod m)). It is easy to see that these bases can also be defined as systems of lirh cycles (2.35) (over the coefficient field $\mathfrak{A}$) with the property that every cycle $z^r \in Z^r(\mathfrak{K}, \mathfrak{A})$ is homologous to a linear combination $\sum a_i z_i^r$, $a_i \in \mathfrak{A}$, of cycles of the system (2.35), i.e., can be written in the form $z^r = u^r + \sum a_i z_i^r$, where $u^r \in H^r(\mathfrak{K}, \mathfrak{A})$.

EXERCISE. Construct examples of 1-dimensional and 2-dimensional $(J, \mathfrak{R})$-bases for the triangulations of the torus, Klein bottle, and projective plane considered in 1.4. For the last two surfaces also construct homology bases (mod 2) (for dimensions 1 and 2).

§2.4. The Euler-Poincaré formula for a finite n-dimensional a-complex. In this subsection we shall take $\mathfrak{A} = J$.

DEFINITION 2.41. The number $\sum_{r=1}^{n} (-1)^r \rho^r$ is called the *Euler characteristic* of the complex $\mathfrak{K}$. [As always, $2\rho^r$ is the number of r-elements of the a-complex $\mathfrak{K}$; this means that in the fundamental case, when $\mathfrak{K}$ is the complex of the oriented elements of a triangulation K, ρ^r is the number of r-simplexes of K.]

THE EULER-POINCARÉ FORMULA is

$$\sum_{r=1}^{n} (-1)^r \rho^r = \sum_{r=0}^{n} (-1)^r \pi^r. \tag{2.4}$$

To prove (2.4) we recall that:

(2.41)
$$\begin{cases} 1^\circ)\ \rho^r \text{ is the rank of the group } L_0{}^r = L_0{}^r(\mathfrak{K}): \\ \qquad\qquad \rho^r = \rho(L_0{}^r); \\ 2^\circ)\ \Delta \text{ is a homomorphism of } L_0{}^r \text{ onto } H_0{}^{r-1}; \text{ consequently, } L_0{}^r/Z_0{}^r \end{cases}$$

is isomorphic to $H_0{}^{r-1}$. Hence they have the same rank:

$$\rho(L_0{}^r/Z_0{}^r) = \rho(H_0{}^{r-1}); \tag{2.42}$$

3°) if $A/B = C$ is true for three groups A, B, C, then

$$\rho A = \rho B + \rho C$$

[Appendix 2, (3.2)]. Hence, by (2.41) and (2.42),

$$\rho^r = \rho(Z_0{}^r) + \rho(H_0{}^{r-1})$$

for $r = 0, 1, \cdots, n$; i.e.,

$$\rho(Z_0{}^r) = \rho^r - \rho(H_0{}^{r-1}). \tag{2.43}$$

Moreover, according to the definitions, $L_0^{-1} = Z_0^{-1} = H_0^{-1}$ is the null group, so that

$$\rho(H_0^{-1}) = 0. \tag{2.44}$$

On the other hand, π^r is the rank of the group $\Delta_0^r = Z_0^r/H_0^r$; hence,

$$\pi^r = \rho(Z_0^r) - \rho(H_0^r) \qquad (0 \leq r \leq n). \tag{2.45}$$

Inserting (2.43) into (2.45), we obtain the formula

$$\pi^r = \rho^r - \rho(H_0^{r-1}) - \rho(H_0^r) \qquad (0 \leq r \leq n), \tag{2.46}$$

which has independent interest.

Since H_0^n is the null group,

$$\rho(H_0^n) = 0, \qquad \pi^n = \rho^n - \rho(H_0^{n-1}). \tag{2.47}$$

Multiplying both sides of (2.46) by $(-1)^r$, summing from $r = 0$ to $r = n$, and keeping (2.44) and (2.47) in mind, we obtain precisely (2.4).

Examples of the application of the Euler-Poincaré formula will be given in the following section.

§3. Pseudomanifolds

In this section we shall take K to be a finite simplicial complex whose oriented simplexes form an a-complex; in particular, K may be thought of as a triangulation or as an open subcomplex of a triangulation (the only two cases which will be considered in the sequel).

§3.1. Pseudomanifolds. In what follows we shall use the terminology defined in VI, 5 (see VI, 5, Defs. 5.23 and 5.24).

DEFINITION 3.11. *A strongly connected n-complex K^n is called an n-dimensional (combinatorial) pseudomanifold if every $(n - 1)$-simplex of K^n is a face of precisely two n-simplexes of K^n.*

REMARK. An unrestricted pseudomanifold is usually referred to as a *closed pseudomanifold.*

In addition to the pseudomanifolds of Def. 3.11, we shall also consider *pseudomanifolds with boundary* which are characterized by

DEFINITION 3.12. A finite unrestricted strongly connected simplicial n-complex K^n is called an *n-dimensional (combinatorial) pseudomanifold with boundary* if every $(n-1)$-simplex of K^n is a face of either one or two n-simplexes of K^n. The subcomplex of K^n consisting of all the faces (proper or not) of the $(n - 1)$-simplexes of K^n which are faces of precisely one n-simplex of K^n is referred to as the *boundary* or *edge* of K^n.

Every closed pseudomanifold is a special case of a pseudomanifold with boundary: the boundary is in this case the empty set.

EXAMPLES OF PSEUDOMANIFOLDS. 1°. A simple closed broken line

$$e_1e_2 \cdots e_{s-1}e_se_1$$

is an example of an unrestricted 1-dimensional pseudomanifold K. Here K consists of the vertices $e_1, e_2, \cdots, e_s$ and the 1-simplexes (e_1e_2), $(e_2e_3), \cdots, (e_{s-1}e_s), (e_se_1)$.

It is easily seen that this example (for $s = 2, 3, \cdots$) exhausts, up to an isomorphism, all the unrestricted 1-dimensional pseudomanifolds.

2°. An arbitrary triangulation of a closed surface (see Chapter III) is an example of a closed 2-dimensional pseudomanifold. In the same way,

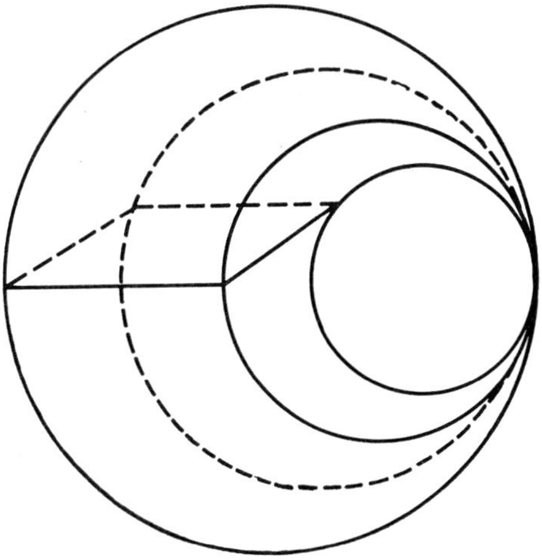

FIG. 121

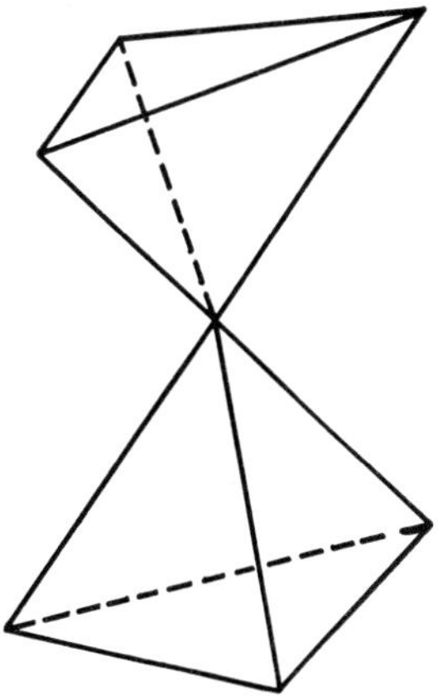

FIG. 122

the triangulations of surfaces with boundary are examples of 2-dimensional pseudomanifolds with boundary.

3°. If we identify two opposite vertices of an octahedron (Fig. 121), we obtain a 2-dimensional curved polyhedron, whose triangulations are examples of 2-dimensional closed pseudomanifolds which are not triangulations of surfaces. The corresponding 3-dimensional figure (considered in some triangulation) is a 3-dimensional pseudomanifold with boundary.

4°. If, on the contrary, we identify two vertices of a pair of tetrahedra (Fig. 122), the result is not even a pseudomanifold with boundary (why?).

5°. An arbitrary subdivision of the complex $|T^n|$, where T^n is an n-simplex or in general an n-dimensional convex polyhedral domain, is an n-dimensional pseudomanifold with boundary (the strong connectedness of $|T^n|$ follows from VI, 5.12 and 5.252). Moreover, the edge is an $(n - 1)$-dimensional closed pseudomanifold.

A complex consisting of a single n-simplex is an n-dimensional pseudomanifold (for $n > 0$ it is obviously not closed).

§3.2. Orientable pseudomanifolds.

DEFINITION 3.21. If T_i^n, T_h^n are two n-simplexes with a common $(n-1)$-face T_j^{n-1}, and t_j^{n-1} is any orientation of T_j^{n-1}, then the orientations t_i^n and t_h^n of T_i^n, T_h^n are said to be *coherent* if

$$(t_i^n : t_j^{n-1}) = -(t_h^n : t_j^{n-1}).$$

This definition is clearly independent of the choice of t_j^{n-1}.

Now let

$$T_1^n, \cdots, T_s^n \tag{3.21}$$

be the n-simplexes of an n-dimensional pseudomanifold K^n. A set of orientations

$$t_1^n, \cdots, t_s^n \tag{3.22}$$

of the simplexes (3.21) is said to be coherent if the orientations t_i^n and t_h^n of an arbitrary pair of simplexes T_i^n, T_h^n of (3.21) having a common $(n-1)$-face are coherent.

DEFINITION 3.22. Let K^n be either an n-dimensional pseudomanifold or pseudomanifold with boundary. K^n is said to be *orientable* if there is a set of coherent orientations of all the n-simplexes of K^n; in the contrary case, K^n is said to be *nonorientable*.

Hence if K^n is a nonorientable pseudomanifold (or a nonorientable pseudomanifold with edge), for every choice of a set of orientations of all the n-simplexes $T_i^n \in K^n$ there is at least one pair of noncoherent orientations t_i^n, t_h^n of two adjacent simplexes T_i^n, T_h^n.

REMARK 1. If K^n is a pseudomanifold with boundary and K^{n-1} is its edge, it is easily seen that K^n is orientable if, and only if, $K^n \setminus K^{n-1}$ is orientable (the latter pseudomanifold is not closed). Hence all questions relating to the orientability of pseudomanifolds with boundary lead immediately to analogous questions for pseudomanifolds without edge (but not closed).

EXAMPLES. 1°. The edge of an n-simplex (that is, the complex

$$K^{n-1} = | T^n | \setminus T^n)$$

is an orientable $(n-1)$-dimensional pseudomanifold. To show this it is enough to choose any orientation t^n of the simplex T^n and orientations t_i^{n-1} of its $(n-1)$-faces such that $(t^n : t_i^{n-1}) = 1$.

2°. The triangulations of the nonorientable closed surfaces (for instance, the projective plane and the Klein bottle) are nonorientable closed pseudomanifolds.

3°. Let us take any triangulation of a nonorientable surface with edge (for instance, the Möbius band) and delete all the boundary elements of the triangulation. The result is a nonorientable pseudomanifold (not closed).

EXERCISE. Prove the assertions in 2° and 3° (see III, 4).

If (3.21) is the collection of all n-simplexes of an orientable pseudomanifold K^n and the orientations (3.22) are coherent, then the orientations $-t_1^n, \cdots, -t_s^n$ are obviously also coherent.

On the other hand, if the orientations t_i^n and t_h^n of two simplexes T_i^n, T_h^n with a common $(n-1)$-face T_j^{n-1} are coherent, then the orientations t_i^n and $-t_h^n$ are noncoherent. Therefore, if a definite orientation t_i^n of any one simplex T_i^n is prescribed, the orientations coherent with t_i^n of all the simplexes T_h^n having a common $(n-1)$-face with T_i^n are uniquely defined. Consequently (because of the strong connectedness of K^n), the coherent orientations of all the remaining simplexes of (3.21) are also determined uniquely.

Hence

3.23. There exist precisely two sets of coherent orientations of all the n-simplexes (3.21) of an n-dimensional orientable pseudomanifold K^n: if (3.22) is one set of coherent orientations of the simplexes (3.21), the second set is

$$-t_1^n, \cdots, -t_s^n.$$

3.24. *Let* (3.21) *be the collection of all n-simplexes of an n-dimensional pseudomanifold* K^n. *The orientations* (3.22) *are coherent if, and only if,*

$$z_1^n = \sum t_i^n \tag{3.24}$$

is a cycle.

This is an immediate consequence of the fact that the value of the chain Δz_1^n on an arbitrary t_j^{n-1} is

$$(t_i^n : t_j^{n-1}) + (t_h^n : t_j^{n-1}),$$

where $|t_i^n|$ and $|t_h^n|$ are the simplexes of K^n which have $|t_j^{n-1}|$ as their common face.

DEFINITION 3.25. If K^n is an orientable pseudomanifold (or an orientable pseudomanifold with boundary K^{n-1}), and (3.22) is a set of coherent orientations of all the n-simplexes of K^n, the cycle $z_1^n = \sum t_i^n$ is called an *orientation of* K^n.

This definition and 3.23 imply

3.26. If K^n is an orientable pseudomanifold or a pseudomanifold with boundary, and (3.22) is a set of coherent orientations of the n-simplexes of K^n, then K^n has *precisely the two orientations* $z_1^n = \sum_i t_i^n$ and $-z_1^n = \sum_i - t_i^n$.

The fundamental theorems on orientability are Theorems 3.27 and 3.28.

THEOREM 3.27. *The group* $\Delta_0^n(K^n) = Z_0^n(K^n)$ *of an orientable n-dimensional pseudomanifold* K^n *is the infinite cyclic group generated by either one of the two orientations of* K^n.

This proposition obviously follows from

3.270. Every n-cycle $z^n \in Z^n(K^n, \mathfrak{A})$ of a pseudomanifold K^n is of the form $z^n = az_1^n$, where z_1^n is any orientation of K^n and $a \in \mathfrak{A}$.

To prove this it suffices to show that every cycle z^n has the same value on all the coherent orientations

$$t_1^n, \cdots, t_s^n.$$

This assertion is a consequence of the strong connectedness of K^n and, to see that it is so, it is enough to show that it holds for two simplexes $|t_i^n|$ and $|t_h^n|$ with a common $(n-1)$-face $|t_j^{n-1}|$. However, since $z^n = \sum a_i t_i^n$ is a cycle,

$$0 = \Delta z^n(t_j^{n-1}) = a_i(t_i^n : t_j^{n-1}) + a_h(t_h^n : t_j^{n-1});$$

and since, in virtue of coherence,

$$(t_h^n : t_j^{n-1}) = -(t_i^n : t_j^{n-1}) \neq 0,$$

it follows that $a_h = a_i$. This completes the proof.

REMARK 2. On the other hand, since, in the notation of 3.270, a chain az_1^n is a cycle, we have proved the following proposition:

3.271. If K^n is an orientable n-dimensional pseudomanifold, the group $\Delta^n(K^n, \mathfrak{A}) = Z^n(K^n, \mathfrak{A})$ is isomorphic to $\mathfrak{A}$.

REMARK 3. Using Theorem 3.27, we may now characterize an orientation of a pseudomanifold in the following terms:

3.250. *An orientation of an n-dimensional orientable pseudomanifold* K^n *is any generator of the group* $\Delta_0^n(K^n) = Z_0^n(K^n)$.

THEOREM 3.28. *If* K^n *is a nonorientable pseudomanifold (with or without boundary),* $Z_0^n(K^n)$ *is the null group.*

We shall prove, indeed, that the existence on K^n of a nonvanishing integral cycle

$$z^n = \textstyle\sum_i a_i t_i^n \tag{3.28}$$

implies both the absence of an edge and the orientability of K^n.

We shall show first that $|a_i| = |a_h|$ for arbitrary i, h if the cycle (3.28) exists. It suffices to prove this equality on the assumption that T_i^n and T_h^n have a common face T_j^{n-1}. But in that case

$$0 = \Delta z^n(t_j^{n-1}) = a_i(t_i^n : t_j^{n-1}) + a_h(t_h^n : t_j^{n-1})$$

for an arbitrary orientation t_j^{n-1} of the common face; hence $a_i = \pm a_h$.

Since $z^n \neq 0$, denoting by a the common value of all the $|a_i|$, we obtain on K^n the integral cycle

$$z_1^n = (1/a)z^n = \sum \varepsilon_i t_i^n, \qquad \varepsilon_i = a_i/a = \pm 1.$$

Suppose that the edge of K^n is nonempty and therefore contains an $(n-1)$-simplex $|t^{n-1}|$. Let $|t_i^n|$ be the unique n-simplex which has $|t^{n-1}|$ as a face. Then

$$(\Delta z_1^n \cdot t^{n-1}) = \varepsilon_i(t_i^n : t^{n-1}) \neq 0,$$

which contradicts the fact that z_1^n is a cycle.

Hence K^n has no edge. In order to prove that K^n is an orientable pseudomanifold, it suffices to set $t'^n_i = \varepsilon_i t_i^n$; it is then immediately obvious that the cycle

$$z_1^n = \sum \varepsilon_i t_i^n = \sum t'^n_i$$

is an orientation of K^n. This proves Theorem 3.28.

§3.3. The groups $\Delta_m^n(K^n)$ of a nonorientable n-dimensional pseudomanifold. Disorienting sequences.

THEOREM 3.31. *The group $\Delta_2^n(K^n) = Z_2^n(K^n)$ is of order 2 for every pseudomanifold K^n.*

Proof. If $z^n = \sum a_i t_i^n \in Z_2^n(K^n)$, then either $a_i = 0$ for all i, or $a_i = 1$ for all i. Indeed, if $a_i = 0$, $a_h = 1$ for two distinct simplexes t_i^n, t_h^n with a common face t_j^{n-1}, then

$$(\Delta z^n \cdot t_j^{n-1}) = 1.$$

Hence $a_i = a_h$ for two simplexes t_i^n, t_h^n with a common face and the same follows for any two simplexes t_i^n, t_h^n from the strong connectedness of K^n.

On the other hand, the chain $z^n = \sum t_i^n$ is a cycle (mod 2) on an arbitrary pseudomanifold K^n, since every simplex $|t_j^{n-1}| \in K^n$ is a face of precisely two simplexes $|t_i^n| \in K^n$; consequently,

$$(\Delta z^n \cdot t_j^{n-1}) = 1 + 1 = 0 \pmod 2.$$

COROLLARY. The nth Betti number (mod 2) of every n-dimensional pseudomanifold is 1.

We shall give still another definition of the orientability of a pseudomanifold which is convenient in certain cases.

DEFINITION 3.32. A sequence of oriented n-simplexes

$$t_1^n, t_2^n, t_3^n, \cdots, t_s^n \tag{3.32}$$

is called a *disorienting sequence* if:

1. The simplexes $|t_i^n|$ and $|t_{i+1}^n|$ have a common $(n-1)$-face $|t_i^{n-1}|$ for $i = 1, 2, \cdots, s-1$; and the simplexes $|t_s^n|$ and $|t_1^n|$ also have a common face $|t_s^{n-1}|$.

2. The orientations t_i^n and t_{i+1}^n $(1 \leq i \leq s - 1)$ are coherent; the orientations t_s^n and t_1^n, however, are noncoherent, so that $(t_1^n : t_s^{n-1}) = (t_s^n : t_s^{n-1})$.

3.33. *In order that an n-dimensional pseudomanifold K^n be nonorientable it is necessary and sufficient that the set of oriented n-simplexes of K^n contain at least one disorienting sequence.*

Indeed, suppose that (3.32) is a disorienting sequence. If $z^n \in Z^n(K^n)$, then it is easy to see that

$$(z^n \cdot t_1^n) = (z^n \cdot t_2^n) = \cdots = (z^n \cdot t_s^n) = a.$$

But then

$$(\Delta z^n \cdot t_s^{n-1}) = 2a$$

and hence $a = 0$. Therefore, every cycle $z^n \in Z^n(K^n)$ is 0 on the simplexes $t_1^n, t_2^n, \cdots, t_s^n$. If K^n were orientable, z^n would define one of the orientations of K^n and z^n would not be 0 on any of the n-simplexes. Hence the existence of a disorienting sequence assures the nonorientability of K^n.

Now let K^n be nonorientable. We shall construct a disorienting sequence as follows:

Let t_1^n be any orientation of the simplex $T_1^n \in K^n$ and suppose that the oriented simplexes $t_1^n, t_2^n, \cdots, t_k^n$ are coherently oriented and that the simplexes $| t_1^n |, \cdots, | t_k^n |$ and their faces form a strongly connected complex, with each simplex after the first adjoining its predecessor in the sequence.

In consequence of the nonorientability of K^n, the simplexes

$$| t_1^n |, \cdots, | t_k^n |$$

cannot exhaust the set of n-simplexes of K^n. Because K^n is strongly connected there is an n-simplex $| t_{k+1}^n |$ in K^n distinct from all the simplexes $| t_1^n |, \cdots, | t_k^n |$ which has a face in common with at least one of

$$| t_1^n |, \cdots, | t_k^n |$$

and such that $| t_1^n |, \cdots, | t_{k+1}^n |$ form a strongly connected complex.

Two cases are possible:

1°. There is an orientation t_{k+1}^n of the simplex $| t_{k+1}^n |$ coherent with the orientations $t_1^n, \cdots, t_k^n$.

2°. Such an orientation t_{k+1}^n does not exist.

In virtue of the nonorientability of K^n, Case 2° will occur sooner or later. Then $| t_{k+1}^n |$ will adjoin (i.e., have a face in common with) at least two previously chosen simplexes, say $| t_i^n |$ and $| t_h^n |$, $i < h$, in such a way that one of the orientations of $| t_{k+1}^n |$, say t_{k+1}^n, is coherent with both t_h^n and $-t_i^n$.

Then, connecting t_i^n and t_h^n with a chain of coherently oriented simplexes $t_i^n, t_{j(1)}^n, t_{j(2)}^n, \cdots, t_{j(s)}^n, t_h^n$, with $j(1), j(2), \cdots, j(s) < k + 1$, we obtain a disorienting sequence

$$t_i^n, t_{j(1)}^n, t_{j(2)}^n, \cdots, t_{j(s)}^n, t_h^n, t_{k+1}^n.$$

This proves 3.33.

3.34. *If K^n is a nonorientable n-dimensional pseudomanifold and m is an odd integer,*

$$\Delta_m^n(K^n) = Z_m^n(K^n)$$

is the null group.

In fact, if $z^n \in Z^n(K^n)$ and (3.32) is a disorienting sequence, then it is easy to see that (see the proof of Theorem 3.270)

$$(z^n \cdot t_1^n) = (z^n \cdot t_2^n) = \cdots = (z^n \cdot t_s^n) = a \in J_m .$$

Hence

$$0 = (\Delta z^n \cdot t_s^{n-1}) = 2a.$$

But since m is odd, $2a \in J_m$ is equal to zero only if $a = 0$. This is what we wished to prove.

Remark. In IX, Theorem 4.43, we shall show that if K^n is a nonorientable n-dimensional pseudomanifold and m is even, then $\Delta_m^n(K^n)$ is the group of order 2. In IX, Theorem 1.7 we shall prove that a pseudomanifold K^n is orientable if, and only if, $\Theta^{n-1}(K^n)$ is the null group and that if K^n is nonorientable, $\Theta^{n-1}(K^n)$ is the group of order 2.

§4. Addenda and examples

§4.1. The Betti groups of the complexes $| T^n |$ and $\dot{T}^n = | T^n | \setminus T^n$. The results of §§2 and 3 enable us to complete the investigation of the groups Δ^r of the complexes $| T^n |$ and $\dot{T}^n = | T^n | \setminus T^n$. For, since $\dot{T}^n$ is an orientable pseudomanifold (3.2, Example 1°) and therefore a connected complex we have, by 1.35, 1.53, and 3.271,

4.11. *The groups $\Delta^0(\dot{T}^n, \mathfrak{A})$ and $\Delta^{n-1}(\dot{T}^n, \mathfrak{A})$ are isomorphic to $\mathfrak{A}$; the groups $\Delta^r(\dot{T}^n, \mathfrak{A})$ are null groups for $0 < r < n - 1$.*

If t^n is any orientation of T^n, Δt^n is a generator of $Z_0^{n-1}(\dot{T}^n)$ and the group $Z_0^{n-1}(\dot{T}^n, \mathfrak{A})$ consists of cycles of the form $a\Delta t^n$, with $a \in \mathfrak{A}$. Since all such cycles are homologous to zero in $| T^n |$ (as boundaries of chains at^n), we have

4.12. *The groups $\Delta^0(| T^n |, \mathfrak{A})$ are isomorphic to $\mathfrak{A}$; the groups $\Delta^r(| T^n |, \mathfrak{A})$ are null groups for $r > 0$.*

Corollary. The 0th and $(n - 1)$st Betti numbers of $\dot{T}^n$ are equal to 1; the remaining Betti numbers of $\dot{T}^n$ are zero. All the Betti numbers of $| T^n |$ (with the exception of the 0th, which is obviously 1) are 0.

§4.2. Surfaces. A triangulation K^2 of a closed (orientable) surface is a 2-dimensional closed (orientable) pseudomanifold. Hence

$$\pi^0(K^2) = 1,$$

while

$$\pi^2(K^2) = 1, \qquad \text{if } K^2 \text{ is orientable}$$

and

$$\pi^2(K^2) = 0, \qquad \text{if } K^2 \text{ is nonorientable.}$$

Let p be the genus of the surface (if the surface is orientable, p is the number of its handles; if the surface is nonorientable, p is 1 less than the number of holes fitted with Möbius bands (see III, Def. 7.11). Then (see III, 7.1)

$$\chi(K^2) = 2 - 2p$$

for orientable surfaces and

$$\chi(K^2) = 1 - p$$

for nonorientable surfaces.

On the other hand, since we always have

$$\chi(K^2) = \pi^0(K^2) - \pi^1(K^2) + \pi^2(K^2),$$

then

$$1 - \pi^1(K^2) + 1 = 2 - 2p,$$

i.e.,

$$\pi^1(K^2) = 2p$$

for orientable surfaces, and

$$1 - \pi^1(K^2) + 0 = 1 - p,$$

i.e.,

$$\pi^1(K^2) = p$$

for nonorientable surfaces.

Hence

4.21. *If K^2 is a triangulation of a closed orientable surface of genus p, then*

$$\pi^0(K^2) = 1, \qquad \pi^1(K^2) = 2p, \qquad \pi^2(K^2) = 1.$$

If K^2 is a triangulation of a closed nonorientable surface of genus p, then

$$\pi^0(K^2) = 1, \qquad \pi^1(K^2) = p, \qquad \pi^2(K^2) = 0.$$

Using this and the expression for the genus of a surface in terms of its connectivity (see III, 7.1) we obtain

4.22. *The connectivity of an orientable surface is equal to its* 1*st Betti number; the connectivity of a nonorientable surface is* 1 *greater than its* 1*st Betti number.*

By means of III, Theorem 7.21 and 4.21 above, we may finally express the fundamental theorem of surface topology as

4.23. *Two closed surfaces are homeomorphic if, and only if, their corresponding Betti numbers are equal.*

§4.3. Simple pseudomanifolds. Elementary triangulations.

DEFINITION 4.31. A combinatorial n-dimensional pseudomanifold Q^n is said to be *simple* if it is orientable and $\Delta_0^r(Q^n)$ $(0 < r < n)$ is the null group.

A triangulation K is said to be *elementary* if it can be represented in the form

$$K = Q \cup K_0, \qquad Q \cap K_0 = 0,$$

where Q is a simple n-dimensional pseudomanifold and K_0 is a triangulation each simplex of which is a proper face of at least one simplex of Q (hence K_0 has dimension $\leq n - 1$).

REMARK. With these conditions, the pseudomanifold Q is obviously an open subcomplex of K.

Let x^n be any orientation of Q and let J_x be the subgroup of $Z_0^{n-1}(K_0)$ consisting of all the cycles $c\Delta x^n$, with c an integer. We note that

$$Z_0^{n-1}(K_0) = \Delta_0^{n-1}(K_0),$$

so that J_x may also be thought of as a subgroup of $\Delta_0^{n-1}(K_0)$.

We may then assert

4.32. *The groups* $\Delta_0^r(K)$ *and* $\Delta_0^r(K_0)$ *are isomorphic for* $r \leq n - 2$; *the group* $\Delta_0^{n-1}(K)$ *is isomorphic to the factor group* $Z_0^{n-1}(K_0)/J_x$; *as for the group* $\Delta_0^n(K)$, *two cases are possible:*

a) $\Delta x^n = 0$; *then* $\Delta_0^n(K) = Z_0^n(K)$ *consists of all the chains* cx^n, c *an integer, that is, it is infinite cyclic.*

b) $\Delta x^n \neq 0$; *then* $\Delta_0^n(K)$ *is the null group.*

Proof. Suppose $r \leq n - 1$. Every homology class $\mathfrak{z}_0^r \in \Delta_0^r(K_0)$ is contained in a homology class $\mathfrak{z}^r = f(\mathfrak{z}_0^r) \in \Delta_0^r(K)$; hence f is a homomorphism of $\Delta_0^r(K_0)$ into $\Delta_0^r(K)$. *The homomorphism* f *maps* $\Delta_0^r(K_0)$ *onto* $\Delta_0^r(K)$. To see this, *it is enough to prove that every cycle* $z^r \in Z_0^r(K)$ *is homologous to a cycle* $z_0^r \in Z_0^r(K_0)$ *in* K.

We shall prove the last assertion.

By hypothesis, there exists a chain $x_Q^{r+1} \in L^{r+1}(Q)$ such that

$$Qz^r = \Delta_Q x_Q^{r+1},$$

whence it follows that (we write Δx_Q^{r+1} instead of $\Delta_K x_Q^{r+1}$)

$$\Delta x_Q^{r+1} = Qz^r + y_0^r, \tag{4.321}$$

where

$$y_0^r = K_0 \Delta x_Q^{r+1} \in L_0^r(K_0).$$

From (4.321) we obtain

$$\Delta x_Q^{r+1} = Qz^r + y_0^r = (Qz^r + K_0 z^r) - (K_0 z^r - y_0^r) = z^r - z_0^r,$$

with

$$\begin{cases} z_0^r = K_0 z^r - y_0^r \in L_0^r(K_0), \\ z_0^r = z^r - \Delta x_Q^{r+1} \in Z_0^r(K), \end{cases}$$

so that $z_0^r \in Z_0^r(K_0)$, and the assertion is proved.

Hence f if a homomorphism of $\Delta_0^r(K_0)$ *onto* $\Delta_0^r(K)$.

The kernel of f is the set of all homology classes $\mathfrak{z}_0^r \in \Delta_0^r(K_0)$ which are contained in the identity of $\Delta_0^r(K)$, that is, which consist of the cycles

$$z_0^r \in Z_0^r(K_0) \cap H_0^r(K). \tag{4.322}$$

We now complete the proof of the theorem in three steps.

To begin with, suppose $r \leq n - 2$. If

$$z_0^r = \Delta x^{r+1}, \qquad x^{r+1} \in L_0^{r+1}(K),$$

then $Qx^{r+1} \in Z_0^{r+1}(Q)$ and consequently

$$Qx^{r+1} = \Delta_Q x_Q^{r+2}, \qquad x_Q^{r+2} \in L_0^{r+2}(Q).$$

Hence

$$\Delta x_Q^{r+2} = Qx^{r+1} + y_0^{r+1}, \qquad y_0^{r+1} \in L_0^{r+1}(K_0).$$

Since $\Delta Q x^{r+1} + \Delta y_0^{r+1} = \Delta\Delta x_Q^{r+2} = 0,$

$$\Delta Q x^{r+1} = -\Delta y_0^{r+1},$$

i.e.,

$$z_0^r = \Delta x^{r+1} = \Delta Q x^{r+1} + \Delta K_0 x^{r+1} = \Delta(K_0 x^{r+1} - y_0^{r+1}) \in H_0^r(K_0).$$

Therefore, $z_0^r \in Z_0^r(K_0) \cap H_0^r(K)$ implies that $z_0^r \in H_0^r(K_0)$, so that the kernel of f is the identity of $\Delta_0^r(K_0)$. This means that *if $r \leq n - 2$, f is an isomorphism.*

Suppose $r = n - 1$. If $z_0^{n-1} \in Z_0^{n-1}(K_0) \cap H_0^{n-1}(K)$, i.e., if $z_0^{n-1} = \Delta y^n$,

$y^n \in L_0{}^n(K)$, then (since K_0 has no n-simplexes) $y^n \in Z_0{}^n(Q)$. Hence $y^n = cx^n$ and $z_0{}^{n-1} = c\Delta x^n \in J_x$. Therefore, the kernel of the homomorphism f of $\Delta_0{}^{n-1}(K_0)$ onto $\Delta_0{}^{n-1}(K)$ is J_x and we have proved that $\Delta_0{}^{n-1}(K_0)/J_x$ is isomorphic to $\Delta_0{}^{n-1}(K)$.

It remains to consider the case $r = n$.

Suppose $z^n \in Z_0{}^n(K)$. Since K_0 has no n-simplexes, $z^n \in L_0{}^n(Q)$ and so $z^n \in Z_0{}^n(Q)$, that is, $z^n = cx^n$, c an integer. But a chain cx^n is a cycle of K if, and only if, $\Delta x^n = 0$. This completes the proof.

Now let $S_\alpha{}^\beta$ be a simplicial mapping of an elementary triangulation $K_\beta = Q_\beta \cup K_{\beta 0}$ onto a triangulation K_α with the further assumption that $S_\alpha{}^\beta$ restricted to Q_β is an isomorphism in the following sense:

1°. Every simplex $T_\beta{}^r \in Q_\beta$ is mapped onto a simplex $S_\alpha{}^\beta T_\beta{}^r \in K_\alpha$ of the same dimension r.

2°. If

$$T_\alpha{}^r = S_\alpha{}^\beta T_\beta{}^r, \qquad T_\beta{}^r \in Q_\beta,$$

then $T_\beta{}^r$ is the unique simplex of K_β mapped by $S_\alpha{}^\beta$ onto $T_\alpha{}^r$.

Remark. Since $S_\alpha{}^\beta$ is a simplicial mapping, the relation *proper face of* between simplexes is of course preserved by $S_\alpha{}^\beta$.

Simplicial mappings $S_\alpha{}^\beta$ satisfying these conditions will be called *admissible.*

4.33. *Let $S_\alpha{}^\beta$ be an admissible simplicial mapping of an elementary triangulation $K_\beta = Q_\beta \cup K_{\beta 0}$ onto a triangulation K_α. Then K_α is an elementary triangulation; moreover, if $Q_\alpha = S_\alpha{}^\beta Q_\beta$, $K_{\alpha 0} = S_\alpha{}^\beta K_{\beta 0}$, then $K_\alpha = Q_\alpha \cup K_{\alpha 0}$, and Q_α, $K_{\alpha 0}$ satisfy* Def. 4.31.

Proposition 4.33 follows from the fact that $Q_\alpha \cap K_{\alpha 0} = 0$ and that $S_\alpha{}^\beta$ obviously induces an isomorphism of the cell complex of all the oriented simplexes of Q_β onto the cell complex of all the oriented simplexes of Q_α.

§4.4. Applications to projective spaces. We shall adopt the following notation. $K_{\beta 0}{}^n$ will designate the second order barycentric subdivision of the boundary S^n of the regular $(n + 1)$-dimensional octahedron E^{n+1} whose vertices are the points v_k, v'_k, $k = 0, 1, \cdots, n$, of R^{n+1}, where $v_k(v'_k)$ has all coordinates equal to zero except the kth, which is equal to 1 (-1) (see IV, 6.2). The triangulation $K_{\beta 0}{}^n$ is symmetric with respect to the origin of coordinates o. We denote by $K_{\alpha 0}{}^n$ the triangulation of the projective space P^n obtained by identifying the simplexes of $K_{\beta 0}{}^n$ symmetric relative to o.

We shall designate the second order barycentric subdivision of the closed n-dimensional octahedron $\bar{E}^n$ (in R^n) by $K_{\beta 1}{}^n$; $K_{\beta 1}{}^n$ is symmetric relative to the origin of coordinates o in R^n. The corresponding triangulation of its boundary S^{n-1} is then $K_{\beta 0}{}^{n-1} \subset K_{\beta 1}{}^n$. Identification of the simplexes of $K_{\beta 0}{}^{n-1}$ symmetric relative to the origin of coordinates o (in R^n) transforms $K_{\beta 1}{}^n$ into a triangulation $K_{\alpha 1}{}^n$ of the projective space P^n. It is clear that $K_{\beta 1}{}^n$ and $K_{\alpha 1}{}^n$ are elementary triangulations and that the identification carrying $K_{\beta 1}{}^n$ into $K_{\alpha 1}{}^n$ is an admissible simplicial mapping $S_\alpha{}^\beta$ of $K_{\beta 1}{}^n$ onto $K_{\alpha 1}{}^n$. We see also that

$$Q_\beta{}^n = K_{\beta 1}{}^n \setminus K_{\beta 0}{}^{n-1} = K_{\alpha 1}{}^n \setminus K_{\alpha 0}{}^{n-1} = Q_\alpha{}^n$$

is a second order barycentric subdivision of the open octahedron.

REMARK. We shall use the following theorem on the invariance of the Betti groups (proved in Chapter X): All triangulations of the same or of homeomorphic polyhedra have isomorphic Betti groups.

THEOREM 4.41. *Suppose $n \geq 1$. The Betti groups of an arbitrary triangulation K^n of the projective n-space are:*

$$\Delta_0{}^r(K^n) = 0, \qquad \textit{if } r \textit{ is even and } 0 < r < n.$$

$$\Delta_0{}^r(K^n) \approx J_2, \qquad \textit{if } r \textit{ is odd and } 0 < r < n.$$

$$\Delta_0{}^n(K^n) = 0, \qquad \textit{if } n \textit{ is even.}$$

$$\Delta_0{}^n(K^n) \approx J, \qquad \textit{if } n \textit{ is odd.}$$

COROLLARY. *The odd-dimensional projective spaces are orientable; the even-dimensional projective spaces are nonorientable.*

LEMMA 4.411. Let $x_\beta{}^n$ be an orientation of the pseudomanifold

$$Q_\beta{}^n = K_{\beta 1}{}^n \setminus K_{\beta 0}{}^{n-1}.$$

Let π be the reflection of $K_{\beta 1}{}^n$ onto itself relative to o. Let

$$T_{\beta i}{}^{n-1} = (e_0 \cdots e_{n-1})$$

be any $(n - 1)$-simplex of $K_{\beta 0}{}^{n-1}$ and set

$$e'_k = \pi e_k \qquad (0 \leq k \leq n - 1),$$

so that $T'_{\beta i}{}^{n-1} = (e'_0 \cdots e'_{n-1})$ is antipodal to the simplex $T_{\beta i}{}^{n-1}$. Finally, let $t_{\beta i}{}^{n-1} = | e_0 \cdots e_{n-1} |$ be the orientation of $T_{\beta i}{}^{n-1}$ such that

$$(\Delta x_\beta{}^n \cdot t_{\beta i}{}^{n-1}) = +1.$$

Then

$$(\Delta x_\beta{}^n \cdot \pi t_{\beta i}{}^{n-1}) = (-1)^n.$$

Indeed, the reflection

$$(4.411) \qquad x'_h = -x_h, \qquad h = 1, \cdots, n,$$

of R^n relative to the origin of coordinates o (which we shall likewise denote by π) has determinant $(-1)^n$; whence, recalling that $K_{\beta 1}{}^n$ is symmetric relative to o, it is easy to deduce that $\pi x_\beta{}^n = (-1)^n x_\beta{}^n$ (here $\pi x_\beta{}^n$ denotes the image of the chain $x_\beta{}^n$ under the simplicial mapping π of $K_{\beta 1}{}^n$ onto itself).

Since π is an isomorphic mapping of the cell complex of the oriented simplexes of $K_{\beta 1}{}^n$ onto itself,

$$(\Delta \pi x_\beta{}^n \cdot \pi t_{\beta i}{}^{n-1}) = (\Delta x_\beta{}^n \cdot t_{\beta i}{}^{n-1}),$$

whence

$$(\Delta x_\beta{}^n \cdot \pi t_{\beta i}{}^{n-1}) = (\Delta(-1)^n \pi x_\beta{}^n \cdot \pi t_{\beta i}{}^{n-1})$$
$$= (-1)^n(\Delta \pi x_\beta{}^n \cdot \pi t_{\beta i}{}^{n-1}) = (-1)^n(\Delta x_\beta{}^n \cdot t_{\beta i}{}^{n-1}) = (-1)^n.$$

This proves the assertion.

4.412. If $t_{\beta i}{}^{n-1}$, $t'_{\beta i}{}^{n-1}$ are orientations of $T_{\beta i}{}^{n-1}$ and $T'_{\beta i}{}^{n-1}$ such that

$$(\Delta x_\beta{}^n \cdot t_{\beta i}{}^{n-1}) = (\Delta x_\beta{}^n \cdot t'_{\beta i}{}^{n-1}) = 1,$$

then

$$S_\alpha{}^\beta t'_{\beta i}{}^{n-1} = (-1)^n S_\alpha{}^\beta t_{\beta i}{}^{n-1}.$$

Indeed, if $t_{\beta i}{}^{n-1} = | e_0 \cdots e_{n-1} |$ and $t'_{\beta i}{}^{n-1} = | e'_{k(0)} \cdots e'_{k(n-1)} |$, then by Lemma 4.411,

$$\operatorname{sgn}\begin{pmatrix} 0 & 1 & \cdots & (n-1) \\ k(0) & k(1) & \cdots & k(n-1) \end{pmatrix} = (-1)^n.$$

But

$$S_\alpha{}^\beta t_{\beta i}{}^{n-1} = | S_\alpha{}^\beta e_0 \cdots S_\alpha{}^\beta e_{n-1} |,$$
$$S_\alpha{}^\beta t'_{\beta i}{}^{n-1} = | S_\alpha{}^\beta e'_{k(0)} \cdots S_\alpha{}^\beta e'_{k(n-1)} |.$$

The assertion follows.

From VII, (8.22) and 4.412 above we obtain

4.413. If $t_{\alpha i}{}^{n-1} = S_\alpha{}^\beta t_{\beta i}{}^{n-1} = (-1)^n S_\alpha{}^\beta t'_{\beta i}{}^{n-1}$ is an oriented simplex of the complex $K_{\alpha 0}{}^{n-1} = S_\alpha{}^\beta K_{\beta 0}{}^{n-1}$ and $x_\alpha{}^n = x_\beta{}^n$ is an orientation of the pseudomanifold $Q_\alpha{}^n = Q_\beta{}^n$, then

$$(\Delta_\alpha x_\alpha{}^n \cdot t_{\alpha i}{}^{n-1}) = (\Delta_\beta x_\beta{}^n \cdot t_{\beta i}{}^{n-1}) + (\Delta_\beta x_\beta{}^n \cdot (-1)^n t'_{\beta i}{}^{n-1})$$
$$= 1 + (-1)^n,$$

i.e., if n is odd, $\Delta_\alpha x_\alpha{}^n = 0$; and if n is even, $\Delta_\alpha x_\alpha{}^n = 2z_\alpha{}^{n-1}$, where

$$z_\alpha{}^{n-1} = \sum_i t_{\alpha i}{}^{n-1}$$

is an orientation of the pseudomanifold $K_{\alpha 0}{}^{n-1}$ (orientable for n even), and where we have written $\Delta_\alpha(\Delta_\beta)$ to denote the boundary in $K_\alpha(K_\beta)$.

Proof of Theorem 4.41. We may now carry out the proof of 4.41 by induction on n. The theorem is true for $n = 1$: $P^1 = \| K^1 \|$ is homeomorphic to a circumference; hence (see 1.4, Example 2), $\Delta_0{}^1(K^1)$ is infinite cyclic.

We shall now suppose the theorem proved for P^{n-1} and, in particular, for the triangulation (see the Remark just before Theorem 4.41) $K_{\alpha 0}{}^{n-1}$ of the polyhedron P^{n-1} and prove it for P^n. We must therefore show that Theorem 4.41 is true for $K^n = K_{\alpha 1}{}^n$. For $r \leq n - 2$, this follows immediately from 4.32 and from the inductive hypothesis. For $r = n - 1$ and $r = n$, the theorem follows from 4.32, 4.413, and the inductive hypothesis (in applying Theorem 4.32 for n even it is necessary to keep in mind that the group J_x consists of cycles $2cz_\alpha{}^{n-1}$, where c is an integer, so that $\Delta_0{}^{n-1}(K_{\alpha 1}{}^n)$, isomorphic to $Z_0{}^{n-1}(K_{\alpha 0})/J_x$ by 4.32, is of order 2).

§5. Simplicial mappings of pseudomanifolds

§5.1. The degree of a mapping. Let K_β and K_α be n-dimensional orientable pseudomanifolds. We recall that the groups

$$\Delta_0{}^n(K_\beta) = Z_0{}^n(K_\beta), \qquad \Delta_0{}^n(K_\alpha) = Z_0{}^n(K_\alpha)$$

are in this case infinite cyclic with generators

$$z_\beta{}^n = \sum_j t_{\beta j}{}^n, \qquad z_\alpha{}^n = \sum_i t_{\alpha i}{}^n,$$

where $\sum_j t_{\beta j}{}^n$, $\sum_i t_{\alpha i}{}^n$ are orientations of K_β and K_α.

In virtue of VII, Theorem 8.31,

$$S_\alpha{}^\beta z_\beta{}^n \in Z_0{}^n(K_\alpha),$$

i.e.,

$$S_\alpha{}^\beta z_\beta{}^n = \gamma z_\alpha{}^n. \tag{5.11}$$

Since $(z_\alpha{}^n \cdot t_{\alpha i}{}^n) = 1$,

$$(S_\alpha{}^\beta z_\beta{}^n \cdot t_{\alpha i}{}^n) = \gamma, \tag{5.12}$$

where γ is an integer (positive, zero, or negative). The integer γ is uniquely determined by the simplicial mapping $S_\alpha{}^\beta$ and the chosen orientations $\sum_j t_{\beta j}{}^n$, $\sum_i t_{\alpha i}{}^n$ of K_β, K_α. It is referred to as the *degree of the simplicial mapping* $S_\alpha{}^\beta$ of the pseudomanifold K_β into the pseudomanifold K_α.

§5.2. The original definition of the degree of a simplicial mapping. Formula (8.22) of VII yields [if $x_\beta{}^r = z_\beta{}^n$, $x_\alpha{}^r = z_\alpha{}^n$, and $(z_\beta{}^n \cdot t_{\beta j}{}^n) = 1$]

$$(S_\alpha{}^\beta z_\beta{}^n \cdot t_{\alpha i}{}^n) = \sum_j (S_\alpha{}^\beta t_{\beta j}{}^n \cdot t_{\alpha i}{}^n),$$

whence, by (5.12),

$$\gamma = \sum_j (S_\alpha{}^\beta t_{\beta j}{}^n \cdot t_{\alpha i}{}^n). \tag{5.21}$$

If we denote by π_i the number of $| t_{\beta j}{}^n |$ for which $S_\alpha{}^\beta t_{\beta j}{}^n = t_{\alpha i}{}^n$, and by ν_i the number of $| t_{\beta j}{}^n |$ such that $S_\alpha{}^\beta t_{\beta j}{}^n = -t_{\alpha i}{}^n$, we arrive at

$$\gamma = \pi_i - \nu_i . \tag{5.22}$$

Hence

5.21. *Let* $z_\alpha{}^n = \sum_i t_{\alpha i}{}^n$, $z_\beta{}^n = \sum_j t_{\beta j}{}^n$ *be orientations of the orientable pseudomanifolds* K_α *and* K_β, *and let* $S_\alpha{}^\beta$ *be a simplicial mapping of* K_β *into* K_α. *Then* $\pi_i - \nu_i$ (with π_i and ν_i *defined as above*) *is independent of i and is equal to the degree of the simplicial mapping* $S_\alpha{}^\beta$.

In this formulation we have the original definition of Brouwer of the degree of a simplicial mapping.

5.22. *If* $S_\alpha{}^\beta K_\beta \neq K_\alpha$ (that is, $S_\alpha{}^\beta K_\beta \subset K_\alpha$), *then the degree of the mapping* $S_\alpha{}^\beta$ *is* 0.

In fact, if $| t_{\alpha i}{}^n |$ is not in $S_\alpha{}^\beta K_\beta$, then $(S_\alpha{}^\beta t_{\beta j}{}^n \cdot t_{\alpha i}{}^n) = 0$ for arbitrary j and $\gamma = 0$ by (5.21).

Remark 1. The degree of a simplicial mapping $S_\alpha{}^\beta$ of an orientable pseudomanifold K_β into an orientable pseudomanifold K_α depends on the choice of the orientations of K_β and K_α. It is easy to see that if the given orientation of any one of the pseudomanifolds K_β or K_α is replaced by its opposite, the degree of the mapping remains unchanged in absolute value, but changes its sign.

Remark 2. Now let K_β and K_α again be n-dimensional pseudomanifolds, but not necessarily orientable. The groups

$$\Delta_2{}^n(K_\beta) = Z_2{}^n(K_\beta), \qquad \Delta_2{}^n(K_\alpha) = Z_2{}^n(K_\alpha)$$

each consist of two elements; the identity and $z_\beta{}^n = \sum T_{\beta j}{}^n$, $z_\alpha{}^n = \sum T_{\alpha i}{}^n$, respectively. Therefore, either $S_\alpha{}^\beta z_\beta{}^n = 0$, or $S_\alpha{}^\beta z_\beta{}^n = z_\alpha{}^n$.

We shall say that the *parity* of $S_\alpha{}^\beta$ is 0 in the first case, and 1 in the second. In the first case, the number of $T_{\beta j}{}^n \in K_\beta$ mapped onto $T_{\alpha i}{}^n$ for a prescribed $T_{\alpha i}{}^n \in K_\alpha$ is even, in the second case it is odd. The parity of a simplicial mapping is sometimes referred to as the "degree (mod 2)" of the mapping.

Remark 3. Examples. Let K_α be a closed m-polygon whose vertices $a_1, \cdots, a_m$ are enumerated in some order, say counterclockwise. Let γ be a natural number and let K_β be a closed γm-polygon with its vertices

$$a_{11}, \cdots, a_{1m}; a_{21}, \cdots, a_{2m}; \cdots; a_{\gamma 1}, \cdots, a_{\gamma m}$$

also indexed counterclockwise. If we orient both 1-dimensional pseudomanifolds K_α and K_β counterclockwise and set

$$S_\alpha{}^\beta a_{ij} = a_j, \qquad i = 1, \cdots, \gamma; \quad j = 1, \cdots, m,$$

we obtain a simplicial mapping $S_\alpha{}^\beta$ of K_β onto K_α whose degree is γ. Reversing the orientation of one of these pseudomanifolds yields a mapping of degree $-\gamma$. In Fig. 123, $m = 3$, $\gamma = 2$.

The simplicial mapping $S_\alpha{}^\beta$ obtained above has (for $n = 1$) the following property: $\pi_i = \gamma$, $\nu_i = 0$ for all n-simplexes $T_{\alpha i}{}^n \in K_\alpha{}^n$ (that is, as a result of the mapping S_α^β each oriented n-simplex of K_α^n is covered by the images of precisely $|\gamma|$ uniquely oriented simplexes of K_β^n, where γ is the degree of S_α^β).

Now let n be a natural number; we shall assume that the n-dimensional orientable pseudomanifolds K_α^n and K_β^n are triangulations in R^{n+1} of polyhedra homeomorphic to the n-sphere S^n and that S_α^β is a simplicial mapping of K_β^n onto K_α^n of degree γ, satisfying the condition formulated above. Let us imbed R^{n+1} in R^{n+2} and choose two points o and o' in

$$R^{n+2} \setminus R^{n+1}$$

on different sides of the plane R^{n+1}. We then consider the cones $<oK_\alpha^n>$, $<o'K_\alpha^n>$; $<oK_\beta^n>$, $<o'K_\beta^n>$. Each pair of cones

$$K_\alpha^{n+1} = <oK_\alpha^n> \cup <o'K_\alpha^n>,$$

$$K_\beta^{n+1} = <oK_\beta^n> \cup <o'K_\beta^n>$$

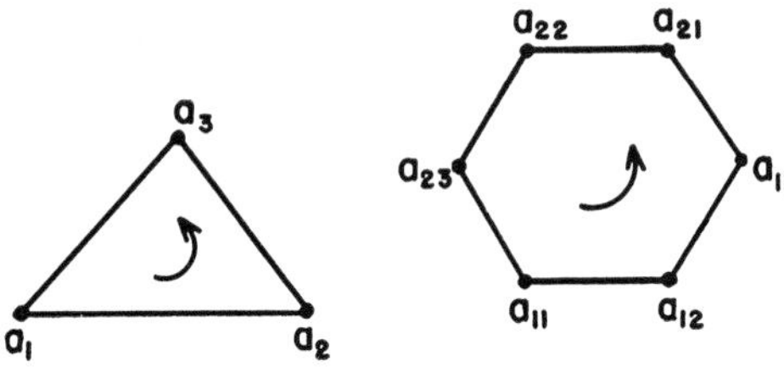

FIG. 123

is an orientable closed $(n + 1)$-dimensional pseudomanifold, and the polyhedron corresponding to each is homeomorphic to S^{n+1}.

Let us now extend the simplicial mapping S_α^β of K_β^n to all of K_β^{n+1} by setting $S_\alpha^\beta(o) = o$, $S_\alpha^\beta(o') = o'$. We obtain a simplicial mapping of K_β^{n+1} onto K_α^{n+1} with the property that $\pi_i = \gamma$, $\nu_i = 0$ for all $T_{\alpha i}^{n+1} \in K_\alpha^{n+1}$. Hence the mapping is of degree γ. Thus we have triangulations K_α^n, K_β^n of polyhedra homeomorphic to the n-sphere and a simplicial mapping of degree γ of K_β^n into K_α^n for arbitrary natural number n and integer $\gamma \neq 0$. We note, finally, that to obtain a mapping of degree zero it suffices to map all of the complex K_β^n onto a single vertex of K_α^n. Hence

5.24. *There exists a simplicial mapping of degree γ of one triangulation of the n-sphere into another for arbitrary natural number n and integer γ.*

REMARK 4. *The case $n = 0$.* A pair of points e_0, e_1 is called a 0-*sphere.* The normal 0-cycles $e_1 - e_0$ and $e_0 - e_1$ are called *orientations of the* 0-*sphere.* Hence a 0-sphere, as is the case with the n-sphere for $n \geq 1$, has precisely two orientations. However, there is no pseudomanifold corresponding to the 0-sphere, since it is not connected.

Let $S_\alpha^0 = \{e_{\alpha 0}, e_{\alpha 1}\}$ and $S_\beta^0 = \{e_{\beta 0}, e_{\beta 1}\}$ be two 0-spheres. The following four mappings exhaust all the possible mappings of S_β^0 into S_α^0:

1) $C_1(e_{\beta 0}) = C_1(e_{\beta 1}) = e_{\alpha 0}$; 2) $C_2(e_{\beta 0}) = C_2(e_{\beta 1}) = e_{\alpha 1}$;

3) $C_3(e_{\beta 0}) = e_{\alpha 0}$, $C_3(e_{\beta 1}) = e_{\alpha 1}$; 4) $C_4(e_{\beta 0}) = e_{\alpha 1}$, $C_4(e_{\beta 1}) = e_{\alpha 0}$.

Let us now choose any orientations z_α^0 and z_β^0 of the spheres S_α^0 and S_β^0, say $z_\alpha^0 = e_{\alpha 1} - e_{\alpha 0}$ and $z_\beta^0 = e_{\beta 1} - e_{\beta 0}$. Then

$$C_1(z_\beta^0) = C_2(z_\beta^0) = 0, \qquad C_3(z_\beta^0) = z_\alpha^0, \qquad C_4(z_\beta^0) = -z_\alpha^0.$$

It is natural to say that the *degree* of the first two mappings is 0; the degree of the third is 1, and the degree of the fourth is -1.

Chapter IX

THE OPERATOR ∇ AND THE GROUPS $\nabla^r(\mathfrak{K}, \mathfrak{A})$. CANONICAL BASES. CALCULATION OF THE GROUPS $\Delta^r(\mathfrak{K}, \mathfrak{A})$ AND $\nabla^r(\mathfrak{K}, \mathfrak{A})$ BY MEANS OF THE GROUPS $\Delta_0^r(\mathfrak{K})$

§1. The operator ∇

§1.1. Definition of the chain ∇x^r. Let $\mathfrak{K}$ be an a-complex. We shall define an $(r+1)$-chain $\nabla_{\mathfrak{K}} x^r$, called the ∇-boundary or upper boundary or coboundary of the chain x^r, for every r-chain $x^r \in L^r(\mathfrak{K}, \mathfrak{A})$. In each pair of opposite cells of $\mathfrak{K}$ denote any one of the cells by t_i and define the value of the chain $\nabla_{\mathfrak{K}} x^r$ on the $(r+1)$-cell t_h^{r+1} in accordance with the formula

$$(1.1)_\nabla \qquad \nabla_{\mathfrak{K}} x^r(t_h^{r+1}) = \textstyle\sum_i (t_h^{r+1} : t_i^r) x^r(t_i^r),$$

where the sum is extended over all t_i^r.

For comparison, we recall the definition

$$(1.1)_\Delta \qquad \Delta_{\mathfrak{K}} x^r(t_j^{r-1}) = \textstyle\sum_i (t_i^r : t_j^{r-1}) x^r(t_i^r).$$

REMARK. If the a-complex $\mathfrak{K}$ is the complex consisting of the oriented elements of a simplicial or polyhedral complex K, we shall write ∇_K instead of $\nabla_{\mathfrak{K}}$. *If there can be no misunderstanding, we shall write simply ∇ in place of $\nabla_{\mathfrak{K}}$ (or ∇_K).*

EXAMPLES. 1°. Let K be the complex consisting of all the edges and vertices of a tetrahedron $(oe_1e_2e_3)$, and consider the 0-chain x^0 which is 1 on the vertex o and 0 on the rest of the vertices of K. The chain ∇x^0 is 1 on the oriented segments $| e_i o |$, $i = 1, 2, 3$ (Fig. 124) and 0 on the remaining 1-elements of K. If we denote ∇x^0 by x^1, then $\nabla x^1 = 0$.

2°. Let the complex K consist of the two triangles $(e_1e_2e_4)$ and $(e_2e_3e_4)$ of Fig. 125, and their sides and vertices. The chain x^1, by definition, is 1 on the oriented simplexes $| e_1e_2 |$, $| e_2e_3 |$, $| e_3e_4 |$, and $| e_4e_1 |$ and 0 on $| e_2e_4 |$. The chain ∇x^1 is 2 on the oriented triangles $| e_1e_2e_4 |$ and $| e_2e_3e_4 |$ (Fig. 125).

3°. Fig. 126 and Fig. 127 show a torus divided into curvilinear rectangles (opposite sides of the square $ABCD$ are to be identified). The complex K consists of the curvilinear rectangles, their sides, and vertices. The chain x^1 is, by definition, 1 on the 1-elements of K (the horizontal and vertical segments) oriented as indicated; then $\nabla x^1 = 0$. The same result is obtained if x^1 is set equal to 1 only on the horizontal segments and 0 on the vertical segments, or conversely.

If we set $x^1 | e_1e_2 | = x^1 | e_2e_3 | = x^1 | e_3e_4 | = x^1 | e_4e_1 | = 1$ (Fig. 128) and x^1 equal to 0 on the remaining segments, then ∇x^1 is the 2-chain which is

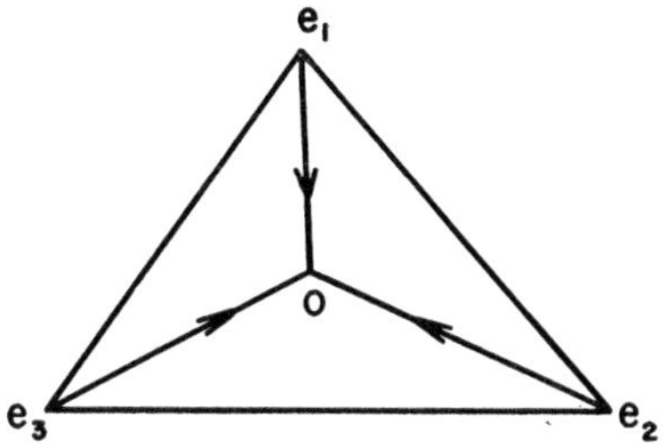

FIG. 124

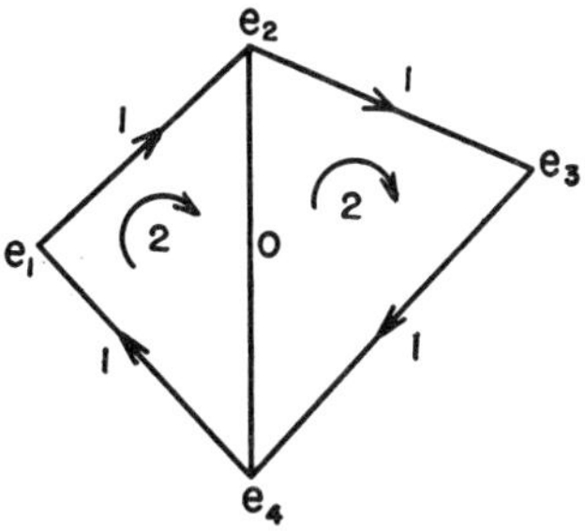

FIG. 125

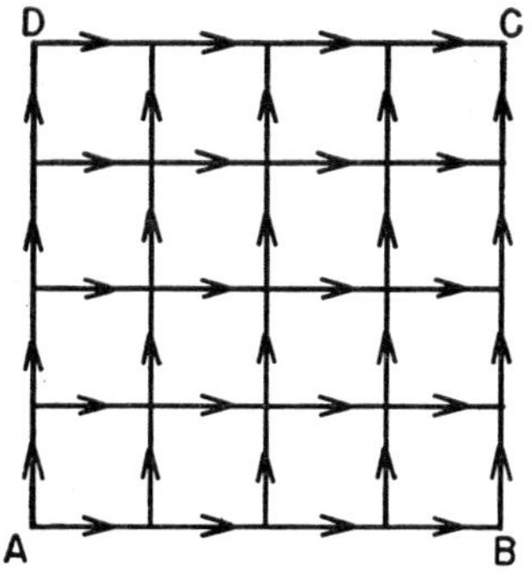

FIG. 126

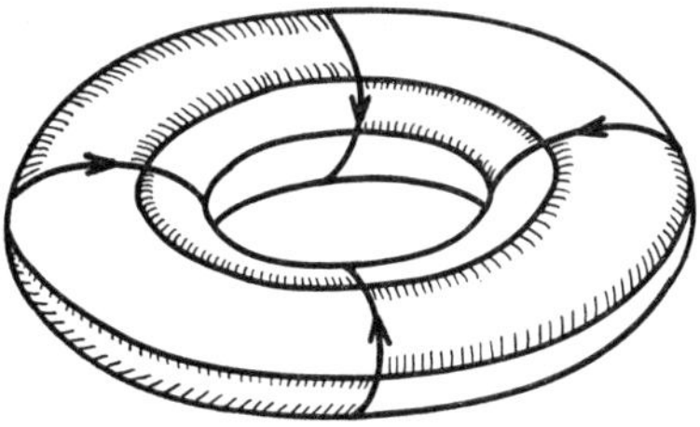

FIG. 127

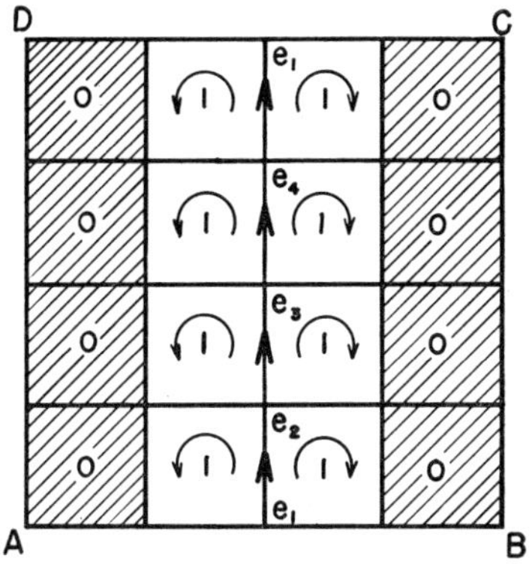

FIG. 128

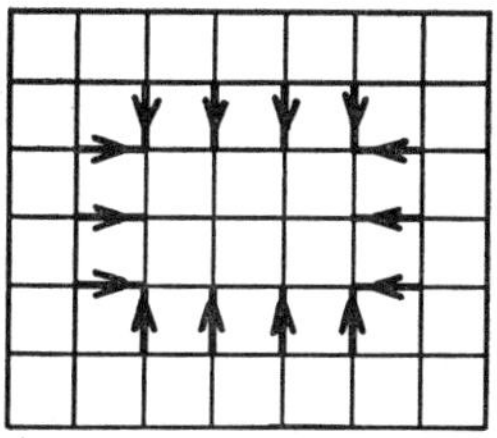

FIG. 129

0 on the hatched squares (Fig. 128) and 1 on the rest of the squares oriented by arrows as indicated.

4°. In Fig. 129 the plane has been divided into congruent squares. The complex K consists of these squares, their sides, and vertices. Let the chain x^1 be 1 on the segments marked with arrows and oriented as indicated by the arrows, and 0 on all the rest of the 1-elements of K. Then $\nabla x^1 = 0$.

§1.2. The chain ∇x^r as a linear form. If the chain x^r is written as a linear form

$$x^r = \sum a_i t_i^r,$$

it is also convenient to write the chain ∇x^r as a linear form. Since ∇x^r has the value $\sum_i (t_h^{r+1} : t_i^r) a_i$ on t_h^{r+1} by Def. $(1.1)_\nabla$, we may write

$$(1.21)_\nabla \qquad \nabla x^r = \sum_h \sum_i (t_h^{r+1} : t_i^r) a_i t_h^{r+1}.$$

For comparison we rewrite the analogous formula for Δx^r:

$$(1.21)_\Delta \qquad \Delta x^r = \sum_j \sum_i (t_i^r : t_j^{r-1}) a_i t_j^{r-1}.$$

In particular, if $x^r = t_k^r$, that is, if $a_i = \delta_{ik}$, then $(1.21)_\nabla$ becomes

$$\nabla t_k^r = \sum_h (t_h^{r+1} : t_k^r) t_h^{r+1}.$$

Comparing this formula with

$$\Delta t_h^{r+1} = \sum_k (t_h^{r+1} : t_k^r) t_k^r,$$

and examining the incidence matrix $\mathfrak{E}^r = \| \varepsilon_{hk}^r \| = \| (t_h^{r+1} : t_k^r) \|$:

$$\begin{array}{c|ccccc}
 & t_1^r & \cdots & t_k^r & \cdots & t_{\rho(r)}^r \\
\hline
t_1^{r+1} & & & & & \\
\vdots & & & \vdots & & \\
t_h^{r+1} & \cdots & & \varepsilon_{hk}^r & & \cdots \\
\vdots & & & \vdots & & \\
t_{\rho(r+1)}^{r+1} & & & & &
\end{array},$$

we obtain the following rule:

Each column of the matrix $\mathfrak{E}^r$ is the ∇-boundary of the simplex t_k^r appearing at the head of the column; each row of $\mathfrak{E}^r$ is the Δ-boundary of the simplex t_h^{r+1} heading the row. In other words, the number $\varepsilon_{hk}^r = (t_h^{r+1} : t_k^r)$ is the value of ∇t_k^r on t_h^{r+1} and also the value of Δt_h^{r+1} on t_k^r.

§1.3. *The operators Δ and ∇ are dual homomorphisms* (see Appendix 2, 5.4) *of the module $L^r(\mathfrak{K})$ into the module $L^{r-1}(\mathfrak{K})$ and of $L^{r-1}(\mathfrak{K})$ into $L^r(\mathfrak{K})$, respectively.*

Indeed, for arbitrary

$$x^r = \sum a_i t_i^r \in L^r(\mathfrak{K})$$

and

$$y^{r-1} = \sum b_j t_j^{r-1} \in L^{r-1}(\mathfrak{K}),$$
$$(\Delta x^r \cdot y^{r-1}) = \sum_j \sum_i (t_i^r : t_j^{r-1}) a_i b_j = \sum_{i,j} (t_i^r : t_j^{r-1}) a_i b_j\,,$$
$$(x^r \cdot \nabla y^{r-1}) = \sum_i a_i \sum_j (t_i^r : t_j^{r-1}) b_j = \sum_{i,j} (t_i^r : t_j^{r-1}) a_i b_j\,,$$

that is,

$$(\Delta x^r \cdot y^{r-1}) = (x^r \cdot \nabla y^{r-1}). \tag{1.31}$$

This completes the proof.

Replacing y^{r-1} by x^r and x^r by y^{r+1} in (1.31), we get $(\Delta y^{r+1} \cdot x^r) = (y^{r+1} \cdot \nabla x^r)$, or

$$(\nabla x^r \cdot y^{r+1}) = (x^r \cdot \Delta y^{r+1}). \tag{1.310}$$

From this we deduce the *fundamental identity*

$$\nabla \nabla x^r = 0 \tag{1.32}$$

for an arbitrary chain x^r of an arbitrary a-complex $\mathfrak{K}$.

We prove this identity by calculating the value of $\nabla\nabla x^r$ on any $t^{r+2} \in \mathfrak{K}$:

$$(\nabla\nabla x^r \cdot t^{r+2}) = (\nabla x^r \cdot \nabla t^{r+2}) = (x^r \cdot \Delta\Delta t^{r+2}) = 0.$$

§1.4. The groups $Z_\nabla{}^r(\mathfrak{K}, \mathfrak{A})$, $H_\nabla{}^r(\mathfrak{K}, \mathfrak{A})$, $\nabla^r(\mathfrak{K}, \mathfrak{A})$. Def. $(1.1)_\nabla$ implies that

$$\nabla(x_1^r \pm x_2^r) = \nabla x_1^r \pm \nabla x_2^r.$$

Consequently, the assignment to every chain x^r of its ∇-boundary yields a homomorphism ∇ of $L^r(\mathfrak{K}, \mathfrak{A})$ into $L^{r+1}(\mathfrak{K}, \mathfrak{A})$. The kernel of this homomorphism is the group $Z_\nabla{}^r(\mathfrak{K}, \mathfrak{A})$ consisting of all the *r-dimensional ∇-cycles* (*r-cocycles*), that is, of all the *r-chains whose ∇-boundary is* 0. The image of $L^r(\mathfrak{K}, \mathfrak{A})$ under ∇ is the subgroup $H_\nabla{}^{r+1}(\mathfrak{K}, \mathfrak{A})$ of $L^{r+1}(\mathfrak{K}, \mathfrak{A})$ of all $(r + 1)$-chains which are ∇-boundaries of r-chains $x^r \in L^r(\mathfrak{K}, \mathfrak{A})$.

A consequence of the fundamental identity (1.32) is:

The coboundary of every chain is a cocycle. Hence $H_\nabla{}^r(\mathfrak{K}, \mathfrak{A})$ is a subgroup of $Z_\nabla{}^r(\mathfrak{K}, \mathfrak{A})$ for arbitrary r.

The elements z^r of $H_\nabla{}^r(\mathfrak{K}, \mathfrak{A})$ are said to be ∇-homologous to zero in $\mathfrak{K}$ over $\mathfrak{A}$ (cohomologous to zero in $\mathfrak{K}$ over $\mathfrak{A}$). We shall write this relation as:

$$z^r \backsim 0 \text{ (in } \mathfrak{K} \text{ over } \mathfrak{A}).$$

If $z_1^r \in Z_\nabla{}^r(\mathfrak{K}, \mathfrak{A})$, $z_2^r \in Z_\nabla{}^r(\mathfrak{K}, \mathfrak{A})$, and $z_1^r - z_2^r \in H_\nabla{}^r(\mathfrak{K}, \mathfrak{A})$, the cocycles z_1^r and z_2^r are said to be ∇-homologous (cohomologous) in $\mathfrak{K}$ over $\mathfrak{A}$:

$$z_1^r \backsim z_2^r \text{ (in } \mathfrak{K} \text{ over } \mathfrak{A}).$$

REMARK 1. The cohomology symbol "$\smile$" possesses all the properties of the homology symbol "$\sim$" enumerated in VII, 6.4.

DEFINITION 1.4. The group

$$\nabla^r(\mathfrak{K}, \mathfrak{A}) = Z_\nabla{}^r(\mathfrak{K}, \mathfrak{A})/H_\nabla{}^r(\mathfrak{K}, \mathfrak{A})$$

is called the *r-dimensional ∇-group* (*r-dimensional cohomology group*), the *r-dimensional upper Betti group* (*rth coBetti group*), or simply the group ∇^r of $\mathfrak{K}$ (over $\mathfrak{A}$).

REMARK 2. The coboundary of every n-chain of an n-complex $\mathfrak{K}^n$ is 0, that is, *in an n-complex all n-chains are cocycles*:

$$Z_\nabla{}^n(\mathfrak{K}^n, \mathfrak{A}) = L^n(\mathfrak{K}^n, \mathfrak{A}).$$

On the other hand, the coboundary of the only element (the identity) of $L^{-1}(\mathfrak{K}, \mathfrak{A})$ is the 0-chain identically equal to zero. Hence $H_\nabla{}^0(\mathfrak{K}, \mathfrak{A})$ is the null group, so that $\nabla^0(\mathfrak{K}, \mathfrak{A}) = Z_\nabla{}^0(\mathfrak{K}, \mathfrak{A})$.

REMARK 3. The group $\nabla^r(\mathfrak{K}, J)$ will be written as $\nabla_0{}^r(\mathfrak{K})$, and the group $\nabla^r(\mathfrak{K}, J_m)$ as $\nabla_m{}^r(\mathfrak{K})$; instead of $\nabla^r(\mathfrak{K}, \mathfrak{A})$ we shall often write $\nabla^r(\mathfrak{K})$, etc.

§1.5. Chains restricted to a subcomplex. Let $x^r \in L^r(\mathfrak{K})$ and let $\mathfrak{M}$ be a closed subcomplex of the complex $\mathfrak{K}$. The fact that $\mathfrak{M}$ is closed implies that

$$(\nabla_{\mathfrak{M}}\mathfrak{M}x^r \cdot t^{r+1}) = (\nabla_{\mathfrak{M}}x^r \cdot t^{r+1}), \qquad (\mathfrak{M}\nabla_{\mathfrak{K}}x^r \cdot t^{r+1}) = (\nabla_{\mathfrak{K}}x^r \cdot t^{r+1})$$

for every $t^{r+1} \in \mathfrak{M}$. Hence

1.51. *For $x^r \in L^r(\mathfrak{K})$ and arbitrary closed subcomplex $\mathfrak{M}$ of $\mathfrak{K}$,*

$$\nabla_{\mathfrak{M}}\mathfrak{M}x^r = \mathfrak{M}\nabla_{\mathfrak{K}}x^r. \tag{1.51}$$

In particular, if $x^r \in Z_\nabla{}^r(\mathfrak{K})$, then $\mathfrak{M}x^r \in Z_\nabla{}^r(\mathfrak{M})$.

§1.6. The groups $\nabla^0(\mathfrak{K}, \mathfrak{A}) = Z_\nabla{}^0(\mathfrak{K}, \mathfrak{A})$. Since $H_\nabla{}^0(\mathfrak{K}, \mathfrak{A})$ is the null group,

$$\nabla^0(\mathfrak{K}, \mathfrak{A}) = Z_\nabla{}^0(\mathfrak{K}, \mathfrak{A}),$$

and to compute $\nabla^0(\mathfrak{K}, \mathfrak{A})$ we must find those 0-chains which are cocycles. We shall restrict ourselves to the case that $\mathfrak{K}$ is the complex of the oriented simplexes of an unrestricted simplicial complex K.

First, let K be a connected complex and z^0 a 0-cocycle of K. We shall prove that z^0 has the same value on all the vertices e_i of K.

Since K is connected, it suffices to show that z^0 has the same value on any two vertices e_0, e_1 which are the endpoints of a segment $(e_0e_1) \in K$. But this assertion follows from the fact that ∇z^0 has the value $z^0(e_1) - z^0(e_0)$ on $|e_0e_1|$. Conversely, if x^0 has the same value on all the vertices, then ∇x^0 has the value

$$z^0(e_{i+1}) - z^0(e_i) = 0$$

on every segment $|e_ie_{i+1}|$, so that $\nabla x^0 = 0$. Hence

1.611. *A* 0-*cocycle of a connected unrestricted simplicial complex* K *may be defined as a* 0-*chain which is constant on* K, *that is, a chain which has the same value on all the vertices of* K.

This at once implies

1.621. If K is a connected unrestricted simplicial complex, $\nabla^0(K, \mathfrak{A})$ is isomorphic to $\mathfrak{A}$.

On the other hand, 1.611 implies

1.61. *A* 0-*cocycle of an arbitrary unrestricted simplicial complex* K *is a* 0-*chain constant on each component of* K.

1.62. *If* K *is an arbitrary unrestricted simplicial complex,* $\nabla^0(K, \mathfrak{A})$ *is the direct sum of* m *copies of* $\mathfrak{A}$, *where* m *is the number of components of* K. *Consequently,* $\nabla^0(K, \mathfrak{A})$ *is isomorphic to* $\Delta^0(K, \mathfrak{A})$.

REMARK. Theorem 4.2 is a generalization of 1.62.

§1.7. The groups $\nabla^n(K^n, J)$ of n-dimensional pseudomanifolds. In this article K^n is an n-dimensional closed pseudomanifold. We shall consider only integral chains.

Let t^n be any oriented n-simplex of K^n. Then the integral chain t^n, as is the case with every n-chain of K^n, is a cocycle.

1.71. *The cocycle* t^n *is not cohomologous to zero in* K^n, i.e., there is no integral $(n - 1)$-chain x^{n-1} of K^n such that

$$\nabla x^{n-1} = t^n.$$

To prove this it is enough to note

1.710. The sum of the coefficients of every n-cocycle z^n, cohomologous to zero on K^n, is even.

Proof. Since every n-cocycle cohomologous to zero on K^n is of the form $\nabla x^{n-1} = \sum a_i \nabla t_i^{n-1}$, where $x^{n-1} \in L^{n-1}(K^n)$, it is enough to prove 1.710 for cocycles of the form ∇t_i^{n-1}. But if $| t_{i(1)}^n |$ and $| t_{i(2)}^n |$ are the two simplexes having $| t_i^{n-1} |$ as a face, and $t_{i(1)}^n$, $t_{i(2)}^n$ are arbitrary orientations of these simplexes, then

$$\nabla t_i^{n-1} = \varepsilon_1 t_{i(1)}^n + \varepsilon_2 t_{i(2)}^n,$$

where $\varepsilon_1, \varepsilon_2 = \pm 1$. Hence $\varepsilon_1 + \varepsilon_2 = 0$ or ± 2. This completes the proof.

If K^n is orientable, 1.710 may be strengthened to

1.720. If $\sum t_i^n$ is any orientation of an orientable n-dimensional pseudomanifold K^n and $z^n = \sum a_i t_i^n$ is a cocycle $\backsim 0$ on K^n, then $\sum a_i = 0$.

In fact,

$$\nabla t_i^{n-1} = t_{i(1)}^n - t_{i(2)}^n$$

for an arbitrary t_i^{n-1}, where $| t_{i(1)}^n |$ and $| t_{i(2)}^n |$ are the simplexes with $| t_i^{n-1} |$ as a face. Hence 1.720 holds for every cocycle of the form ∇t_i^{n-1} and thus for every cocycle of the form ∇x^{n-1}.

1.720 implies

1.72. If $|t^n| \in K^n$ and K^n is orientable, then $at^n \backsim 0$ on K^n only if $a = 0$.

Now let K^n be an arbitrary n-dimensional pseudomanifold.

1.73. If t_i^n and t_h^n are two oriented n-simplexes of K^n,

$$t_i^n \backsim \pm t_h^n \quad \text{in} \quad K^n.$$

In virtue of the strong connectedness of K^n it is enough to prove 1.73 for two simplexes $|t_i^n|$ and $|t_h^n|$ having a common $(n - 1)$-face $|t_i^{n-1}|$. But in that case

$$\nabla t_j^{n-1} = (t_i^n : t_j^{n-1})t_i^n + (t_h^n : t_j^{n-1})t_h^n. \qquad (1.73)$$

Since

$$(t_i^n : t_j^{n-1}) = \pm 1, \qquad (t_h^n : t_j^{n-1}) = \pm 1,$$

$t_i^n \pm t_h^h \backsim 0$. This proves the assertion.

REMARK. If t_i^n and t_h^n are coherently oriented, $(t_i^n : t_j^{n-1}) = -(t_h^n : t_j^{n-1})$ and 1.73 becomes $\nabla t_j^{n-1} = \pm(t_i^n - t_h^n)$, that is, $t_i^n \backsim t_h^n$. It follows that

1.74. If K^n is nonorientable,

$$2t_i^n \backsim 0$$

for an arbitrary oriented simplex t_i^n.

Proof. If $t_1^n, t_2^n, \cdots, t_s^n$ is any disorienting sequence of K^n,

$$t_1^n \backsim t_2^n \backsim \cdots \backsim t_s^n \backsim -t_1^n,$$

i.e.,

$$t_1^n \backsim -t_1^n \quad \text{and} \quad 2t_1^n \backsim 0.$$

Since $t_i^n \backsim \pm t_1^1$ for every $|t_i^n| \in K^n$, $2t_i^n \backsim 0$. This is what we wished to prove.

A further consequence of 1.73 is

1.75. If z^n is an arbitrary integral n-cocycle of K^n,

$$z^n \backsim at^n,$$

where $|t^n|$ is an arbitrary n-simplex of K^n and a is an integer.

From 1.75, 1.72, and 1.74 we obtain the

FUNDAMENTAL THEOREM 1.7. *Let K^n be an n-dimensional pseudomanifold; if K^n is orientable, $\nabla_0{}^n(K^n)$ is infinite cyclic; if K^n is nonorientable, $\nabla_0{}^n(K^n)$ is of order* 2.

We shall derive still another property of closed pseudomanifolds from the propositions proved in this subsection.

1.76. *If K_0 is a closed proper subcomplex of an n-dimensional closed pseudomanifold K^n, then $\nabla_0{}^n(K_0)$ is the null group.*

Proof. Let t^n be an oriented n-simplex of K_0. It is enough to prove that the n-cocycle t^n is cohomologous to zero in K_0.

Let $|t_1^n| \in K^n \setminus K_0$; let t_1^n be any one of its orientations. In virtue of 1.73,

$$t^n \backsim \varepsilon t_1^n \quad \text{in} \quad K^n, \qquad \varepsilon = \pm 1.$$

Hence there is a chain $x^{n-1} \varepsilon L_0^{n-1}(K^n)$ such that

$$\nabla_n x^{n-1} = t^n - \varepsilon t_1^n,$$

where ∇_n denotes the coboundary in K^n. Since K_0 is a closed subcomplex of K^n, we have, by Theorem 1.51,

$$\nabla_0 K_0 x^{n-1} = K_0 \nabla_n x^{n-1} = K_0 t^n - K_0 \varepsilon t_1^n = t^n,$$

where ∇_0 denotes the coboundary in K_0. This means that

$$t^n \backsim 0 \quad \text{in} \quad K_0,$$

which was to be proved.

§2. Bases of the modules $L_0^r(\mathfrak{K})$

§2.1. Preliminary remarks. From now on till the end of this chapter we shall assume that $\mathfrak{K}$ is a finite n-dimensional a-complex.

The matrices of the homomorphisms Δ relative to the initial bases

$$t_1^r, \cdots, t_{\rho(r)}^r, \qquad r = 0, 1, \cdots, n,$$

of the modules $L_0^r(\mathfrak{K})$ (see Appendix 2, 2.7 and 5.2) are the incidence matrices

$$\mathfrak{E}^{r-1} = \| \varepsilon_{ij}^{r-1} \|, \qquad \varepsilon_{ij}^{r-1} = (t_i^r : t_j^{r-1}).$$

Let $X^r = \{x_1^r, \cdots, x_{\rho(r)}^r\}$,

(2.11) $$x_h^r = \sum_i a_{hi}^r t_i^r, \qquad h = 1, 2, \cdots, \rho^r,$$

be any basis of $L_0^r(\mathfrak{K})$ and let

(2.12) $$\Delta x_h^r = \sum_k \eta_{hk}^{r-1} x_k^{r-1}, \qquad h = 1, 2, \cdots, \rho^r.$$

Then every boundary

(2.13) $$\Delta x^r = u^{r-1}, \qquad x^r = \sum a_h x_h^r$$

is a linear combination with coefficients a_h of the boundaries (2.12). Hence, if $\mathfrak{A}^r$ is the matrix $\| a_{hi}^r \|$, the matrix $\mathfrak{H}^{r-1} = \| \eta_{kh}^{r-1} \|$ is given by

$$\mathfrak{H}^{r-1} = \mathfrak{A}^r \mathfrak{E}^{r-1} (\mathfrak{A}^{r-1})^{-1}.$$

§2.2. Dual bases of $L_0^r(\mathfrak{K})$. Let

$$(2.21)\qquad \begin{cases} W^r = \{w_1^r, \cdots, w_{\rho(r)}^r\}, \\ \overline{W}^r = \{\bar{w}_1^r, \cdots, \bar{w}_{\rho(r)}^r\}; \end{cases}$$

$$(2.22)\qquad \begin{cases} W^{r-1} = \{w_1^{r-1}, \cdots, w_{\rho(r-1)}^{r-1}\}, \\ \overline{W}^{r-1} = \{\bar{w}_1^{r-1}, \cdots, \bar{w}_{\rho(r-1)}^{r-1}\} \end{cases}$$

be two pairs of dual bases (Appendix 2, 5.3) of $L_0^r(\mathfrak{K})$ and $L_0^{r-1}(\mathfrak{K})$, respectively. Let $\mathfrak{H}^{r-1} = \| \eta_{ij}^{r-1} \|$ be the matrix of the homomorphism

$$(2.23)\qquad \Delta w_i^r = \textstyle\sum_j \eta_{ij}^{r-1} w_j^{r-1}$$

of $L_0^r(\mathfrak{K})$ into $L_0^{r-1}(\mathfrak{K})$ relative to the bases W^r, W^{r-1}. Since the homomorphisms Δ and ∇ are dual,

$$(2.24)\qquad \nabla \bar{w}_j^{r-1} = \textstyle\sum_i \eta_{ij}^{r-1} \bar{w}_i^r$$

by Appendix 2, Theorem 5.4.

Hence if the rows of $\mathfrak{H}^{r-1}$ are the Δ-boundaries of the elements of the basis W^r heading these rows in the table

$$\begin{array}{c|ccccc}
 & w_1^{r-1} & \cdots & w_j^{r-1} & \cdots & w_{\rho(r-1)}^{r-1} \\ \hline
w_1^r & & & \cdot & & \\
\cdot & & & \cdot & & \\
\cdot & & & \cdot & & \\
\cdot & & & \cdot & & \\
w_i^r & \cdots & \cdots & \eta_{ij}^{r-1} & \cdots & \cdots \\
\cdot & & & \cdot & & \\
\cdot & & & \cdot & & \\
\cdot & & & \cdot & & \\
w_{\rho(r)}^r & & & \cdot & &
\end{array},$$

the columns of $\mathfrak{H}^{r-1}$ are the coboundaries of the elements of the basis $\overline{W}^{r-1}$ appearing at the head of the columns in the table

$$\begin{array}{c|ccccc}
 & \bar{w}_1^{r-1} & \cdots & \bar{w}_j^{r-1} & \cdots & \bar{w}_{\rho(r-1)}^{r-1} \\ \hline
\bar{w}_1^r & & & \cdot & & \\
\cdot & & & \cdot & & \\
\cdot & & & \cdot & & \\
\cdot & & & \cdot & & \\
\bar{w}_i^r & \cdots & \cdots & \eta_{ij}^{r-1} & \cdots & \cdots \\
\cdot & & & \cdot & & \\
\cdot & & & \cdot & & \\
\cdot & & & \cdot & & \\
\bar{w}_{\rho(r)}^r & & & \cdot & &
\end{array}.$$

Hence η_{ij}^{r-1} is the coefficient of w_j^{r-1} in Δw_i^r; it is also the coefficient of $\bar{w}_i^r$ in $\nabla \bar{w}_j^{r-1}$.

In accordance with the Remark following Theorem 5.4 of Appendix 2, η_{ij}^{r-1} may also be interpreted as the jth contravariant component of $\Delta w_i^r \in L_0^{r-1}(\mathfrak{K})$ relative to the basis W^{r-1} and also as the ith covariant component of $\nabla \bar{w}_j^{r-1}$ relative to the basis $\overline{W}^r$.

§2.3. The elements of the group $L^r(\mathfrak{K}, \mathfrak{A})$ expressed in terms of a basis of the module $L_0^r(\mathfrak{K})$.

THEOREM 2.3. *Let*

$$X^r = \{x_1^r, \cdots, x_{\rho(r)}^r\}$$

be a basis of $L_0^r(\mathfrak{K})$, *and let* $\mathfrak{A}$ *be an arbitrary Abelian group. Then every chain* $x^r \in L^r(\mathfrak{K}, \mathfrak{A})$ *may be uniquely written in the form*

$$x^r = \sum_i a_i x_i^r,$$

with $a_i \in \mathfrak{A}$.

Proof. If

$$x_i^r = \sum_j \alpha_{ij} t_j^r, \qquad t_j^r = \sum_k b_{jk} x_k^r,$$

the matrices $\| \alpha_{ij} \|$ and $\| b_{jk} \|$ are unimodular and inverse to each other. Hence

$$\sum_j \alpha_{ij} b_{jk} = \delta_{ik} = \begin{cases} 0 \ (i \neq k), \\ 1 \ (i = k). \end{cases}$$

Every chain $x^r = \sum_j b_j t_j^r \in L^r(\mathfrak{K}, \mathfrak{A})$, $b_j \in \mathfrak{A}$, is of the form

$$\sum_{j,k} b_j b_{jk} x_k^r = \sum_k a_k x_k^r, \qquad a_k = \sum_j b_j b_{jk}.$$

To prove uniqueness it is enough to show that $\sum_i a_i x_i^r = 0$ implies that $a_i = 0$. This is easily done. If

$$\sum_i a_i x_i^r = 0,$$

then

$$\sum_{i,j} a_i \alpha_{ij} t_j^r = 0,$$

and therefore $\sum_i a_i \alpha_{ij} = 0$ (for arbitrary j). Multiplying by b_{jk} and summing over j, we get

$$\sum_{i,j} a_i \alpha_{ij} b_{jk} = \sum_i a_i \delta_{ik} = a_k = 0 \text{ (for arbitrary } k).$$

This completes the proof.

§3. Canonical systems of bases. The groups $\nabla_0^r(\mathfrak{K})$

§3.1. Preliminary remarks. In this section, for brevity, we shall write L^r for $L^r(\mathfrak{K}, J)$; and $Z_\Delta^r, H_\Delta^r, \hat{H}^r, Z_\nabla^r, \cdots$ for $Z_\Delta^r(\mathfrak{K}, J), H_\Delta^r(\mathfrak{K}, J), \hat{H}^r(\mathfrak{K}, J), Z_\nabla^r(\mathfrak{K}, J), \cdots$.

3.11. *The group Z_Δ^r is a division closed subgroup of L^r.*

In fact, for $a \in J$, $a \neq 0$,

$$x^r \in L^r, \qquad ax^r \in Z_\Delta^r$$

imply that

$$\Delta ax^r = a\Delta x^r = 0.$$

Hence

$$\Delta x^r = 0$$

and

$$x^r \in Z_\Delta^r.$$

Since L^r/Z_Δ^r is isomorphic to H_Δ^{r-1}, Theorem 2.61 of Appendix 2 implies

3.12. *The group L^r is a direct sum*

$$L^r = Y^r + Z_\Delta^r, \tag{3.12}$$

where the subgroup Y^r of L^r is isomorphic to H_Δ^{r-1}.

§3.2. Canonical bases of the groups Z_Δ^r. Setting

$$X = Z_\Delta^r, \qquad U = H_\Delta^r$$

in Theorem 2.52 of Appendix 2, we obtain

3.21. *The group Z_Δ^r has a basis*

$$z_1^r, \cdots, z_{\pi(r)}^r; \qquad u_1^r, \cdots, u_{\tau(r)}^r; \qquad v_1^r, \cdots, v_{\sigma(r)}^r, \tag{3.21}$$

with the following properties:

a) *$v_1^r, \cdots, v_{\sigma(r)}^r$ are elements of H_Δ^r; the order of the element u_i^r, $i = 1, 2, \cdots, \tau^r$, relative to H_Δ^r is a natural number $\theta_i^r > 1$, with*

$$\theta_i^r \equiv 0 \ (\text{mod } \theta_{i+1}^r).$$

b) *The cycles $\theta_1^r u_1^r, \cdots, \theta_{\tau(r)}^r u_{\tau(r)}^r; v_1^r, \cdots, v_{\sigma(r)}^r$ form a basis for H_Δ^r.*

c) *The cycles $u_1^r, \cdots, u_{\tau(r)}^r; v_1^r, \cdots, v_{\sigma(r)}^r$ form a basis for $\hat{H}^r$.*

DEFINITION 3.22. A basis (3.21) of Z_Δ^r with properties a), b), c) is called a *canonical basis of Z_Δ^r*.

3.23. *The numbers θ_i^r are the rth torsion numbers of $\mathfrak{K}$ and hence are uniquely determined; π^r is the rth Betti number of $\mathfrak{K}$.*

Proof. Denote by Z_h^r, U_i^r, V_j^r the infinite cyclic groups generated by

z_h^r, u_i^r, v_j^r, respectively. Then we have the following decompositions into direct sums:

$$(3.22)\qquad \begin{cases} Z_\Delta^r = \sum_{h=1}^{\pi(r)} Z_h^r + \sum_{i=1}^{\tau(r)} U_i^r + \sum_{j=1}^{\sigma(r)} V_j^r, \\ H_\Delta^r = \qquad\qquad \sum_{i=1}^{\tau(r)} \theta_i^r U_i^r + \sum_{j=1}^{\sigma(r)} V_j^r. \end{cases}$$

Consequently (Appendix 2, Theorem 1.52)

$$(3.23)\qquad \Delta_0^r(\mathfrak{K}) \approx \sum_{h=1}^{\pi(r)} Z_h^r + \sum_{i=1}^{\tau(r)} (U_i^r/\theta_i^r U_i^r),$$

where $\approx$ denotes isomorphism. The groups $U_i^r/\theta_i^r U_i^r$ are finite cyclic of order θ_i^r, while the groups Z_h^r are infinite cyclic; the number π^r is the rank of $\Delta_0^r(\mathfrak{K})$, that is, the rth Betti number of $\mathfrak{K}$. The direct sum $\sum (U_i^r/\theta_i^r U_i^r)$ consists of all the elements of finite order of $\Delta_0^r(\mathfrak{K})$ and hence coincides (up to an isomorphism) with the rth torsion group $\Theta^r(\mathfrak{K})$ of $\mathfrak{K}$:

$$(3.24)\qquad \Theta^r(\mathfrak{K}) = \sum_{i=1}^{\tau(r)} (U_i^r/\theta_i^r U_i^r).$$

Since the finite group $\Theta^r(\mathfrak{K})$ is the direct sum of τ^r cyclic subgroups of order θ_i^r and $\theta_i^r \equiv 0 \pmod{\theta_{i+1}^r}$, the numbers θ_i^r and τ^r are uniquely determined by $\Theta^r(\mathfrak{K})$: the numbers θ_i^r are the torsion numbers of $\Theta^r(\mathfrak{K})$ (see Appendix 2, 4.3). Hence they are also the torsion numbers of $\Delta_0^r(\mathfrak{K})$. This is what we wished to prove.

§3.3. Canonical homology bases. Let x^r be an arbitrary integral cycle. Then, since $Z_\Delta^r = \sum Z_h^r + \sum U_i^r + \sum V_j^r$, x^r can be uniquely represented as a linear combination

$$\sum a_h z_h^r + \sum b_i u_i^r + \sum c_j v_j^r.$$

Since

$$\{\theta_1^r u_1^r, \cdots, \theta_{\tau(r)}^r u_{\tau(r)}^r, v_1^r, \cdots, v_{\sigma(r)}^r\}$$

is a basis of H_Δ^r, $x^r = \sum a_h z_h^r + \sum b_i u_i^r + \sum c_j v_j^r$ is contained in H_Δ^r if, and only if,

$$\sum a_h z_h^r + \sum b_i u_i^r + \sum c_j v_j^r = \sum d_i \theta_i^r u_i^r + \sum e_j v_j^r.$$

This condition, in view of the linear independence of the chains z_h^r, u_i^r, v_j^r, is satisfied only if

$$a_h = 0, \qquad b_i = d_i \theta_i^r, \qquad c_j = e_j .$$

Hence

3.31. *Every integral cycle $x^r \in Z_\Delta^r$ is uniquely representable in the form*

$$(3.31)\qquad x^r = \sum a_h z_h^r + \sum b_i u_i^r + \sum c_j v_j^r.$$

A cycle x^r is homologous to zero if, and only if,

$$a_h = 0, \qquad b_i \equiv 0 \pmod{\theta_i^r}$$

in (3.31).

It follows that

3.32. *Every integral cycle is homologous to a cycle*

$$\sum a_h z_h^r + \sum b_i u_i^r,$$

where the coefficients a_h are uniquely determined and the coefficients b_i are determined up to multiples of the corresponding θ_i^r.

COROLLARY. The elements $z_1^r, \cdots, z_{\pi(r)}^r$ of a canonical basis of Z_Δ^r form an r-dimensional $(J, \mathfrak{R})$-basis of $\mathfrak{K}$.

The converse follows from Appendix 2, Theorem 2.63:

3.33. *Every r-dimensional $(J, \mathfrak{R})$-basis of $\mathfrak{K}$ can be extended to a canonical basis of Z_Δ^r.*

§3.4. A system of canonical bases of the groups L^r. Suppose that a definite canonical basis (3.21) of $Z^r(\mathfrak{K})$ has been chosen for each r, $r = 0, \cdots, n$. The isomorphism Δ between L^r/Z_Δ^r and H_Δ^{r-1} associates with the basis

$$\theta_1^{r-1} u_1^{r-1}, \cdots, \theta_{\tau(r-1)}^{r-1} u_{\tau(r-1)}^{r-1}; \qquad v_1^{r-1}, \cdots, v_{\sigma(r-1)}^{r-1}$$

of the group H_Δ^{r-1} the basis

$$\mathfrak{x}_1^r, \cdots, \mathfrak{x}_{\tau(r-1)}^r; \qquad \mathfrak{y}_1^r, \cdots, \mathfrak{y}_{\sigma(r-1)}^r$$

of the group L^r/Z_Δ^r, where

$$\Delta x_i^r = \theta_i^{r-1} u_i^{r-1}, \qquad \Delta y_j^r = v_j^{r-1}$$

for arbitrary $x_i^r \in \mathfrak{x}_i^r$, $y_j^r \in \mathfrak{y}_j^r$. Let us choose fixed chains $x_i^r \in \mathfrak{x}_i^r$, $y_j^r \in \mathfrak{y}_j^r$. In view of Appendix 2, Theorem 2.64, the chains

$$x_1^r, \cdots, x_{\tau(r-1)}^r; \quad y_1^r, \cdots, y_{\sigma(r-1)}^r; \quad z_1^r, \cdots, z_{\pi(r)}^r;$$
$$u_1^r, \cdots, u_{\tau(r)}^r; \quad v_1^r, \cdots, v_{\sigma(r)}^r$$

form a basis for L^r.

We have the following fundamental result:

THEOREM AND DEFINITION 3.4. *Let $\mathfrak{K}$ be a finite n-dimensional a-complex. It is possible, for every $r = 0, 1, 2, \cdots, n$, to construct a basis*

$$(3.4) \quad x_1^r, \cdots, x_{\tau(r-1)}^r; \quad y_1^r, \cdots, y_{\sigma(r-1)}^r;$$
$$z_1^r, \cdots, z_{\pi(r)}^r; \quad u_1^r, \cdots, u_{\tau(r)}^r; \quad v_1^r, \cdots, v_{\sigma(r)}^r$$

for L^r, with the following properties:

a) *the chains $z_1^r, \cdots, z_{\pi(r)}^r$; $u_1^r, \cdots, u_{\tau(r)}^r$; $v_1^r, \cdots, v_{\sigma(r)}^r$ are cycles and form a basis for Z_Δ^r;*

b) *the cycles* $u_1^r, \cdots, u_{\tau(r)}^r; v_1^r, \cdots, v_{\sigma(r)}$ *form a basis of* $\hat{H}_\Delta^r$;

c) *for every* $i = 1, \cdots, \tau^r$, *the order of* u_i^r *relative to* H_Δ^r *is a natural number* $\theta_i^r > 1$, *where* $\theta_i^r \equiv 0$ (*mod* θ_{i+1}^r) *and the cycles* $\theta_1^r u_1^r, \cdots, \theta_{\tau(r)}^r u_{\tau(r)}^r$; $v_1^r, \cdots, v_{\sigma(r)}^r$ *form a basis for* H_Δ^r;

d) $\Delta x_k^r = \theta_k^{r-1} u_k^{r-1}$, $\Delta y_i^r = v_i^{r-1}$ (*so that the chains* $x_1^r, \cdots, x_{\tau(r-1)}^r$ *and* $y_1^r, \cdots, y_{\sigma(r-1)}^r$ *are not cycles*).

The bases (3.4), *constructed for every* $r = 0, 1, \cdots, n$ *and satisfying* a)–d) *for every* r, *form, by definition, a system of canonical bases of* L^r.

(3.41)

	$x_1^{r-1}\cdots x_{\tau(r-2)}^{r-1}$	$y_1^{r-1}\cdots y_{\sigma(r-2)}^{-1}$	$z_1^{r-1}\cdots z_{\pi(r-1)}^{r-1}$	$u_1^{r-1}\cdots u_{\tau(r-1)}^{r-1}$	$v_1^{r-1}\cdots v_{\sigma(r-1)}^{r-1}$
$x_1^r \; \vdots \; x_{\tau(r-1)}^r$	0	0	0	$\theta_1^{r-1} \; \cdot \; \cdot \; \cdot \; \theta_{\tau(r-1)}^{r-1}$	0
$y_1^r \; \vdots \; y_{\sigma(r-1)}^r$	0	0	0	0	$1 \; \cdot \; \cdot \; \cdot \; 1$
$z_1^r \; \vdots \; z_{\pi(r)}^r$	0	0	0	0	0
$u_1^r \; \vdots \; u_{\tau(r)}^r$	0	0	0	0	0
$v_1^r \; \vdots \; v_{\sigma(r)}^r$	0	0	0	0	0

Remark. We recall that π^r is the rth Betti number and the θ_i^r are the rth torsion numbers of $\mathfrak{R}$.

Conditions a)–d) enable us to write the matrix of the homomorphism Δ of L^r into L^{r-1} relative to the canonical bases (3.4). This matrix is given in Table (3.41).

Each row in the matrix is the Δ-boundary of the chain heading the row.

The matrix of the homomorphism Δ of L^{r+1} into L^r is given by Table (3.42).

Example 1 (see VII, 4.2, Examples 1–3 and VIII, 1.4, Exercise 4). Let K be the complex consisting of the three sides and the three vertices of a triangle $(e_1e_2e_3)$. We construct a system of canonical bases for K as

(3.42)

	$x_1^r \cdots x_{\tau(r-1)}^r$	$y_1^r \cdots y_{\sigma(r-1)}^r$	$z_1^r \cdots z_{\pi(r)}^r$	$u_1^r \cdots u_{\tau(r)}^r$	$v_1^r \cdots v_{\sigma(r)}^r$
$x_1^{r+1} \vdots x_{\tau(r)}^{r+1}$	0	0	0	$\theta_1^r \ \ddots \ \theta_{\tau(r)}^r$	0
$y_1^{r+1} \vdots y_{\sigma(r)}^{r+1}$	0	0	0	0	$1 \ \ddots \ 1$
$z_1^{r+1} \vdots z_{\pi(r+1)}^{r+1}$	0	0	0	0	0
$u_1^{r+1} \vdots u_{\tau(r+1)}^{r+1}$	0	0	0	0	0
$v_1^{r+1} \vdots v_{\sigma(r+1)}^{r+1}$	0	0	0	0	0

follows:

$$v_1^0 = e_3 - e_2, \qquad v_2^0 = e_1 - e_3, \qquad z_1^0 = e_3,$$

$$y_1^1 = t_1^1 = |\, e_2e_3 \,|, \quad y_2^1 = t_2^1 = |\, e_3e_1 \,|, \quad z_1^1 = t_1^1 + t_2^1 + t_3^1, \quad t_3^1 = |\, e_2e_1 \,|.$$

EXAMPLE 2 (see the references given for Example 1). $\mathfrak{K}$ is the a-complex consisting of the elements

$$\pm t_1^2, \qquad \pm t_1^1, \qquad \pm t_2^1, \qquad \pm t_1^0,$$

with

$$(t_1^2 : t_1^1) = (t_1^2 : t_2^1) = 0, \qquad (t_1^1 : t_1^0) = (t_2^1 : t_1^0) = 0.$$

A system of canonical bases is

$$z_1^0 = t_1^0, \qquad z_1^1 = t_1^1, \qquad z_2^1 = t_2^1, \qquad z_1^2 = t_1^2.$$

EXAMPLE 3 (see the references given for Example 1). $\mathfrak{K}$ is an a-complex with the elements

$$\pm t_1^2, \qquad \pm t_1^1, \qquad t_1^0$$

whose incidence numbers are

$$(t_1^2 : t_1^1) = 2, \qquad (t_1^1 : t_1^0) = 0.$$

A system of canonical bases is

$$z_1^0 = t_1^0, \qquad u_1^1 = t_1^1, \qquad x_1^2 = t_1^2.$$

§3.5. A system of ∇-bases for $\mathfrak{K}$; the groups $\nabla_0^r(\mathfrak{K})$. Let

$$W^r = \{x_1^r, \cdots, x_{\tau(r-1)}^r;\ y_1^r, \cdots, y_{\sigma(r-1)}^r;\ z_1^r, \cdots, z_{\pi(r)}^r;\ u_1^r, \cdots, u_{\tau(r)}^r;\ v_1^r, \cdots, v_{\sigma(r)}^r\} \tag{3.51}$$

$(r = 0, 1, \cdots, n)$ be a canonical system of bases for $\mathfrak{K}$.

A system of bases for L^r, $r = 0, 1, \cdots, n$, dual to the bases (3.51) is called *a system of ∇-bases or cohomology bases of the complex* $\mathfrak{K}$. The elements of these bases dual to the elements

$$x_1^r, \cdots, x_{\tau(r-1)}^r; \quad y_1^r, \cdots, y_{\sigma(r-1)}^r; \quad z_1^r, \cdots, z_{\pi(r)}^r; \quad u_1^{(r)}, \cdots, u_{\tau(r)}^r; \quad v_1^r, \cdots, v_{\sigma(r)}^r$$

are denoted by

$$\bar{u}_1^r, \cdots, \bar{u}_{\tau(r-1)}^r; \quad \bar{v}_1^r, \cdots, \bar{v}_{\sigma(r-1)}^r; \quad \bar{z}_1^r, \cdots, \bar{z}_{\pi(r)}^r; \quad \bar{x}_1^r, \cdots, \bar{x}_{\tau(r)}^r; \quad \bar{y}_1^r, \cdots, \bar{y}_{\sigma(r)}^r, \tag{3.52}$$

respectively.

The basis (3.52) itself we denote by $\overline{W}^r$. Recalling that the homomorphism

(3.53^r)

	$\bar{u}_1^{r-1} \cdots \bar{u}_{\tau(r-2)}^{r-1}$	$\bar{v}_1^{r-1} \cdots \bar{v}_{\sigma(r-2)}^{r-1}$	$\bar{z}_1^{r-1} \cdots \bar{z}_{\pi(r-1)}^{r-1}$	$\bar{x}_1^{r-1} \cdots \bar{x}_{\tau(r-1)}^{r-1}$	$\bar{y}_1^{r-1} \cdots \bar{y}_{\sigma(r-1)}^{r-1}$
$\bar{u}_1^r \ \vdots \ \bar{u}_{\tau(r-1)}^r$	0	0	0	$\theta_1^{r-1} \ \ddots \ \theta_{\tau(r-1)}^{r-1}$	0
$\bar{v}_1^r \ \vdots \ \bar{v}_{\sigma(r-1)}^r$	0	0	0	0	$1 \ \ddots \ 1$
$\bar{z}_1^r \ \vdots \ \bar{z}_{\pi(r)}^r$	0	0	0	0	0
$\bar{x}_1^r \ \vdots \ \bar{x}_{\tau(r)}^r$	0	0	0	0	0
$\bar{y}_1^r \ \vdots \ \bar{y}_{\sigma(r)}^r$	0	0	0	0	0

(3.53^{r+1})

	$\bar{u}_1^r \cdots \bar{u}^r_{\tau(r-1)}$	$\bar{v}_1^r \cdots \bar{v}^r_{\sigma(r-1)}$	$\bar{z}_1^r \cdots \bar{z}^r_{\pi(r)}$	$\bar{x}_1^r \cdots \bar{x}^r_{\tau(r)}$	$\bar{y}_1^r \cdots \bar{y}^r_{\tau(r)}$
$\bar{u}_1^{r+1} \vdots \bar{u}^{r+1}_{\tau(r)}$	0	0	0	$\theta_1^r \ddots \theta^r_{\tau(r)}$	0
$\bar{v}_1^{r+1} \vdots \bar{v}^{r+1}_{\sigma(r)}$	0	0	0	0	$1 \ddots 1$
$\bar{z}_1^{r+1} \vdots \bar{z}^{r+1}_{\pi(r+1)}$	0	0	0	0	0
$\bar{x}_1^{r+1} \vdots \bar{x}^{r+1}_{\tau(r+1)}$	0	0	0	0	0
$\bar{y}_1^{r+1} \vdots \bar{y}^{r+1}_{\sigma(r+1)}$	0	0	0	0	0

Δ of L^r into L^{r-1} (of L^{r+1} into L^r) is represented by Table (3.41) [(3.42)], and applying the rule of 2.2, we see that the homomorphism ∇ of L^{r-1} into L^r (of L^r into L^{r+1}) is represented by Table (3.53^r) [(3.53^{r+1})]. Each column in these tables represents the ∇-boundary of the element at the head of the column.

It is therefore clear that the chains $\bar{u}_i^r$, $\bar{v}_j^r$, and $\bar{z}_h^r$ are cocycles, but that the chains $\bar{x}_k^r$ and $\bar{y}_l^r$ are not cocycles:

$$(3.54) \qquad \begin{cases} \nabla \bar{x}_k^r = \theta_k^r \bar{u}_k^{r+1}, & \nabla \bar{y}_l^r = \bar{v}_l^{r+1}, \\ \nabla \bar{u}_i^r = \nabla \bar{v}_j^r = \nabla \bar{z}_k^r = 0. \end{cases}$$

Examples of Cohomology Bases. We shall consider the examples of canonical bases constructed in 3.4, and find corresponding cohomology bases. For Example 1 we obtain

$$\left.\begin{matrix} \bar{y}_1^0 = -e_2, & \bar{y}_2^0 = e_1, & \bar{z}_1^0 = e_1 + e_2 + e_3 \\ \bar{v}_1^1 = t_1^1 - t_3^1, & \bar{v}_2^1 = t_2^1 - t_3^1, & \bar{z}_1^1 = t_3^1 \end{matrix}\right\} \begin{matrix} \nabla \bar{y}_1^0 = \bar{v}_1^1 \\ \nabla \bar{y}_2^0 = \bar{v}_2^1. \end{matrix}$$

In Example 2 of 3.4 the system of canonical bases is self-dual, that is, it coincides with the corresponding system of cohomology bases.

In Example 3, we have the system of cohomology bases

$$\bar{z}_1^0 = t_1^0, \qquad \bar{x}_1^1 = t_1^1, \qquad \bar{u}_1^2 = t_1^2,$$

where

$$\nabla \bar{x}_1^1 = 2\bar{u}_1^2.$$

Since (3.52) is a basis for L^r, every chain $x^r \in L^r$ can be uniquely written in the form

$$x^r = \sum a_i \bar{u}_i^r + \sum b_j \bar{v}_j^r + \sum c_h \bar{z}_h^r + \sum d_k \bar{x}_k^r + \sum e_l \bar{y}_l^r,$$

with

$$(3.55)\quad \begin{aligned} \nabla x^r &= \sum a_i \nabla \bar{u}_i^r + \sum b_j \nabla \bar{v}_j^r + \sum c_h \nabla \bar{z}_h^r + \sum d_k \nabla \bar{x}_k^r + \sum e_l \nabla \bar{y}_l^r \\ &= \sum d_k \nabla \bar{x}_k^r + \sum e_l \nabla \bar{y}_l^r = \sum_1^{\tau(r)} d_k \theta_k^r \bar{u}_k^{r+1} + \sum_1^{\sigma(r)} e_l \bar{v}_l^{r+1}. \end{aligned}$$

Hence x^r is a cocycle if, and only if,

$$d_k = 0, \qquad e_l = 0 \begin{cases} k = 1, 2, \cdots, \tau^r, \\ l = 1, 2, \cdots, \sigma^r \end{cases}$$

in (3.55).

In other words, the elements

$$\bar{u}_i^r, \bar{v}_j^r, \bar{z}_h^r \begin{cases} i = 1, 2, \cdots, \tau^{r-1}, \\ j = 1, 2, \cdots, \sigma^{r-1}, \\ h = 1, 2, \cdots, \pi^r \end{cases}$$

of (3.52) form a basis for Z_∇^r.

Moreover, since $\overline{W}^r$ and $\overline{W}^{r-1}$ are bases of the groups L^r and L^{r-1}, respectively, every coboundary

$$\nabla x^{r-1} = x^r$$

is a linear combination of coboundaries

$$\nabla \bar{x}_i^{r-1} = \theta_i^{r-1} \bar{u}_i^r,$$
$$\nabla \bar{y}_j^{r-1} = \bar{v}_j^r.$$

But this means that a cocycle

$$(3.56)\qquad x^r = \sum a_i \bar{u}_i^r + \sum b_j \bar{v}_j^r + \sum c_h \bar{z}_h^r$$

is in H_∇^r if, and only if,

$$c_h = 0, \qquad a_i \equiv 0 \ (\mathrm{mod}\ \theta_i^{r-1})$$
$$\text{for all } h = 1, 2, \cdots, \pi^r; \qquad i = 1, 2, \cdots, \tau^{r-1}$$

in (3.56). Hence the cocycles

$$\theta_1^{r-1} \bar{u}_1^r, \cdots, \theta_{\tau(r-1)}^{r-1} \bar{u}_{\tau(r-1)}^r; \qquad \bar{v}_1^r, \cdots, \bar{v}_{\sigma(r-1)}^r$$

form a basis for H_∇^r. Therefore, if

$$\bar{X}_k^r,\ \bar{Y}_l^r,\ \bar{Z}_h^r, \bar{U}_i^r,\ \bar{V}_j^r$$

are, respectively, the infinite cyclic subgroups of L^r generated by the elements

$$\bar{x}_k^r,\ \bar{y}_l^r,\ \bar{z}_h^r,\ \bar{u}_i^r,\ \bar{v}_j^r,$$

we have the following decompositions into direct sums:

$$L^r = \sum_{k=1}^{\tau(r)} \bar{X}_k^r + \sum_{l=1}^{\sigma(r)} \bar{Y}_l^r + \sum_{h=1}^{\pi(r)} \bar{Z}_h^r + \sum_{i=1}^{\tau(r-1)} \bar{U}_i^r + \sum_{j=1}^{\sigma(r-1)} \bar{V}_j^r,$$

$$Z_\nabla^r = \sum_{h=1}^{\pi(r)} \bar{Z}_h^r + \sum_{i=1}^{\tau(r-1)} \bar{U}_i^r + \sum_{j=1}^{\sigma(r-1)} \bar{V}_j^r,$$

$$H_\nabla^r = \sum_{i=1}^{\tau(r-1)} \theta_i^{r-1}\bar{U}_i^r + \sum_{j=1}^{\sigma(r-1)} \bar{V}_j^r.$$

Hence (see Appendix 2, Theorem 1.52)

$$\nabla_0^r = Z_\nabla^r/H_\nabla^r = \sum_{h=1}^{\pi(r)} \bar{Z}_h^r + \sum_{i=1}^{\tau(r-1)} (\bar{U}_i^r/\theta_i^{r-1}\bar{U}_i^r).$$

Here $\bar{U}_i^r/\theta_i^{r-1}\bar{U}_i^r$ is the infinite cyclic group of order θ_i^{r-1}, so that, by (3.23), $\sum_{i=1}^{\tau(r-1)} (\bar{U}_i^r/\theta_i^{r-1}\bar{U}_i^r)$ is isomorphic to the $(r - 1)$st torsion group $\Theta^{r-1}(\mathfrak{K})$.

The groups $\bar{Z}_h^r$ are infinite cyclic. Hence the group $\sum_{h=1}^{\pi(r)} \bar{Z}_h^r$ is obviously isomorphic to $\Delta_{00}^r(\mathfrak{K})$.

Consequently:

THEOREM 3.51. *The group $\nabla_0^r(\mathfrak{K})$ is isomorphic to the direct sum of the $(r - 1)$st torsion group $\Theta^{r-1}(\mathfrak{K})$ and the free Abelian group of rank $\pi^r(\mathfrak{K})$.*

Let K^n be an orientable pseudomanifold. Then, by Theorem 1.7, $\nabla_0^n(K^n)$ is infinite cyclic, so that $\pi^n(K^n) = 1$ and $\Theta^{n-1}(K^n)$ is the null group.

But if K^n is a nonorientable pseudomanifold, then, by Theorem 1.7, $\nabla_0^n(K^n)$ is of order 2; consequently, $\pi^n(K^n) = 0$ and $\Theta^{n-1}(K^n)$ is of order 2.

Hence

Theorem 3.52. *An orientable n-dimensional pseudomanifold is $(n - 1)$-torsion free; the $(n - 1)$st torsion group of an n-dimensional nonorientable pseudomanifold is the group of order 2.*

Moreover, we have from Theorems 3.51 and 1.76:

3.53. *If K_0 is a proper closed subcomplex of an n-dimensional closed pseudomanifold K, then K_0 is $(n - 1)$-torsion free.*

For, by Theorem 3.51, $\nabla_0^n(K_0)$ is isomorphic to the direct sum of $\Theta^{n-1}(K_0)$ and $\Delta_{00}^n(K_0)$. But by Theorem 1.76, $\nabla_0^n(K_0)$ is the null group. Hence both $\Theta^{n-1}(K_0)$ and $\Delta_{00}^n(K_0)$ are null groups.

EXAMPLES OF GROUPS $\nabla(\mathfrak{K})$. Let us return to the examples at the end of 3.4. In the first two examples the ∇-groups are isomorphic to the ∇-groups of the same dimensions. In the third example ∇_0^1 is the null group, while ∇_0^2 is of order 2.

EXERCISE. Determine the ∇-groups of the various complexes considered in VIII, 1.4 both by a direct calculation and by application of Theorem 3.51.

§4. Calculation of the groups $\Delta^r(\mathfrak{K}, \mathfrak{A})$ and $\nabla^r(\mathfrak{K}, \mathfrak{A})$ by means of the groups $\Delta_0{}^r(\mathfrak{K})$

§4.1. Calculation of the groups $\Delta^r(\mathfrak{K}, \mathfrak{A})$. Let

$$(4.11)\quad x_1{}^r, \cdots, x_{\tau(r-1)}{}^r; \qquad y_1{}^r, \cdots, y_{\sigma(r-1)}{}^r;$$
$$z_1{}^r, \cdots, z_{\pi(r)}{}^r; \qquad u_1{}^r, \cdots, u_{\tau(r)}{}^r; \qquad v_1{}^r, \cdots, v_{\sigma(r)}{}^r,$$
$$\Delta x_k{}^r = \theta_k{}^{r-1} u_k{}^{r-1}, \qquad \Delta y_l{}^r = v_l{}^{r-1}; \qquad r = 1, 2, \cdots, n,$$

be a canonical system of bases for L^r, and let $x^r \in L^r$. The representation

$$(4.12)\quad x^r = \sum a_k x_k{}^r + \sum b_l y_l{}^r + \sum c_h z_h{}^r + \sum d_i u_i{}^r + \sum e_j v_j{}^r,$$
$$a_k \in \mathfrak{A}, \qquad b_l \in \mathfrak{A}, \qquad c_h \in \mathfrak{A}, \qquad d_i \in \mathfrak{A}, \qquad e_j \in \mathfrak{A}$$

is unique by Theorem 2.3. The boundary

$$\Delta x^r = \sum a_k \Delta x_k{}^r + \sum b_l \Delta y_l{}^r + \sum c_h \Delta z_h{}^r + \sum d_i \Delta u_i{}^r + \sum e_j \Delta v_j{}^r$$
$$= \sum \theta_k{}^{r-1} a_k u_k{}^{r-1} + \sum b_l v_l{}^{r-1}$$

is zero if, and only if,

$$(4.13)\quad \begin{cases} \theta_k{}^{r-1} a_k = 0, & k = 1, 2, \cdots, \tau^{r-1}, \\ b_l = 0, & l = 1, 2, \cdots, \sigma^{r-1}. \end{cases}$$

Hence, in order that $x^r \in L^r(\mathfrak{K}, \mathfrak{A})$ be a cycle, it is necessary and sufficient that the coefficients a_k and b_l of (4.12) satisfy (4.13).

We may, therefore, say that

All cycles $z^r \in Z_\Delta{}^r(\mathfrak{K}, \mathfrak{A})$, and only cycles, are representable in the form

$$(4.130)\quad z^r = \sum a_k x_k{}^r + \sum c_h z_h{}^r + \sum d_i u_i{}^r + \sum e_j v_j{}^r$$

where c_h, d_i, e_j are arbitrary elements of $\mathfrak{A}$ and every a_k is an arbitrary element of the group ${}_{\theta_k^{r-1}}\mathfrak{A}$ (we recall that ${}_m\mathfrak{A}$ is the subgroup of $\mathfrak{A}$ consisting of all $a \in \mathfrak{A}$ such that $ma = 0$). In the sequel we shall write this group as $\mathfrak{A}[\theta_k{}^{r-1}]$. If θ appears without subscripts or superscripts, we shall write ${}_\theta\mathfrak{A}$.

REMARK. If all the $a_k = 0$ in (4.130), the cycle z^r is said to be a cycle of the *first kind*. In the contrary case, z^r is called a cycle of the *second kind*.

Let us now consider the direct sum

$$(4.14)\quad \mathfrak{S} = \sum_{k=1}^{\tau(r-1)} \mathfrak{A}[\theta_k{}^{r-1}] + \sum_{h=1}^{\pi(r)} \mathfrak{A}_h + \sum_{i=1}^{\tau(r)} \mathfrak{A}_i + \sum_{i=1}^{\sigma(r)} \mathfrak{A}_j,$$

where $\mathfrak{A}_h = \mathfrak{A}_i = \mathfrak{A}_j = \mathfrak{A}$. A unique cycle

$$\sum a_k x_k{}^r + \sum c_h z_h{}^r + \sum d_i u_i{}^r + \sum e_j v_j{}^r$$

corresponds to each element

$$(4.15) \quad (a_1, \cdots, a_{\tau(r-1)}; c_1, \cdots, c_{\pi(r)}; d_1, \cdots, d_{\tau(r)}; e_1, \cdots, e_{\sigma(r)})$$

of the group $\mathfrak{S}$. Conversely, we have just seen that a unique element (4.15) of the group $\mathfrak{S}$ corresponds to each cycle (4.12) in virtue of (4.13). Hence: *The group $Z_\Delta{}^r(\mathfrak{K}, \mathfrak{A})$ is isomorphic to the group* (4.14).

Now let

$$(4.12') \qquad x^r = \sum a_k x_k{}^r + \sum c_h z_h{}^r + \sum d_i u_i{}^r + \sum e_j v_j{}^r$$

be a cycle and let

$$x^{r+1} = \sum a_i' x_i{}^{r+1} + \sum b_j' y_j{}^{r+1} + \sum c_h' z_h{}^{r+1} + \sum d_k' u_k{}^{r+1} + \sum e_l' v_l{}^{r+1}$$

be a chain. We shall find necessary and sufficient conditions that $\Delta x^{r+1} = x^r$. We have

$$\begin{aligned}\Delta x^{r+1} &= \sum a_i' \Delta x_i{}^{r+1} + \sum b_j' \Delta y_j{}^{r+1} + \sum c_h' \Delta z_h{}^{r+1} + \sum d_k' \Delta u_k{}^{r+1} \\ &\quad + \sum e_l' \Delta v_l{}^{r+1} \\ &= \sum \theta_i{}^r a_i' u_i{}^r + \sum b_j' v_j{}^r.\end{aligned}$$

In order that $\Delta x^{r+1} = x^r$, it is obviously necessary and sufficient that

$$(4.13') \qquad \begin{aligned} d_i &= \theta_i{}^r a_i', & i &= 1, 2, \cdots, \tau^r, \\ e_j &= b_j', & j &= 1, 2, \cdots, \sigma^r, \\ a_k &= 0, & k &= 1, 2, \cdots, \tau^{r-1}, \\ c_h &= 0, & h &= 1, 2, \cdots, \pi^r. \end{aligned}$$

Therefore,

The cycles homologous to zero, i.e., chains $x^r \in H_\Delta{}^r(\mathfrak{K}, \mathfrak{A})$, are uniquely representable in the form

$$x^r = \sum_{i=1}^{\tau(r)} d_i u_i{}^r + \sum_{j=1}^{\sigma(r)} e_j v_j{}^r,$$

where $d_i \in \theta_i{}^r\mathfrak{A}$, $e_j \in \mathfrak{A}$; and conversely.

Consider now the direct sum

$$(4.14') \qquad \mathfrak{S}' = \sum_{i=1}^{\tau(r)} \theta_i{}^r \mathfrak{A} + \sum_{j=1}^{\sigma(r)} \mathfrak{A}_j, \qquad \mathfrak{A}_j = \mathfrak{A}.$$

To each element

$$(4.16) \qquad (d_1, \cdots, d_{\tau(r)}; e_1, \cdots, e_{\sigma(r)}), \qquad d_i = \theta_i{}^r a_i'$$

($a_i' \in \mathfrak{A}$ arbitrary) of the group $\mathfrak{S}'$ there corresponds the chain

$$x^r = \sum d_i u_i{}^r + \sum e_j v_j{}^r$$

which is the boundary of the chain

$$x^{r+1} = \sum a_i' x_i^{r+1} + \sum b_j' y_j^{r+1} + \sum c_h' z_h^{r+1} + \sum d_k' u_k^{r+1} + \sum e_l' v_l^{r+1},$$

with a_i', b_j' given by (4.13′) and c_h', d_k', e_l', completely arbitrary.

Hence a unique element x^r of $H_\Delta{}^r(\mathfrak{K}, \mathfrak{A})$ corresponds to each element (4.16) of $\mathfrak{S}'$. Conversely, if $x^r \in H_\Delta{}^r(\mathfrak{K}, \mathfrak{A})$ and $x^r = \Delta x^{r+1}$, then (4.12) satisfies (4.13′) and the element

$$(d_1, \cdots, d_{\tau(r)}; e_1, \cdots, e_{\sigma(r)})$$

of $\mathfrak{S}'$ corresponds to the cycle x^r. Hence

The group $H_\Delta{}^r(\mathfrak{K}, \mathfrak{A})$ is isomorphic to the group $\mathfrak{S}'$ defined in (4.14′).

From this it is easy to derive

THEOREM 4.1. *The group $\Delta^r(\mathfrak{K}, \mathfrak{A})$ is isomorphic to the direct sum*

$$\sum_{i=1}^{\tau(r-1)} \mathfrak{A}_i[\theta_i^{r-1}] + \sum_{i=1}^{\pi(r)} \mathfrak{A}_i + \sum_{i=1}^{\tau(r)} \mathfrak{A}_i/\theta_i^r \mathfrak{A}_i, \qquad \mathfrak{A}_i = \mathfrak{A}.$$

Theorem 4.1 follows from the definition $\Delta^r(\mathfrak{K}, \mathfrak{A}) = Z_\Delta{}^r(\mathfrak{K}, \mathfrak{A})/H_\Delta{}^r(\mathfrak{K}, \mathfrak{A})$, from Theorem 1.52 of Appendix 2, and from the fact that the groups $Z_\Delta(\mathfrak{K}, \mathfrak{A})$ and $H_\Delta{}^r(\mathfrak{K}, \mathfrak{A})$ are isomorphic to the groups (4.14), (4.14′), respectively.

§4.2. Calculation of the groups $\nabla^r(\mathfrak{K}, \mathfrak{A})$. The calculation of these groups is effected by means of reasoning similar to that immediately preceding. However, instead of starting with canonical Δ-bases, it is now necessary to begin with ∇-bases.

If $x^r \in L^r(\mathfrak{K}, \mathfrak{A})$ is arbitrary, then

$$x^r = \sum_{k=1}^{\tau(r)} a_k \bar{x}_k^r + \sum_{l=1}^{\sigma(r)} b_l \bar{y}_l^r + \sum_{h=1}^{\pi(r)} c_h \bar{z}_h^r + \sum_{i=1}^{\tau(r-1)} d_i \bar{u}_i^r + \sum_{j=1}^{\sigma(r-1)} e_j \bar{v}_j^r, \tag{4.21}$$

where the representation is in terms of the elements of a cohomology basis. Hence, by (3.54),

$$\nabla x^r = \sum_{k=1}^{\tau(r)} a_k \Delta \bar{x}_k^r + \sum_{l=1}^{\sigma(r)} b_l \nabla \bar{y}_l^r = \sum_{k=1}^{\tau(r)} \theta_k^r a_k \bar{u}_k^{r+1} + \sum_{l=1}^{\sigma(r)} b_l \bar{v}_l^{r+1}.$$

Therefore, $\nabla x^r = 0$ if, and only if,

$$\begin{cases} \theta_k^r a_k = 0, & k = 1, 2, \cdots, \tau^r, \\ b_l = 0, & l = 1, 2, \cdots, \sigma^r. \end{cases} \tag{4.22}$$

Hence, as in 4.1, we arrive at

4.21. *The group $Z_\nabla{}^r(\mathfrak{K}, \mathfrak{A})$ is isomorphic to the direct sum*

$$\sum_{k=1}^{\tau(r)} \mathfrak{A}[\theta_k^r] + \sum_{h=1}^{\pi(r)} \mathfrak{A}_h + \sum_{i=1}^{\tau(r-1)} \mathfrak{A}_i + \sum_{j=1}^{\sigma(r-1)} \mathfrak{A}_j,$$

where $\mathfrak{A}_h$, $\mathfrak{A}_i$, $\mathfrak{A}_j$, are all equal to $\mathfrak{A}$.

Given a cocycle (4.21) and a chain

$$x^{r-1} = \sum a_i'\bar{x}_i^{r-1} + \sum b_j'\bar{y}_j^{r-1} + \sum c_h'\bar{z}_h^{r-1} + \sum d_k'\bar{u}_k^{r-1} + \sum e_l'\bar{v}_l^{r-1},$$

then

$$\nabla x^{r-1} = \sum a_i'\nabla\bar{x}_i^{r-1} + \sum b_j'\nabla\bar{y}_j^{r-1} = \sum \theta_i^{r-1}a_i'\bar{u}_i^r + \sum b_j'\bar{v}_j^r .$$

Consequently $x^r = \nabla x^{r-1}$ if, and only if,

$$\text{(4.220)} \qquad \begin{cases} d_i = \theta_i^{r-1}a_i', \\ e_j = b_j', \\ a_k = b_l = c_h = 0. \end{cases}$$

As in 4.1, it follows that

4.22. *The group $H_\nabla^r(\mathfrak{K}, \mathfrak{A})$ is isomorphic to the direct sum*

$$\sum_{i=1}^{\tau(r-1)} \theta_i^{r-1}\mathfrak{A} + \sum_{j=1}^{\sigma(r-1)} \mathfrak{A}_j, \qquad \mathfrak{A}_j = \mathfrak{A}.$$

The following theorem now follows from the definition $\nabla^r(\mathfrak{K}, \mathfrak{A}) = Z_\nabla^r(\mathfrak{K}, \mathfrak{A})/H_\nabla^r(\mathfrak{K}, \mathfrak{A})$, from 4.21, 4.22, and from Appendix 2, Theorem 1.52:

THEOREM 4.2. *The group $\nabla^r(\mathfrak{K}, \mathfrak{A})$ is isomorphic to the direct sum*

$$\sum_{k=1}^{\tau(r)} \mathfrak{A}[\theta_k^r] + \sum^{\pi(r)} \mathfrak{A} + \sum_{i=1}^{\tau(r-1)} (\mathfrak{A}/\theta_i^{r-1}\mathfrak{A}),$$

where $\sum^{\pi(r)} \mathfrak{A}$ is the direct sum of π^r terms each equal to $\mathfrak{A}$.

§4.3. The coefficient domains J, $\mathfrak{R}$, $\mathfrak{R}_1$. We shall consider several special cases of Theorems 4.1 and 4.2.

a) $\mathfrak{A} = J$. Since ${}_\theta J$ is the null group for $\theta > 1$, while the group $J/\theta J$ is the cyclic group of order θ, Theorem 4.1 gives for $\mathfrak{A} = J$ the already known result (see (3.23), (3.24)):

$$\Delta_0^r(\mathfrak{K}) = \sum^{\pi(r)} J + \sum_{i=1}^{\tau(r)} J(\theta_i^r) = \sum^{\pi(r)} J + \Theta^r(\mathfrak{K}),$$

where $J(\theta_i^r) = J_{\theta_i^r} = J/\theta_i^r J$. We shall use this notation in the sequel, whenever θ appears with subscripts or superscripts.

For $\mathfrak{A} = J$, Theorem 4.2 yields

$$\Delta_0^r(\mathfrak{K}) = \sum^{\pi(r)} J + \sum_{i=1}^{\tau(r-1)} J(\theta_i^{r-1}) = \sum^{\pi(r)} J + \Theta^{r-1}(\mathfrak{K}).$$

This result was obtained in Theorem 3.51.

b) $\mathfrak{A} = \mathfrak{R}$. Since ${}_\theta\mathfrak{R}$ is the null group and $\theta\mathfrak{R} = \mathfrak{R}$ for θ an arbitrary natural number, it follows that

$$\Delta^r(\mathfrak{K}, \mathfrak{R}) = \sum^{\pi(r)} \mathfrak{R},$$

$$\nabla^r(\mathfrak{K}, \mathfrak{R}) = \sum^{\pi(r)} \mathfrak{R}.$$

Hence

The groups $\Delta^r(\mathfrak{K}, \mathfrak{R})$ and $\nabla^r(\mathfrak{K}, \mathfrak{R})$ are isomorphic and completely determine the rth Betti number of $\mathfrak{K}$.

REMARK. Theorems 4.1 and 4.2 imply that if $\mathfrak{K}$ is both r- and $(r-1)$-torsion free, then the groups $\Delta^r(\mathfrak{K}, \mathfrak{A})$, $\nabla^r(\mathfrak{K}, \mathfrak{A})$ are direct sums of π^r groups isomorphic to the group $\mathfrak{A}$ for arbitrary coefficient domain $\mathfrak{A}$. Consequently, they completely determine a unique invariant: the rth Betti number of $\mathfrak{K}$.

c) $\mathfrak{A} = \mathfrak{R}_1$. For arbitrary natural number θ,

$$_\theta\mathfrak{R}_1 \approx J(\theta), \qquad \theta\mathfrak{R}_1 = \mathfrak{R}_1 .$$

Therefore

$$(4.31) \quad \begin{cases} \Delta^r(\mathfrak{K}, \mathfrak{R}_1) = \sum_{k=1}^{\tau(r-1)} J(\theta_k^{r-1}) + \sum^{\pi(r)} \mathfrak{R}_1 = \Theta^{r-1}(\mathfrak{K}) + \sum^{\pi(r)} \mathfrak{R}_1 , \\ \nabla^r(\mathfrak{K}, \mathfrak{R}_1) = \sum_{k=1}^{\tau(r)} J(\theta_k^r) + \sum^{\pi(r)} \mathfrak{R}_1 = \Theta^r(\mathfrak{K}) + \sum^{\pi(r)} \mathfrak{R}_1 . \end{cases}$$

§4.4. The groups $\Delta_m^r(\mathfrak{K})$ and $\nabla_m^r(\mathfrak{K})$ (see Appendix 2, 4.5, Remark 2). Let $\mathfrak{A} = J_m$. For arbitrary $\theta > 1$,

$$_\theta J_m = (J_\theta)_m = J_{(m,\theta)} ,$$

$$J_m/\theta J_m = (J_m)_\theta = J_{(m,\theta)} .$$

Hence [writing $\nabla_m^r(\mathfrak{K})$ instead of $\nabla^r(\mathfrak{K}, J_m)$]

$$(4.41) \quad \begin{aligned} \Delta_m^r(\mathfrak{K}) &= \sum_{i=1}^{\tau(r-1)} J_{(m,\alpha)} + \sum^{\pi(r)} J_m + \sum_{i=1}^{\tau(r)} J_{(m,\beta)} , \\ \nabla_m^r(\mathfrak{K}) &= \sum_{i=1}^{\tau(r)} J_{(m,\beta)} + \sum^{\pi(r)} J_m + \sum_{i=1}^{\tau(r-1)} J_{(m,\alpha)} , \end{aligned}$$

where $\alpha = \theta_i^{r-1}$, $\beta = \theta_i^r$. Then, in particular,

$$(4.410) \qquad \Delta_m^r(\mathfrak{K}) = \nabla_m^r(\mathfrak{K}).$$

We formulate (4.41) as

THEOREM 4.41. *The groups $\Delta_m^r(\mathfrak{K})$ and $\nabla_m^r(\mathfrak{K})$ are isomorphic to the direct sum of π^r cyclic groups of order m, τ^{r-1} cyclic groups of order (m, θ_i^{r-1}) $(i = 1, \cdots, \tau^{r-1})$, and τ^r cyclic groups of order (m, θ_i^r) $(i = 1, \cdots, \tau^r)$.*

The following remarks supplement this result:

Since $J_{(m,\theta)} = (J_\theta)_m$ and $\sum_{i=1}^{\tau(r)} J_\beta = \Theta^r(\mathfrak{K})$,

$$\sum_{i=1}^{\tau(r)} J_{(m,\beta)} = \sum_{i=1}^{\tau(r)} (J_\beta)_m = \Big(\sum_{i=1}^{\tau(r)} J_\beta\Big)_m = (\Theta^r(\mathfrak{K}))_m ;$$
$$\sum_{i=1}^{\tau(r-1)} J_{(m,\alpha)} = (\Theta^{r-1}(\mathfrak{K}))_m .$$

Hence

$$(4.411) \quad \begin{aligned} \Delta_m^r(\mathfrak{K}) &= \sum_{i=1}^{\tau(r-1)} J_{(m,\alpha)} + \sum_{i=1}^{\tau(r)} J_{(m,\beta)} + \sum^{\pi(r)} J_m \\ &= (\Theta^{r-1}(\mathfrak{K}))_m + (\Theta^r(\mathfrak{K}))_m + \sum^{\pi(r)} J_m . \end{aligned}$$

Finally, noting that

$$(\Theta^r(\mathfrak{K}))_m + \sum^{\pi(r)} J_m = \Big(\Theta^r(\mathfrak{K}) + \sum^{\pi(r)} J\Big)_m = (\Delta^r(\mathfrak{K}))_m ,$$

we may write

(4.412) $$\Delta_m^{\ r}(\mathfrak{K}) = \nabla_m^{\ r}(\mathfrak{K}) = (\Theta^{r-1}(\mathfrak{K}))_m + (\Delta^r(\mathfrak{K}))_m .$$

REMARK 1. Let us return to (4.41) and assume that m is divisible by all the θ_i^{r-1} and $\theta_i^{\ i}$ [for this it is enough, in particular, to have m divisible by the orders of the groups $\Theta^{r-1}(K)$ and $\Theta^r(K)$]. Then

$$(m, \theta_i^{r-1}) = \theta_i^{r-1}, \qquad (m, \theta_i^{\ r}) = \theta_i^{\ r}$$

and

$$\sum_{i=1}^{\tau(r-1)} J_{(m,\alpha)} = \sum_{i=1}^{\tau(r-1)} J_\alpha = \Theta^{r-1}(\mathfrak{K}),$$
$$\sum_{i=1}^{\tau(r)} J_{(m,\beta)} = \sum_{i=1}^{\tau(r-1)} J_\beta = \Theta^r(\mathfrak{K}).$$

Hence

THEOREM 4.42. *If m is divisible by all the $\theta_i^{\ r}$ and θ_i^{r-1} [in particular, if m is divisible by the orders of the groups $\Theta^r(\mathfrak{K})$ and $\Theta^{r-1}(\mathfrak{K})$],* then

$$\nabla_m^{\ r}(\mathfrak{K}) = \nabla_m^{\ r}(\mathfrak{K}) = \Theta^{r-1}(\mathfrak{K}) + \Theta^r(\mathfrak{K}) + \sum^{\pi(r)} J_m .$$

3.52 and 4.42 imply

4.43. *If K^n is a nonorientable pseudomanifold and m is even, then $\Delta_m^{\ n}(K^n) = \nabla_m^{\ n}(K^n)$ is the group of order 2.*

REMARK 2. If m is prime and θ is an arbitrary natural number,

$$(m, \theta) = m, \quad \text{if} \quad \theta \equiv 0 \pmod{m},$$

$$(m, \theta) = 1, \quad \text{if} \quad \theta \not\equiv 0 \pmod{m}.$$

For arbitrary r denote the number of torsion numbers $\theta_i^{\ r}$ which are divisible by m by $\delta_m^{\ r}$.

Then 4.41 implies

4.44. *For m prime the groups $\Delta_m^{\ r}(\mathfrak{K})$ and $\nabla_m^{\ r}(\mathfrak{K})$ are direct sums of $\pi^r + \delta_m^{\ r} + \delta_m^{\ r-1}$ cyclic groups of order m, so that the rth Betti number* (mod m) *is*

(4.42) $$\pi_m^{\ r} = \pi^r + \delta_m^{\ r} + \delta_m^{\ r-1}.$$

Multiplying both sides of (4.42) by $(-1)^r$ and summing over r from $r = 0$ to $r = n$, and noting that $\delta_m^{\ 0} = \delta_m^{\ n} = 0$, we obtain

$$\sum_{r=0}^{n} (-1)^r \pi_m^{\ r} = \sum_{r=0}^{n} (-1)^r \pi^r = \sum_{r=0}^{n} (-1)^r \rho^r.$$

The formula

(4.43) $$\sum_{r=0}^{n} (-1)^r \pi_m^{\ r} = \sum_{r=0}^{n} (-1)^r \rho^r$$

is called the *Euler-Poincaré formula* (mod m), m a prime.

§4.5. Integral chains and homologies (mod m).

DEFINITION 4.51. Let $x^r = \sum a_i t_i^{\ r} \in L^r$ be an integral chain.

Set

$$\mathfrak{B}_m x^r = \sum a_i{}^m t_i{}^r,$$

where $a_i{}^m$ is the residue of a_i (mod m). If $\mathfrak{B}_m x^r \in Z_\Delta{}^r(\mathfrak{K}, J_m)$, we shall say that x^r is a *cycle* (mod m).

DEFINITION 4.52. We say that an integral cycle $x^r \in Z_\Delta{}^r$ is *homologous to zero* (mod m), $x^r \sim 0$ (mod m), if

$$\mathfrak{B}_m x^r \in H_\Delta{}^r(\mathfrak{K}, J_m).$$

THEOREM 4.53. *An integral cycle* $x^r \sim 0$ (mod m) *if, and only if, there exists* $y^r \in Z_\Delta{}^r$ *such that*

$$x^r - my^r \in H_\Delta{}^r.$$

Necessity. If

$$\mathfrak{B}_m x^r \in H_\Delta{}^r(\mathfrak{K}, J_m),$$

there is a chain $x_{(m)}{}^{r+1} \in Z_\Delta{}^r(\mathfrak{K}, J_m)$ for which

$$\mathfrak{B}_m x^r = \Delta x_{(m)}{}^{r+1}.$$

Now suppose that x^{r+1} is any integral chain such that

$$\mathfrak{B}_m x^{r+1} = x_{(m)}{}^{r+1}.$$

Then

$$\mathfrak{B}_m x^r = \Delta x_{(m)}{}^{r+1} = \Delta \mathfrak{B}_m x^{r+1} = \mathfrak{B}_m \Delta x^{r+1},$$

that is,

$$\mathfrak{B}_m(x^r - \Delta x^{r+1}) = 0;$$

$$x^r - \Delta x^{r+1} = my^r; \qquad y^r \in L^r.$$

Since x^{r+1} and x^r are integral cycles, my^r, and hence y^r, is an integral cycle and

$$x^r - my^r = \Delta x^{r+1} \in H_\Delta{}^r.$$

This proves necessity.

Sufficiency. If

$$x^r - my^r = \Delta x^{r+1},$$

then

$$\mathfrak{B}_m x^r = \mathfrak{B}_m \Delta x^{r+1} = \Delta \mathfrak{B}_m x^{r+1},$$

and the sufficiency is proved.

THEOREM 4.54. *Let*

$$x^r = \sum c_h z_h^r + \sum a_i u_i^r + \sum b_j v_j^r \tag{4.541}$$

be an integral cycle. Then $x^r \sim 0$ (mod m) *if, and only if,*

$$c_h \equiv 0 \pmod m; \qquad a_i \equiv 0 \pmod{(m, \theta_i^r)} \tag{4.542}$$

for all h *and* i *in* (4.541).

For, if $x^r \sim 0$ (mod m), then by the preceding theorem

$$x^r = \Delta x^{r+1} + m y^r, \qquad y^r \in Z_\Delta r.$$

Let

$$y^r = \sum c_h^* z_h^r + \sum a_i^* u_i^r + \sum b_j^* v_j^r,$$

$$x^{r+1} = \sum a_i' x_i^{r+1} + \sum b_j' y_j^{r+1} + \sum c_h' z_h^{r+1} + \sum d_k' u_k^{r+1} + \sum e_l' v_l^{r+1}.$$

Then

$$\Delta x^{r+1} = \sum a_i' \Delta x_i^{r+1} + \sum b_j' \Delta y_j^{r+1} = \sum a_i' \theta_i^r u_i^r + \sum b_j' v_j^r.$$

Hence

$$\sum c_h z_h^r + \sum a_i u_i^r + \sum b_j v_j^r$$
$$= \sum m c_h^* z_h^r + \sum (m a_i^* + \theta_i^r a_i') u_i^r + \sum (m b_j^* + b_j') v_j^r,$$

that is,

$$c_h = m c_h^*; \qquad a_i = m a_i^* + \theta_i^r a_i'; \qquad b_j = m b_j^* + b_j'. \tag{4.543}$$

The first two equalities of (4.543) imply (4.542) and prove the necessity of this condition.

Now suppose that (4.542) is satisfied. Then

$$c_h = m\gamma_h, \qquad a_i = (m, \theta_i^r) a_i'' = \alpha_i m + \beta_i \theta_i^r$$

for integers γ_h, α_i, β_i, and (4.541) becomes

$$x^r = m \sum \gamma_h z_h^r + m \sum \alpha_i u_i^r + \sum \beta_i \theta_i^r u_i^r + \sum b_j v_j^r$$
$$= m\left(\sum \gamma_h z_h^r + \sum \alpha_i u_i^r\right) + \Delta\left(\sum \beta_i x_i^{r+1} + \sum b_j y_j^{r+1}\right),$$

that is,

$$x^r \sim 0 \pmod m.$$

THEOREM 4.55. *If an integral cycle* x^r *is not homologous to zero (over the integers), it cannot be homologous to zero* (mod m) *for any* m.

We shall prove a stronger proposition:

4.551. *If* $x^r \notin H_\Delta^r$, *but* $x^r \in \hat{H}_\Delta^r$, *then* $x^r \not\sim 0$ (mod m) *for any natural* *number m divisible by the order of the group $\Theta^r(\mathfrak{K})$. If both $x \notin H_\Delta{}^r$ and $x \notin \hat{H}_\Delta{}^r$, then $x \nsim 0 \pmod{m}$ for every sufficiently large m.*

Indeed, let $x \in \hat{H}_\Delta{}^r$ and suppose that m is divisible by the order of $\Theta^r(\mathfrak{K})$, that is, by all the $\theta_i{}^r$. Then $(m, \theta_i{}^r) = \theta_i{}^r$. Hence if $x^r \sim 0 \pmod{m}$, it follows from (4.542) that

$$a_i \equiv 0 \pmod{\theta_i{}^r}, \quad \text{i.e.,} \quad a_i = \theta_i{}^r a_i'',$$

a_i'' an integer. Then (4.541) becomes

$$\begin{aligned} x^r &= \sum c_h z_h{}^r + \sum a_i'' \Delta x_i{}^{r+1} + \sum b_j \Delta y_j{}^{r+1} \\ &= \sum c_h z_h{}^r + \Delta\left(\sum a_i'' x_i{}^{r+1} + \sum b_j y_j{}^{r+1}\right). \end{aligned} \tag{4.543}$$

But $x^r \in \hat{H}_\Delta{}^r$ implies that all the c_h are zero, so that

$$x^r = \Delta\left(\sum a_i'' x_i{}^{r+1} + \sum b_j y_j{}^{r+1}\right) \in H_\Delta{}^r.$$

Now suppose $x^r \notin \hat{H}_\Delta{}^r$. Then in (4.541) not all the c_i vanish. If, for instance, $c_1 \neq 0$, taking $m > |c_1|$ we see that the first condition of (4.542) cannot be satisfied. Hence $x^r \nsim 0 \pmod{m}$.

§5. Calculation of the groups $\Delta^r(\mathfrak{K}, \mathfrak{A})$ and $\nabla^r(\mathfrak{K}, \mathfrak{A})$ by means of the groups $\Delta^r(\mathfrak{K}, \mathfrak{R}_1)$ and $\nabla_m{}^r(\mathfrak{K})$

§5.1. In the preceding section we derived formulas for the calculation of the groups $\Delta^r(\mathfrak{K}, \mathfrak{A})$ and $\nabla^r(\mathfrak{K}, \mathfrak{A})$ over an arbitrary coefficient domain $\mathfrak{A}$ from the groups $\Delta_0{}^r(\mathfrak{K})$ and $\Delta_0{}^{r-1}(\mathfrak{K})$. However, it is sometimes expedient to reduce the calculation of $\Delta^r(\mathfrak{K}, \mathfrak{A})$ and $\nabla^r(\mathfrak{K}, \mathfrak{A})$ to the groups $\Delta^r(\mathfrak{K}, \mathfrak{R}_1)$ [or $\nabla^r(\mathfrak{K}, \mathfrak{R}_1)$] and particularly to $\Delta_m{}^r(\mathfrak{K})$ instead of $\Delta_0{}^r(\mathfrak{K})$. To achieve this it is enough to reduce $\Delta_0{}^r(\mathfrak{K})$ to these groups.

THEOREM 5.1 (the proof given here is due to H. Hopf). *If the groups $\Delta^r(\mathfrak{K}, \mathfrak{R}_1)$ are known for all $r = 0, 1, \cdots$, so are the groups $\Delta_0{}^r(\mathfrak{K})$, $r = 0, 1, \cdots$, and hence the groups $\Delta^r(\mathfrak{K}, \mathfrak{A})$ and $\nabla^r(\mathfrak{K}, \mathfrak{A})$ for arbitrary $\mathfrak{A}$. Furthermore, the group $\Delta_0{}^r(\mathfrak{K})$ may be calculated from $\nabla^r(\mathfrak{K}, \mathfrak{R}_1)$.*

Both assertions of Theorem 5.1 follow from (4.31) and the following lemma:

LEMMA 5.11. *If a group $\mathfrak{G}$ is the direct sum of a finite group $\mathfrak{T}$ and of a finite number of subgroups $\mathfrak{G}_1, \cdots, \mathfrak{G}_p$ of $\mathfrak{G}$, each of which is isomorphic to $\mathfrak{R}_1$, then both p and $\mathfrak{T}$ are uniquely determined (the latter up to an isomorphism).*

Proof of the Lemma. Suppose

$$\mathfrak{G} = \mathfrak{T} + \mathfrak{G}_1 \cdots + \mathfrak{G}_p. \tag{5.11}$$

We shall prove first that the subgroup

$$\mathfrak{S} = \mathfrak{G}_1 + \cdots + \mathfrak{G}_p$$

consists of precisely those elements x of $\mathfrak{G}$ which satisfy the following condition:

For every natural number m there exists $y \in \mathfrak{G}$ such that $my = x$.

Indeed, if $x \in \mathfrak{S}$ and $m \geq 1$, then

$$x = x_1 + \cdots + x_p, \qquad x_i \in \mathfrak{G}_i .$$

But there is a $y_i \in \mathfrak{G}_i$ such that $my_i = x_i$, and putting

$$y = y_1 + \cdots + y_p ,$$

it follows that $my = x$.

Conversely, suppose that for every $x \in \mathfrak{G}$ and $m \geq 1$ there exists a $y \in \mathfrak{G}$ (with y depending on m) such that $my = x$. Choose m so that it is the order of $\mathfrak{T}$.

If

$$x = my, \qquad y = t + s, \qquad t \in \mathfrak{T}, \qquad s \in \mathfrak{S},$$

then

$$x = my = mt + ms \in \mathfrak{S},$$

and this proves the assertion.

Hence the elements of $\mathfrak{S}$ are independent of the decomposition (5.11), that is, $\mathfrak{S}$ always consists of the same elements of $\mathfrak{G}$ no matter how the latter is decomposed into a direct sum. It follows that the group $\mathfrak{T}$ is uniquely determined up to an isomorphism: it is isomorphic to $\mathfrak{G}/\mathfrak{S}$. It remains to be shown that the number p in

$$\mathfrak{S} = \mathfrak{G}_1 + \cdots + \mathfrak{G}_p ,$$

where all the $\mathfrak{G}_i$ are isomorphic to $\mathfrak{R}_1$, is uniquely determined. To prove this it is enough to show that p is defined by the equation $2^p = N$, where N is the number of elements x of $\mathfrak{S}$ such that $2x = 0$.

If $x = x_1 + \cdots + x_p$ $(x_i \in \mathfrak{G}_i)$ satisfies this condition, then $2x_i = 0$ for every i. But there are precisely two elements x_i in $\mathfrak{R}_1$ for which $2x_i = 0$, namely, $x_i = 0$ and $x_i = \frac{1}{2}$; hence the number of elements $x \in \mathfrak{S}$ for which $2x = 0$ is indeed equal to 2^p.

This proves Lemma 5.11 and hence Theorem 5.1.

§5.2. THEOREM 5.2. *If $\mathfrak{K}$ is a complex, there exists a natural number $m > 1$ with the following property: if the groups $\Delta_m{}^r(\mathfrak{K})$ are known for all $r = 0, 1, 2, \cdots$, then all the groups $\Delta_0{}^r(\mathfrak{K})$ are also known and consequently also all the groups $\Delta^r(\mathfrak{K}, \mathfrak{A})$, $\nabla^r(\mathfrak{K}, \mathfrak{A})$ for every $\mathfrak{A}$.*

Proof. Let m_0 be a natural number divisible by the orders of all the groups $\Theta^r(\mathfrak{K})$ (if some $\Theta^r(\mathfrak{K})$ is the null group, its order is 1).

We shall show that Theorem 5.2 holds for arbitrary natural number $m = km_0$, where k is an integer ≥ 2.

Indeed, by Theorem 4.42,

$$\Delta_m^{\ r}(\mathfrak{K}) = \Theta^r(\mathfrak{K}) + \Theta^{r-1}(\mathfrak{K}) + \sum\nolimits^{\pi(r)} J_m . \tag{5.21}$$

Hence, recalling that $m_0\Theta^r(\mathfrak{K})$ and $m_0\Theta^{r-1}(\mathfrak{K})$ are null groups, we obtain

$$m_0\Delta_m^{\ r}(\mathfrak{K}) = m_0\Theta^r(\mathfrak{K}) + m_0\Theta^{r-1}(\mathfrak{K}) + \sum\nolimits^{\pi(r)} m_0J_m = \sum\nolimits^{\pi(r)} m_0J_m$$

for the subgroups $m_0\Delta_m^{\ r}(\mathfrak{K}) \subseteq \Delta_m^{\ r}(\mathfrak{K})$. But

$$m_0J_m = J_k ,$$

where $k = m/m_0$, so that

$$m_0\Delta_m^{\ r}(K) = \sum\nolimits^{\pi(r)} J_k .$$

Hence π^r, the rth Betti number of $\mathfrak{K}$, may be defined as the number of cyclic groups of order $k = m/m_0$ which appear in the direct sum decomposition of the group $m_0\Delta_m^{\ r}(\mathfrak{K})$; hence π^r is known. We must now determine the groups $\Theta^r(\mathfrak{K})$.

$\Theta^0(\mathfrak{K})$ is the null group. Suppose that the group $\Theta^{r-1}(\mathfrak{K})$ has been determined up to an isomorphism. Then, since π^r is known, the group $\Theta^{r-1}(\mathfrak{K}) + \sum^{\pi(r)} J_m$ is also known up to an isomorphism. But then, by (5.21), the group

$$\Theta^r(\mathfrak{K}) = \Delta_m^{\ r}(\mathfrak{K})/(\Theta^{r-1}(\mathfrak{K}) + \sum\nolimits^{\pi(r)} J_m)$$

is also defined up to an isomorphism. Hence, the groups $\Theta^r(\mathfrak{K})$ are determined.

§6. The homomorphism $\bar{S}_\beta^{\ \alpha}$ of $L^r(K_\alpha, \mathfrak{A})$ into $L^r(K_\beta, \mathfrak{A})$ induced by a simplicial mapping $S_\alpha^{\ \beta}$ of a complex K_β into a complex K_α

§6.1. Definition of the homomorphism $\bar{S}_\beta^{\ \alpha}$. Let K_α and K_β be unrestricted simplicial complexes and suppose that $S_\alpha^{\ \beta}$ is a simplicial mapping of K_β into K_α.

We define the homomorphism $\bar{S}_\beta^{\ \alpha}$ of $L^r(K_\alpha, \mathfrak{A})$ into $L^r(K_\beta, \mathfrak{A})$ as follows: for an arbitrary chain $x_\alpha^{\ r} \in L^r(K_\alpha, \mathfrak{A})$ and an arbitrary oriented simplex $t_{\beta j}^{\ r}$ of K_β we set

$$(\bar{S}_\beta^{\ \alpha}x_\alpha^{\ r} \cdot t_{\beta j}^{\ r}) = (x_\alpha^{\ r} \cdot S_\alpha^{\ \beta}t_{\beta j}^{\ r}), \tag{6.1}$$

that is, the value of the chain $\bar{S}_\beta^{\ \alpha}x_\alpha^{\ r}$ on $t_{\beta j}^{\ r}$ is, by definition, the value of the chain $x_\alpha^{\ r}$ on the simplex $S_\alpha^{\ \beta}t_{\beta j}^{\ r}$, if $S_\alpha^{\ \beta}t_{\beta j}^{\ r} \neq 0$, and is 0, if $S_\alpha^{\ \beta}t_{\beta j}^{\ r} = 0$.

If $\mathfrak{A} = J$ or $\mathfrak{A} = J_m$ and $x_\beta^{\ r} = \sum a_jt_{\beta j}^{\ r} \in L^r(K_\beta, \mathfrak{A})$, then by (6.1)

$$\begin{aligned}(\bar{S}_\beta^{\ \alpha}x_\alpha^{\ r} \cdot x_\beta^{\ r}) &= (\bar{S}_\beta^{\ \alpha}x_\alpha^{\ r} \cdot \sum a_jt_{\beta j}^{\ r}) = \sum a_j(\bar{S}_\beta^{\ \alpha}x_\alpha^{\ r} \cdot t_{\beta j}^{\ r}) \\ &= \sum a_j(x_\alpha^{\ r} \cdot S_\alpha^{\ \beta}t_{\beta j}) = \sum (x_\alpha^{\ r} \cdot S_\alpha^{\ \beta}a_jt_{\beta j}^{\ r}) \\ &= (x_\alpha^{\ r} \cdot \sum a_jS_\alpha^{\ \beta}t_{\beta j}^{\ r}) = (x_\alpha^{\ r} \cdot S_\alpha^{\ \beta}x_\beta^{\ r}).\end{aligned}$$

Hence

6.11. *The homomorphisms* $\bar{S}_\beta^{\ \alpha}$ *and* $S_\alpha^{\ \beta}$ *are dual.*

§6.2. Commutativity of the operators ∇ and $\bar{S}_\beta^{\ \alpha}$. We shall prove the identity

$$\nabla \bar{S}_\beta^{\ \alpha} x_\alpha^{\ r} = \bar{S}_\beta^{\ \alpha} \nabla x_\alpha^{\ r} \tag{6.2}$$

for any chain $x_\alpha^{\ r} \in L^r(K_\alpha, \mathfrak{A})$. This identity is analogous to VII, (8.3).

In virtue of (6.1) and the fact that $(\nabla x^r \cdot y^{r+1}) = (x^r \cdot \Delta y^{r+1})$ [(1.310)], we have

$$(\bar{S}_\beta^{\ \alpha} \nabla x_\alpha^{\ r} \cdot t_{\beta j}^{\ r+1}) = (\nabla x_\alpha^{\ r} \cdot S_\alpha^{\ \beta} t_{\beta j}^{\ r+1}) = (x_\alpha^{\ r} \cdot \Delta S_\alpha^{\ \beta} t_{\beta j}^{\ r+1}),$$

$$(\nabla \bar{S}_\beta^{\ \alpha} x_\alpha^{\ r} \cdot t_{\beta j}^{\ r+1}) = (\bar{S}_\beta^{\ \alpha} x_\alpha^{\ r} \cdot \Delta t_{\beta j}^{\ r+1}) = (x_\alpha^{\ r} \cdot S_\alpha^{\ \beta} \Delta t_{\beta j}^{\ r+1}).$$

Since $\Delta S_\alpha^{\ \beta} t_{\beta j}^{\ r+1} = S_\alpha^{\ \beta} \Delta t_{\beta j}^{\ r+1}$, the assertion is proved.

The reader will find examples of homomorphisms in X, 3.2, which may be read at this time.

§6.3. THEOREM 6.3. *Let K_β and K_α be finite unrestricted simplicial complexes and let $S_\alpha^{\ \beta}$, $S'^{\ \beta}_\alpha$ be two simplicial mappings of K_β into K_α. If $S_\alpha^{\ \beta}$ and $S'^{\ \beta}_\alpha$ induce the same homomorphism of the group $\Delta_m^{\ r}(K_\beta)$ into $\Delta_m^{\ r}(K_\alpha)$ for arbitrary dimension r and arbitrary m = 0, 2, 3, $\cdots$, then the homomorphisms $\bar{S}_\beta^{\ \alpha}$ and $\bar{S}'^{\ \alpha}_\beta$ of $\nabla^r(K_\alpha, \mathfrak{A})$ into $\nabla^r(K_\beta, \mathfrak{A})$ also coincide for arbitrary coefficient domain $\mathfrak{A}$ and arbitrary dimension r.*

Proof. Let

$$W_\beta^{\ r} = \{x_{\beta i}^{\ r}, y_{\beta i}^{\ r}, z_{\beta i}^{\ r}, u_{\beta i}^{\ r}, v_{\beta i}^{\ r}\}, \qquad W_\alpha^{\ r} = \{x_{\alpha h}^{\ r}, y_{\alpha h}^{\ r}, z_{\alpha h}^{\ r}, u_{\alpha h}^{\ r}, v_{\alpha h}^{\ r}\}$$

be a canonical system of bases and

$$\overline{W}_\beta^{\ r} = \{\bar{u}_{\beta i}^{\ r}, \bar{v}_{\beta i}^{\ r}, \bar{z}_{\beta i}^{\ r}, \bar{x}_{\beta i}^{\ r}, \bar{y}_{\beta i}^{\ r}\}, \qquad \overline{W}_\alpha^{\ r} = \{\bar{u}_{\alpha h}^{\ r}, \bar{v}_{\alpha h}^{\ r}, \bar{z}_{\alpha h}^{\ r}, \bar{x}_{\alpha h}^{\ r}, \bar{y}_{\alpha h}^{\ r}\}$$

a dual system for K_β and K_α.

Let us consider the matrices of the homomorphisms $S_\alpha^{\ \beta}$ and $S'^{\ \beta}_\alpha$ of $L_0^{\ r}(K_\beta)$ into $L_0^{\ r}(K_\alpha)$ relative to the bases $W_\beta^{\ r}$ and $W_\alpha^{\ r}$; every row of a matrix consists of the coefficients of the linear combination of the elements of $W_\alpha^{\ r}$ into which $S_\alpha^{\ \beta}$ maps the element of $W_\beta^{\ r}$ heading the given row.

Since $S_\alpha^{\ \beta}$ maps cycles into cycles and bounding cycles into bounding cycles, $S_\alpha^{\ \beta} z_{\beta i}^{\ r}$ is a linear combination of the elements $z_{\alpha h}^{\ r}$, $u_{\alpha h}^{\ r}$, $v_{\alpha h}^{\ r}$; $S_\alpha^{\ \beta} u_{\beta i}^{\ r}$ is a linear combination of the elements $u_{\alpha h}^{\ r}$, $v_{\alpha h}^{\ r}$; and $S_\alpha^{\ \beta} v_{\beta i}^{\ r}$ is a linear combination of $\theta_{\alpha h}^{\ r} u_{\alpha h}^{\ r}$, $v_{\alpha h}^{\ r}$. The same is true, of course, for $S'^{\ \beta}_\alpha$.

Hence the matrix of $S_\alpha^{\ \beta}$ has the form shown in Table (6.3) (only the elements in which we are interested at the moment are written out explicitly). The matrix of $S'^{\ \beta}_\alpha$ has the same structure; we shall not exhibit it, but merely note that its elements will be denoted by the same letters as the corresponding elements of the matrix of $S_\alpha^{\ \beta}$, marked, however, with primes.

(6.3)

	$x_{\alpha h}^r$	$y_{\alpha h}^r$	$z_{\alpha h}^r$	$u_{\alpha h}^r$	$v_{\alpha h}^r$
$x_{\beta i}^r$	a_{ih}		d_{ih}	e_{ih}	
$y_{\beta i}^r$				f_{ih}	
$z_{\beta i}^r$	0	0	c_{ih}	p_{ih}	
$u_{\beta i}^r$	0	0	0	b_{ih}	
$v_{\beta i}^r$	0	0	0	$\theta_{\alpha h}^r g_{ih}$	

Since the homomorphisms S_α^β and S'^β_α induce the same homomorphism of $\Delta_0^r(K_\beta)$ into $\Delta_0^r(K_\alpha)$,

$$S_\alpha^\beta z_{\beta i}^r \sim S'^\beta_\alpha z_{\beta i}^r \quad \text{in} \quad K_\alpha$$

for arbitrary $i = 1, 2, \cdots, \pi_\beta^r$; that is,

$$S_\alpha^\beta z_{\beta i}^r - S'^\beta_\alpha z_{\beta i}^r$$

is a linear combination solely of the cycles $\theta_{\alpha h}^r u_{\alpha h}^r$ and $v_{\alpha h}^r$. Hence

$$c_{ih} = c_{ih}' \tag{6.31}$$

for arbitrary $i = 1, 2, \cdots, \pi_\beta^r, h = 1, 2, \cdots, \pi_\alpha^r$.

Furthermore,

$$\Delta x_{\beta i}^r = \theta_{\beta i}^{r-1} u_{\beta i}^{r-1},$$

so that $x_{\beta i}^r$ is a cycle (mod $m_i = \theta_{\beta i}^{r-1}$). Since S_α^β and S'^β_α induce identical homomorphisms of $\Delta_m^r(K_\beta)$ into $\Delta_m^r(K_\alpha)$ for arbitrary m and in particular for $m = m_i$, if the operator $\mathfrak{B}_m$ has the same meaning as in Def. 4.51, we get

$$\mathfrak{B}_{m_i}(S_\alpha^\beta x_{\beta i}^r - S'^\beta_\alpha x_{\beta i}^r) \in H_\Delta^r(K_\alpha, J_{m_i}).$$

Therefore, by the last two equations of (4.13′),

$$\mathfrak{B}_{m_i} a_{ih} - \mathfrak{B}_{m_i} a_{ih}' = 0,$$

$$\mathfrak{B}_{m_i} d_{ih} - \mathfrak{B}_{m_i} d_{ih}' = 0,$$

that is,

$$a_{ih} - a_{ih}' \equiv 0 \ (\text{mod } \theta_{\beta i}^{r-1}), \tag{6.32}$$

$$d_{ih} - d_{ih}' \equiv 0 \ (\text{mod } \theta_{\beta i}^{r-1}); \tag{6.33}$$

or, for suitable integers μ_{hi}, ν_{ih}:

$$a_{ih} - a_{ih}' = \mu_{ih}\theta_{\beta i}^{r-1}, \qquad i = 1, \cdots, \tau_\beta^{r-1}, h = 1, \cdots, \tau_\alpha^{r-1};$$

$$d_{ih} - d_{ih}' = \nu_{ih}\theta_{\beta i}^{r-1}, \qquad i = 1, \cdots, \tau_\beta^{r-1}, h = 1, \cdots, \pi_\alpha^r.$$

Let us now consider the homomorphism $\bar{S}_\beta{}^\alpha$ relative to the bases $\overline{W}_\alpha{}^r$, $\overline{W}_\beta{}^r$. Since $\bar{S}_\beta{}^\alpha$ and $S_\alpha{}^\beta$ are dual, applying Theorem 5.4 of Appendix 2 we obtain

$$\bar{S}_\beta{}^\alpha \bar{u}_{\alpha h}{}^r = \sum_i a_{ih} \bar{u}_{\beta i}{}^r + \text{a linear combination of the } \bar{v}_{\beta i}{}^r,$$

$$\bar{S}_\beta{}^\alpha \bar{z}_{\alpha h}{}^r = \sum_i d_{ih} \bar{u}_{\beta i}{}^r + \sum_i c_{ih} \bar{z}_{\beta i}{}^r + \text{a linear combination of the } \bar{v}_{\beta i}{}^r,$$

and analogously

$$\bar{S}'_\beta{}^\alpha \bar{u}_{\alpha h}{}^r = \sum_i a_{ih}' \bar{u}_{\beta i}{}^r + \text{a linear combination of the } \bar{v}_{\beta i}{}^r,$$

$$\bar{S}'_\beta{}^\alpha \bar{z}_{\alpha h}{}^r = \sum_i d_{ih}' \bar{u}_{\beta i}{}^r + \sum_i c_{ih}' \bar{z}_{\beta i}{}^r + \text{a linear combination of the } \bar{v}_{\beta i}{}^r.$$

[It is suggested that the matrix (6.3) be rewritten with the left-hand column $x_{\beta i}{}^r$, $y_{\beta i}{}^r$, $z_{\beta i}{}^r$, $u_{\beta i}{}^r$, $v_{\beta i}{}^r$ replaced by $\bar{u}_{\beta i}{}^r$, $\bar{v}_{\beta i}{}^r$, $\bar{z}_{\beta i}{}^r$, $\bar{x}_{\beta i}{}^r$, $\bar{y}_{\beta i}{}^r$, respectively, and the upper row $x_{\alpha h}{}^r$, $y_{\alpha h}{}^r$, $z_{\alpha h}{}^r$, $u_{\alpha h}{}^r$, $v_{\alpha h}{}^r$, by $\bar{u}_{\alpha h}{}^r$, $\bar{v}_{\alpha h}{}^r$, $\bar{z}_{\alpha h}{}^r$, $\bar{x}_{\alpha h}{}^r$, $\bar{y}_{\alpha h}{}^r$, respectively.]

Recalling that $\nabla \bar{x}_{\beta i}{}^{r-1} = \theta_{\beta i}{}^{r-1} \bar{u}_{\beta i}{}^r$ and $\nabla \bar{y}_{\beta i}{}^{r-1} = \bar{v}_{\beta i}{}^r$, we get from (6.31), (6.32), (6.33):

$$\begin{aligned}\bar{S}_\beta{}^\alpha \bar{u}_{\alpha h}{}^r - \bar{S}'_\beta{}^\alpha \bar{u}_{\alpha h}{}^r &= \sum_i \mu_{ih} \theta_{\beta i}{}^{r-1} \bar{u}_{\beta i}{}^r + \text{a linear combination of} \\ \text{the } \bar{v}_{\beta i}{}^r &= \sum_i \mu_{ih} \nabla \bar{x}_{\beta i}{}^{r-1} + \text{a linear combination of the } \nabla \bar{y}_{\beta i}{}^{r-1},\end{aligned} \tag{6.34}$$

$$\begin{aligned}\bar{S}_\beta{}^\alpha \bar{z}_{\alpha h}{}^r - \bar{S}'_\beta{}^\alpha \bar{z}_{\alpha h}{}^r &= \sum_i \nu_{ih} \theta_{\beta i}{}^{r-1} \bar{u}_{\beta i}{}^{r-1} + \text{a linear combination of} \\ \text{the } \bar{v}_{\beta i}{}^r &= \sum_i \nu_{ih} \nabla \bar{x}_{\beta i}{}^{r-1} + \text{a linear combination of the } \nabla \bar{y}_{\beta i}{}^{r-1},\end{aligned} \tag{6.35}$$

or

$$\begin{cases} \bar{S}_\beta{}^\alpha \bar{u}_{\alpha h}{}^r \backsim \bar{S}'_\beta{}^\alpha \bar{u}_{\alpha h}{}^r \quad \text{in} \quad K_\beta \qquad \text{for } h = 1, \cdots, \tau_\alpha{}^{r-1}, \\ \bar{S}_\beta{}^\alpha \bar{z}_{\alpha h}{}^r \backsim \bar{S}'_\beta{}^\alpha \bar{z}_{\alpha h}{}^r \quad \text{in} \quad K_\beta \qquad \text{for } h = 1, \cdots, \pi_\alpha{}^r. \end{cases} \tag{6.36}$$

Since every integral r-cocycle of K_α is cohomologous to a linear combination of cocycles $\bar{u}_{\alpha h}{}^r$, $\bar{z}_{\alpha h}{}^r$ and since $\bar{S}_\beta{}^\alpha$, $\bar{S}'_\beta{}^\alpha$ preserve cohomologies, it follows that

$$\bar{S}_\beta{}^\alpha \bar{z}_\alpha{}^r \backsim \bar{S}'_\beta{}^\alpha \bar{z}_\alpha{}^r \quad \text{in} \quad K_\beta$$

for *arbitrary* $\bar{z}_\alpha{}^r \in Z_\nabla{}^r(K_\beta, J)$. This means that $\bar{S}_\beta{}^\alpha$ and $\bar{S}'_\beta{}^\alpha$ induce identical homomorphisms of $\nabla_0{}^r(K_\alpha)$ into $\nabla_0{}^r(K_\beta)$.

To show that the simplicial mappings $S_\alpha{}^\beta$ and $S'_\alpha{}^\beta$ induce identical homomorphisms of the groups $\nabla^r(K_\alpha, \mathfrak{A})$ into $\nabla^r(K_\beta, \mathfrak{A})$ over an *arbitrary coefficient domain* $\mathfrak{A}$, we note that according to 4.2 every cocycle $\bar{z}_\alpha{}^r \in Z_\nabla{}^r(K_\alpha, \mathfrak{A})$ has a unique representation

$$\bar{z}_\alpha{}^r = \sum \mathfrak{a}_h \bar{x}_{\alpha h}{}^r + \sum \mathfrak{c}_h \bar{z}_{\alpha h}{}^r + \sum \mathfrak{d}_h \bar{u}_{\alpha h}{}^r + \sum \mathfrak{e}_h \bar{v}_{\alpha h}{}^r,$$

where $\mathfrak{a}_h, \mathfrak{c}_h, \mathfrak{d}_h, \mathfrak{e}_h \in \mathfrak{A}$ and $\mathfrak{a}_h$ satisfies the condition

$$\theta_{\alpha h}{}^r \mathfrak{a}_h = 0. \tag{6.37}$$

In virtue of (6.36) we obviously need merely prove that

$$\bar{S}_\beta{}^\alpha \mathfrak{a}_h \bar{x}_{\alpha h}{}^r \backsim \bar{S}'_\beta{}^\alpha \mathfrak{a}_h \bar{x}_{\alpha h}{}^r \qquad (\text{in } K_\beta \text{ over } \mathfrak{A}). \tag{6.38}$$

To this end, we return to Table (6.3) and write

$$\bar{S}_\beta{}^\alpha \bar{x}_{\alpha h}{}^r = \sum_i e_{ih}\bar{u}_{\beta i}{}^r + \sum_i f_{ih}\bar{v}_{\beta i}{}^r$$
$$+ \sum_i p_{ih}\bar{z}_{\beta i}{}^r + \sum_i b_{ih}\bar{x}_{\beta i}{}^r + \sum_i \theta_{\alpha h}{}^r g_{ih}\bar{y}_{\beta i}{}^r.$$

Since

$$S_\alpha{}^\beta z_{\beta i}{}^r \sim S'_\alpha{}^\beta z_{\beta i}{}^r \qquad (\text{in } K_\alpha \text{ over } J),$$

$$p_{ih} \equiv p_{ih}' \pmod{\theta_{\alpha h}{}^r}. \tag{6.39$_p$}$$

Since

$$S_\alpha{}^\beta u_{\beta i}{}^r \sim S'_\alpha{}^\beta u_{\beta i}{}^r \qquad (\text{in } K_\alpha \text{ over } J),$$

$$b_{ih} \equiv b_{ih}' \pmod{\theta_{\alpha h}{}^r}. \tag{6.39$_b$}$$

Setting $e_{ih} - e_{ih}' = A_{ih}$ and keeping (6.37) in mind, we obtain

$$\bar{S}_\beta{}^\alpha \mathfrak{a}_h \bar{x}_{\alpha h}{}^r - \bar{S}'_\beta{}^\alpha \mathfrak{a}_h \bar{x}_{\alpha h}{}^r = \sum_i A_{ih}\mathfrak{a}_h \bar{u}_{\beta i}{}^r + \sum_i (f_{ih} - f_{ih}')\mathfrak{a}_h \bar{v}_{\beta i}{}^r$$

from $(6.39)_p$ and $(6.39)_b$. Since $\bar{v}_{\beta i}{}^r \backsim 0$ (in K_β over J), the proof reduces to showing that

$$\sum_i A_{ih}\mathfrak{a}_h \bar{u}_{\beta i}{}^r \backsim 0 \qquad (\text{in } K_\beta \text{ over } \mathfrak{A}). \tag{6.39}$$

To this end we recall that $x_{\beta i}{}^r$ is a cycle (mod $m_i = \theta_{\beta i}{}^{r-1}$) and that

$$\mathfrak{B}_{m_i}(S_\alpha{}^\beta x_{\beta i}{}^r - S'_\alpha{}^\beta x_{\beta i}{}^r) \in H_\Delta{}^r(K_\alpha, J_{m_i}),$$

$S_\alpha{}^\beta x_{\beta i}{}^r - S'_\alpha{}^\beta x_{\beta i}{}^r = \sum A_{ih} u_{\alpha h}{}^r +$ a linear combination of the chains $x_{\alpha h}{}^r$, $y_{\alpha h}{}^r$, $z_{\alpha h}{}^r$, $v_{\alpha h}{}^r$. According to the first equation of (4.13′),

$$\mathfrak{B}_{m_i} A_{ih} = \theta_{\alpha h}{}^r \mu_{ih}$$

for some $\mu_{ih} \in J_{m_i}$ or

$$A_{ih} \equiv m_{ih}\theta_{\alpha h}{}^r \pmod{m_i}$$

for some integer m_{ih}, or finally

$$A_{ih} = m_{ih}\theta_{\alpha h}{}^r + n_{ih}\theta_{\beta i}{}^{r-1}$$

for suitable integers m_{ih}, n_{ih}. Hence

$$\sum_i A_{ih}\mathfrak{a}_h \bar{u}_{\beta i}{}^r = \sum_i m_{ih}\theta_{\alpha h}{}^r \mathfrak{a}_h \bar{u}_{\beta i}{}^r + \sum_i n_{ih}\theta_{\beta i}{}^{r-1}\mathfrak{a}_h \bar{u}_{\beta i}{}^r.$$

Applying (6.37),

$$\sum_i A_{ih}\mathfrak{a}_h\bar{u}_{\beta i}{}^r = \sum_i n_{ih}\mathfrak{a}_h\theta_{\beta i}{}^{r-1}\bar{u}_{\beta i}{}^r \backsim 0.$$

This completes the proof.

Theorem 6.3 and VII, Theorem 9.4 yield the following result, required in Chapter XIV:

Corollary to Theorem 6.3. Let $S_\alpha{}^\beta$ and $S'{}_\alpha{}^\beta$ be simplicial mappings of K_β into K_α, where both K_β and K_α are finite unrestricted simplicial complexes. Suppose that for every simplex $T_\beta \in K_\beta$ there exists a simplex $T_\alpha \in K_\alpha$ having both $S_\alpha{}^\beta T_\beta$ and $S'{}_\alpha{}^\beta T_\beta$ as faces. Then $S_\alpha{}^\beta$ and $S'{}_\alpha{}^\beta$ induce identical homomorphisms $\bar{S}_\beta{}^\alpha = \bar{S}'{}_\beta{}^\alpha$ of the groups $\nabla^r(K_\alpha, \mathfrak{A})$ into $\nabla^r(K_\beta, \mathfrak{A})$.

Chapter X

INVARIANCE OF THE BETTI GROUPS

In this chapter we shall prove that the Betti groups of all the triangulations of homeomorphic polyhedra are isomorphic.

In §1 we discuss our method of attacking the proof of this theorem, which is finally realized in §6.

§2 contains a proof of the invariance of the Betti groups under barycentric subdivisions of triangulations.

§§3 and 5 are devoted to a discussion of normal and canonical displacements, as well as to ϵ-displacements in compacta and polyhedra, which form the basis of a considerable part of this and the next chapter.

In §4 we treat certain auxiliary concepts, which are related to those developed in §3, but are not required until Chapter XIV; consequently, this section may be read in conjunction with Chapter XIV.

§6 contains a proof of the fundamental invariance theorem based on §§2, 3, and 5.

In §7 we prove the invariance of the pseudomanifolds, that is, we prove that if a triangulation of a polyhedron is a combinatorial pseudomanifold, then every triangulation of the polyhedron (or of any polyhedron homeomorphic to the given polyhedron) has the same property.

§1. Formulation of the invariance theorems

§1.1. Definition of the numbers $b^r(\Phi)$. Let Φ be a compactum and r an integer ≥ 0.

DEFINITION 1.1 We denote by $b^r(\Phi)$ the least integer k with the following property:

For every open covering ω of Φ there exists a closed refinement α of ω with nerve K_α such that $\pi^r(K_\alpha) = k$. [By a closed refinement of ω we mean a closed covering α of Φ with the property that every element of α is contained in some element of ω (see I, Def. 8.11). For the definition of nerve see IV, 2.1.]

If there is no integer k with this property, we set $b^r(\Phi) = \infty$.

If $b^r(\Phi) = \infty$, for every k there exists an open covering ω of Φ such that for every closed refinement α of ω, $\pi^r(K_\alpha) > k$.

The following theorem is a consequence of I, Theorem 8.35:

1.11. *The number $b^r(\Phi)$ is the least integer k satisfying the condition: for every $\epsilon > 0$ there exists a closed ϵ-covering of Φ such that the rth Betti number of its nerve is k.*

REMARK. Let m be a prime. The numbers $b_m{}^r(\Phi)$ are defined in complete analogy with the numbers $b^r(\Phi)$ by considering the Betti numbers (mod m) of the nerves K_α of closed coverings α of Φ.

Clearly, if the compacta Φ and Φ' are homeomorphic, then

$$b^r(\Phi) = b^r(\Phi'), \qquad b_m^r(\Phi) = b_m^r(\Phi'),$$

that is, both sets of numbers are *topological invariants.*

§1.2. Definition of the groups $\mathfrak{G}^r(\Phi)$.

ALGEBRAIC LEMMA. *If A and B are Abelian groups with a finite number of generators and A is isomorphic to a subgroup of B, while B is isomorphic to a subgroup of A, then A and B are isomorphic.*

For a proof see Appendix 2, Theorem 4.41.

We shall now define a set (perhaps empty) $\mathfrak{G}^r(\Phi)$ of Abelian groups for every compactum Φ and integer $r \geq 0$.

The set $\mathfrak{G}^r(\Phi)$ consists, by definition, of all Abelian groups $\mathfrak{B}$ (defined up to an isomorphism) with the following property:

For every open covering ω of Φ there exists a closed refinement α of ω such that $\mathfrak{B}$ is isomorphic to the group $\Delta_0^r(K_\alpha)$, where K_α is the nerve of α.

In consequence of I, 8.35, we may now say that

$\mathfrak{G}^r(\Phi)$ can be defined as the set of all groups $\mathfrak{B}$ with the following property: For every $\epsilon > 0$ there exists a closed ϵ-covering of Φ such that $\Delta_0^r(K_\alpha)$ is isomorphic to $\mathfrak{B}$.

A group $\mathfrak{B}_0 \in \mathfrak{G}^r(\Phi)$ is said to be a *minimal group of the system* $\mathfrak{G}^r(\Phi)$ if every group $\mathfrak{B} \in \mathfrak{G}^r(\Phi)$ contains a subgroup isomorphic to $\mathfrak{B}_0$.

The system $\mathfrak{G}^r(\Phi)$ may contain no minimal group in the above sense; but if $\mathfrak{G}^r(\Phi)$ does contain a minimal group, it contains just one (according to the Algebraic Lemma).

DEFINITION 1.2. *The minimal group (if it exists) of the system $\mathfrak{G}^r(\Phi)$ is denoted by $\mathfrak{B}^r(\Phi)$.*

REMARK. Clearly, if $\mathfrak{B}^r(\Phi)$ exists, so does $\mathfrak{B}^r(\Phi')$ for every Φ' homeomorphic to Φ, and $\mathfrak{B}^r(\Phi)$ is isomorphic to $\mathfrak{B}^r(\Phi')$. Hence the groups $\mathfrak{B}^r(\Phi)$ are topological invariants.

§1.3. Formulation of the invariance theorems. The basic purpose of this chapter is to prove the following theorems (and incidentally to give another proof of the Pflastersatz):

THEOREM 1.31. *The equalities*

$$b^r(\Phi) = \pi^r(K), \qquad b_m^r(\Phi) = \pi_m^r(K)$$

hold for every triangulation K of a polyhedron Φ.

THEOREM 1.32. *If Φ is a polyhedron and K is a triangulation of Φ, the group $\mathfrak{B}^r(\Phi)$ exists and is isomorphic to $\Delta_0^r(K)$.*

In virtue of the topological invariance of $b^r(\Phi)$ and $\mathfrak{B}^r(\Phi)$, these theorems imply

THEOREM 1.33. *If K and K' are, respectively, triangulations of two homeo-*

morphic polyhedra Φ *and* Φ', *the groups* $\Delta_0^r(K)$ *and* $\Delta_0^r(K')$ *are isomorphic and* $\pi^r(K) = \pi^r(K')$.

[The equality of $\pi^r(K)$ and $\pi^r(K')$ is an obvious consequence of the isomorphism of the groups $\Delta_0^r(K)$ and $\Delta_0^r(K')$.]

Theorem 1.33 is called the *invariance theorem for the Betti groups* and was first proved in 1916 by Alexander. The term *invariance* in the name of the theorem is to be taken in the sense that any two triangulations of two homeomorphic polyhedra have isomorphic Betti groups. This invariance may also, of course, be understood in the sense that any two triangulations of the *same* polyhedron have isomorphic Betti groups. Indeed, in accordance with IV, 6.1, Remark 2, we may also formulate Theorems 1.31–1.33 in the following way:

1.34. *The groups* $\Delta_0^r(K)$ *are isomorphic for all topological triangulations of a topological polyhedron* Φ (see IV, 6.1), *since they are all isomorphic to* $\mathfrak{B}^r(\Phi)$, *and the numbers* $\pi^r(K)$ *are all equal to* $b^r(\Phi)$.

Since the groups $\Delta_0^r(K)$ completely determine the groups $\Delta^r(K, \mathfrak{A})$ and $\nabla^r(K, \mathfrak{A})$ over an arbitrary coefficient domain $\mathfrak{A}$ (see IX, 4.1 and 4.2), 1.33 implies the

General Invariance Theorem for the Betti Groups 1.35. *If* K *and* K' *are triangulations of two homeomorphic polyhedra* Φ *and* Φ', *respectively, then the groups* $\Delta^r(K, \mathfrak{A})$ $[\nabla^r(K, \mathfrak{A})]$ *are isomorphic to* $\Delta^r(K', \mathfrak{A})$ $[\nabla^r(K', \mathfrak{A})]$ *for arbitrary* $r \geq 0$ *and arbitrary coefficient domain* $\mathfrak{A}$.

§2. Subdivision of chains. Fundamental systems of subcomplexes and chains. Invariance of the Δ- and ∇-groups under elementary and barycentric subdivisions

§2.1. The isomorphism s_β^α. Throughout this section K_α, K_β will denote triangulations or open subcomplexes of triangulations.

Let K_β be a subdivision of a complex K_α. Choose a definite orientation t_α^r for every element T_α^r of K_α. We shall call $T_\beta^r \in K_\beta$ a *principal element* if T_β^r is contained (set-theoretically) in some $T_\alpha^r \in K_\alpha$ of the same dimension r. Of the two orientations of a principal element $T_\beta^r \in K_\beta$ we denote by t_β^r that which is coherent with the orientation t_α^r of the carrier T_α^r of T_β^r. If T_β^r is not a principal element, t_β^r will denote either one of its orientations. Moreover, for brevity, we shall now write L_α^r, L_β^r, Z_α^r, Z_β^r, Δ_α^r, Δ_β^r, etc. in place of $L^r(K_\alpha, \mathfrak{A})$, $L^r(K_\beta, \mathfrak{A})$, $Z_\Delta^r(K_\alpha, \mathfrak{A})$, $Z_\Delta^r(K_\beta, \mathfrak{A})$, $\Delta^r(K_\alpha, \mathfrak{A})$, $\Delta^r(K_\beta, \mathfrak{A})$, etc. The elements of $L_\alpha^r(L_\beta^r)$ are denoted by $x_\alpha^r(x_\beta^r)$, and additional indices are used whenever necessary.

Definition 2.11. For every chain x_α^r we shall define a chain x_β^r as follows:

a) if T_β^r is a principal element contained in T_α^r, then

$$(x_\beta^r \cdot t_\beta^r) = (x_\alpha^r \cdot t_\alpha^r);$$

b) if T_β^r is not a principal element, then

$$(x_\beta^r \cdot t_\beta^r) = 0.$$

We denote the chain x_β^r by $s_\beta^\alpha x_\alpha^r$ and call it the *subdivision of the chain* x_α^r *in* K_β.

REMARK. If $x_\alpha^r = t_\alpha^r$ is a monomial chain, then obviously

$$s_\beta^\alpha t_\alpha^r = \sum t_{\beta j}^r, \qquad T_{\beta j}^r \subseteq T_\alpha^r. \tag{2.111}$$

For an arbitrary chain $x_\alpha^r = \sum_i a_{\alpha i} t_{\alpha i}^r$,

$$s_\beta^\alpha x_\alpha^r = \sum_i a_{\alpha i} s_\beta^\alpha t_{\alpha i}^r. \tag{2.112}$$

Assigning the chain $s_\beta^\alpha x_\alpha^r$ to every chain $x_\alpha^r \in L_\alpha^r$, we obtain a homomorphism of L_α^r into L_β^r; in fact, it is easy to see that

$$s_\beta^\alpha(x_\alpha^r \pm y_\alpha^r) = s_\beta^\alpha x_\alpha^r \pm s_\beta^\alpha y_\alpha^r.$$

Moreover, if $x_\alpha^r \neq 0$, then $s_\beta^\alpha x_\alpha^r \neq 0$. Hence s_β^α *is an isomorphism of* L_α^r *into* L_β^r.

THEOREM 2.12.

$$\Delta s_\beta^\alpha x_\alpha^r = s_\beta^\alpha \Delta x_\alpha^r. \tag{2.12}$$

Proof. In consequence of (2.112) it is enough to show that

$$\Delta s_\beta^\alpha t_\alpha^r = s_\beta^\alpha \Delta t_\alpha^r.$$

If T_β^{r-1} is arbitrary and is contained in T_α^r, there exist precisely two elements $T_{\beta h}^r$ and $T_{\beta k}^r$ of K_β having T_β^{r-1} as a face. Since the orientations $t_{\beta h}^r$ and $t_{\beta k}^r$ are coherent,

$$(t_{\beta h}^r : t_\beta^{r-1}) = -(t_{\beta k}^r : t_\beta^{r-1})$$

for an arbitrary choice of t_β^{r-1}. Hence by (2.111) and the definition of Δ,

$$\begin{aligned}(\Delta s_\beta^\alpha t_\alpha^r \cdot t_\beta^{r-1}) = (\Delta \textstyle\sum t_{\beta j}^r \cdot t_\beta^{r-1}) &= (\Delta t_{\beta h}^r \cdot t_\beta^{r-1}) + (\Delta t_{\beta k}^r \cdot t_\beta^{r-1}) \\ &= (t_{\beta h}^r : t_\beta^{r-1}) + (t_{\beta k}^r : t_\beta^{r-1}) = 0.\end{aligned}$$

Since $T_\beta^{r-1} \subset T_\alpha^r$ is not a principal element, $s_\beta^\alpha \Delta t_\alpha^r$ is zero on t_β^{r-1}. Therefore, if $T_\beta^{r-1} \subset T_\alpha^r$, both $\Delta s_\beta^\alpha t_\alpha^r$ and $s_\beta^\alpha \Delta t_\alpha^r$ vanish on t_β^{r-1}.

Now suppose that T_β^{r-1} is contained in a face $T_\alpha^{r-1} \in K_\alpha$ of $T_\alpha^r \in K_\alpha$. Then only one simplex $T_\beta^r \subseteq T_\alpha^r$ has T_β^{r-1} as a face. Let us denote by $R^r(R^{r-1})$ the carrying space of the simplex $T_\alpha^r(T_\alpha^{r-1})$, and by C the affine mapping of R^r onto itself which carries the oriented simplex t_α^r into the oriented simplex t_β^r. We shall denote the restriction of C to R^{r-1} by C'. Since T_α^r and $T_\beta^r = CT_\alpha^r$ lie on the same side of the plane $R^{r-1} = CR^{r-1}$, putting $\varepsilon = +1$ in Theorem 1.521 of Appendix 1, we obtain

$$\operatorname{sgn} C \cdot \operatorname{sgn} C' = +1.$$

Since the orientations t_α^r and t_β^r are coherent, sgn $C = +1$. Hence sgn $C' = +1$, so that

$$Ct_\alpha^{r-1} = C't_\alpha^{r-1} = t_\beta^{r-1}$$

and

$$(2.121) \qquad (t_\beta^r : t_\beta^{r-1}) = (Ct_\alpha^r : Ct_\alpha^{r-1}) = (t_\alpha^r : t_\alpha^{r-1}).$$

But $(t_\alpha^r : t_\alpha^{r-1})$ is the value of Δt_α^r on t_α^{r-1} and it is also the value of $s_\beta^\alpha \Delta t_\alpha^r$ on t_β^{r-1}. On the other hand, $\Delta s_\beta^\alpha t_\alpha^r$ and Δt_β^r have the same value $(t_\beta^r : t_\beta^{r-1})$ on t_{β}^{r-1}. In view of (2.121), this proves Theorem 2.12.

§2.2. Fundamental systems of subcomplexes of a complex K. As we indicated at the beginning of 2.1, K, K_α, K_β will denote triangulations or open subcomplexes of triangulations in this section.

DEFINITION 2.21. A system of subcomplexes $U_i^r \subset K$ of various dimensions (from zero to the dimension n of K) satisfying the following conditions will be called a *fundamental system of subcomplexes of* K:

1°. All the U_i^r are disjoint simple pseudomanifolds and their union is K. (An r-dimensional pseudomanifold U^r is said to be *simple* if it is orientable and if all its Δ^p-groups are null groups for $0 < p < r$.)

2°. For every U_i^r the complex B_i^{r-1} consisting of all the simplexes $T \in K$ which do not belong to U_i^r and which are faces of at least one simplex of U_i^r is an $(r-1)$-complex and is the union of pseudomanifolds U_j^s, $0 \leq s \leq r - 1$.

REMARK. Setting $U_j^s < U_i^r$ if $U_i^s \subseteq B_i^{r-1}$, we see that the set of all U_i^r is partially ordered. We shall denote it by K. If we assign to every $U_i^r \in \mathsf{K}$ a dimension number which is the same as its dimension as a subcomplex of K, we can then look upon K as an abstract complex in the sense of IV, 1.7. We shall make use of this remark in the sequel.

We shall designate the set of all $U_j^s < U_i^r$ by B_i^{r-1}. It is clear that B_i^{r-1} is a closed subcomplex of K; the union of all the complexes U_j^s which are elements of B_i^{r-1} is B_i^{r-1}.

From 2° it follows easily that:

2.22. The union of the subcomplexes U_i^r which are elements of a subcomplex K_0 of K is a closed subcomplex K_0 of K if, and only if, K_0 is a closed subcomplex of K.

From 2.22 we obtain an analogous proposition for open subcomplexes.

2.23. The set of all U_i^s, $s \leq r$, is a closed subcomplex of K denoted by K^r. Hence $\mathsf{K} = \mathsf{K}^n$. The union of all the elements of the complex K^r (that is, the union of all the complexes U_i^s which are elements of K^r) is a closed subcomplex of K denoted by K^r.

We shall require the following easily proved theorem in the sequel:

2.24. If $U_k^{r-1} < U_i^r$, U_k^{r-1} is an open subcomplex of B_i^{r-1}.

For, since U_k^{r-1}, because of its dimension, is not less than any element of $\mathbf{B}_i^{r-1}$, the set of all elements of $\mathbf{B}_i^{r-1}$ different from U_k^{r-1} is a closed subcomplex $\mathbf{K}_0$ of $\mathbf{B}_i^{r-1}$ and the union of all the elements of $\mathbf{K}_0$ is a closed subcomplex K_0 of B_i^{r-1}. Therefore, $U_k^{r-1} = B_i^{r-1} \setminus K_0$ is an open subcomplex of B_i^{r-1}. This completes the proof.

Similarly,

2.25. Every U_i^r is an open subcomplex of K^r.

EXAMPLES OF FUNDAMENTAL SYSTEMS OF SUBCOMPLEXES.

1. Let K be the triangulation of the torus discussed in VIII, 1.4, Example 5 (Fig. 116). Denote by U_1^2 the subcomplex of K consisting of the elements of K inside the square $ABCD$ (Fig. 116), and let $U_1^1(U_2^1)$ be the subcomplex of K made up of the elements of K on the open segment $AB = DC$ (on $AD = BC$). Finally, let U_1^0 consist of the vertex $A = B = C = D$. The complexes U_1^2, U_1^1, U_2^1, U_1^0 form a fundamental system of subcomplexes of K.

EXERCISE. Prove the last assertion of Example 1 by using arguments analogous to those of VIII, 1.4, Example 5.

2. Let K be the triangulation of the projective plane of VIII, 1.4, Example 7 (Fig. 116; the second diagonal of the square is not shown in this figure). Let U_1^2 be the subcomplex of K consisting of all the elements of K inside the square $ABCD$; let U_1^1 be the subcomplex made up of the segments $| t_1^1 | = | t_7^1 |$, $| t_2^1 | = | t_8^1 |$, $| t_3^1 | = | t_9^1 |$, $| t_4^1 | = | t_{10}^1 |$, $| t_5^1 | = | t_{11}^1 |$, $| t_6^1 | = | t_{12}^1 |$ and of the vertices of these segments except for $A = C$; and let U_1^0 consist of the vertex $A = C$. The complexes U_1^2, U_1^1, U_1^0 form a fundamental system of subcomplexes of K.

3. We state the following general theorem [we shall need only the special case 2.26 (see below)]:

2.261. *Suppose that K_1 is a subdivision of a complex K. If T_i^r, $r = 1, 2, \cdots, m$; $i = 1, 2, \cdots, \rho^r$, are the elements of K and U_i^r is the subcomplex of K_1 consisting of the elements whose carrier is T_i^r, then the complexes U_i^r, $r = 1, 2, \cdots, m$; $i = 1, 2, \cdots, \rho^r$, form a fundamental system of subcomplexes of K_1.*

Theorem 2.261 may be proved most simply by the methods of 6.2, once the invariance of the Betti groups has been established (which depends only on Theorem 2.26). The proof of Theorem 2.261 for the case when K_1 is a *barycentric* subdivision of K is left to the reader as an exercise. A direct proof of 2.261 in its general form may be found in Alexandroff-Hopf [A–H; Chapter VI, §3] (see Bibliography at the end of Vol. 1).

Theorem 2.261 makes the value of a fundamental system of subcomplexes quite clear. As we pointed out above, we shall not use this theorem in all its generality; but a special case of the theorem, dealing with elementary subdivisions, is essential for the proof of the invariance of the Betti groups. We shall now state and prove this special case.

2.26. *Let* K_1 *be an elementary subdivision* (see IV, 4.3) *of a complex* K *relative to a simplex* $T \in K$. *Denote by* K *the set of all subcomplexes* U_i^r *of* K_1 *which are subdivisions of simplexes* $T_i^r \in K$. (Each of the complexes U_i^r consists either of just one nonsubdivided simplex of K, which then appears in K_1 in unchanged form, or of an elementary subdivision of a simplex of K.) *The set* K *is a fundamental system of subcomplexes of* K.

Proof. In consequence of VII, Theorem 9.22 it is enough to show that an elementary subdivision K_β of a simplex T_α^n is an orientable pseudomanifold. It is not hard to show by induction over the dimension number of T_α^n that K_β is a strongly connected complex, each of whose $(n - 1)$-elements is a face of precisely two n-simplexes, that is, that K_β is a pseudomanifold. An orientation of K_β is obtained by taking an orientation t_α^n of T_α^n and constructing the chain $s_\beta^\alpha t_\alpha^n$, as in 2.1. The chain $s_\beta^\alpha t_\alpha^n$ is then obviously an n-cycle of K_β.

§2.3. Fundamental systems of chains.

Definition 2.31. A system of integral chains $u_{\beta i}^r$ (of various dimensions) of a complex K_β is said to be a *fundamental system of chains of* K_β if the following conditions are satisfied:

1°. $\Delta_\beta u_{\beta i}^r$ *is a linear combination (with integral coefficients) of the chains* $u_{\beta j}^{r-1}$ (Δ_β is the boundary operator in K_β).

2°. *The chains* $u_{\beta i}^r$ *are linearly independent.*

3°. *Each homology class* $\mathfrak{z}_\beta^r \in \Delta_0^r(K_\beta)$ *contains cycles which are linear combinations of the chains* $u_{\beta i}^r$.

4°. *If* $x_\beta^r \in H_\Delta^r(K_\beta, J)$ *and* x_β^r *is a linear combination of the chains* $u_{\beta i}^r$:

$$x_\beta^r = \textstyle\sum_i b_i u_{\beta i}^r,$$

there exists a linear combination

$$x_\beta^{r+1} = \textstyle\sum_h a_h u_{\beta h}^{r+1}$$

such that

$$\Delta_\beta x_\beta^{r+1} = x_\beta^r.$$

The importance of the concept of a fundamental system of chains is based on Theorems 2.32 and 2.35, which follow.

Theorem 2.32. *Let* $U_{\beta i}^r$ *be a fundamental system of subcomplexes of an* n*-complex* K_β. *If* $u_{\beta i}^r$ *is a definite orientation of the pseudomanifold* $U_{\beta i}^r$, *the system of all chains* $u_{\beta i}^r$ *is a fundamental system of chains of* K_β.

Proof. We shall retain the notation of 2.2, at the same time inserting the index β to obtain $U_{\beta i}^r$, $B_{\beta i}^r$, K_β^r, etc.

We shall show that the chains $u_{\beta i}^r$ satisfy all the Conditions 1°–4° of Def. 2.31.

Condition 1°. Since $u_{\beta i}^r$ is a cycle on the orientable pseudomanifold $U_{\beta i}^r$, $\Delta_\beta u_{\beta i}^r$ can be different from zero only on $(r - 1)$-simplexes of the complex $B_{\beta i}^{r-1} = | U_{\beta i}^r | \setminus U_{\beta i}^r$. But $B_{\beta i}^{r-1}$ is a union of terms $U_{\beta j}^s$ of dimensions $s \leq r - 1$. Consequently, if $\Delta_\beta u_{\beta i}^r$ does not vanish on a simplex $t_{\beta k}^{r-1}$, then

$$| t_{\beta k}^{r-1} | \in U_{\beta j}^{r-1} \subset B_{\beta i}^{r-1}.$$

Since $U_{\beta j}^{r-1}$ (by 2.24) is an open subcomplex of $B_{\beta i}^{r-1}$ (which contains the cycle $\Delta_\beta u_{\beta i}^r$), $U_{\beta j}^{r-1} \Delta_\beta u_{\beta i}^r$ is a cycle of the pseudomanifold $U_{\beta j}^{r-1}$, that is, it is of the form $c_{ij} u_{\beta j}^{r-1}$, c_{ij} an integer.

Since $\Delta_\beta u_{\beta i}^r$ is the sum of its restrictions to distinct subcomplexes $U_{\beta j}^{r-1} \subset B_{\beta i}^{r-1}$, $\Delta_\beta u_{\beta i}^r = \sum c_{ij} u_{\beta j}^{r-1}$. This proves that Condition 1° is satisfied.

Condition 2° is an immediate consequence of the fact that the chain $u_{\beta i}^r$ is contained in $U_{\beta i}^r$ and that two distinct pseudomanifolds $U_{\beta i}^r$ are disjoint (as complexes).

To show that 3° and 4° are also satisfied, we shall prove two lemmas.

LEMMA 2.33. *Let K_β^r be the union of all the $U_{\beta j}^s$, $s \leq r$. If a chain $x_\beta^r \in L_0^r(K_\beta)$ is on K_β^r and its boundary is on K_β^{r-1}, then x_β^r is a linear combination of the chains $u_{\beta i}^r$.*

Indeed, by 2.25, $U_{\beta i}^r$ is an open subcomplex of K_β^r; therefore, in view of the fact that $\Delta_\beta x_\beta^r$ is on K_β^{r-1}, we may write

$$\Delta_{ir} U_{\beta i}^r x_\beta^r = U_{\beta i}^r \Delta_\beta x_\beta^r = 0,$$

where Δ_{ir} is the boundary in $U_{\beta i}^r$ and Δ_β is the boundary in K_β. Hence $U_{\beta i}^r x_\beta^r$ is a cycle on the oriented pseudomanifold $U_{\beta i}^r$, so that $U_{\beta i}^r x_\beta^r = c_i u_{\beta i}^r$, and

$$x_\beta^r = \textstyle\sum_i U_{\beta i}^r x_\beta^r = \sum_i c_i u_{\beta i}^r.$$

COROLLARY. Every r-cycle on K_β^r is of the form $\sum_i c_i u_{\beta i}^r$.
In particular, every n-cycle of K_β is of the form $\sum_i c_i u_{\beta i}^n$.

LEMMA 2.34. *Let*

$$x_\beta^r \in L_0^r(K_\beta), \qquad \Delta_\beta x_\beta^r \in Z_0^{r-1}(K_\beta^{r-1}).$$

Denote by $q = q(x_\beta^r)$ the maximum integer such that $U_{\beta i}^q x_\beta^r \neq 0$ for some $U_{\beta i}^q \in \mathsf{K}$. Then, if $q > r$, there exists a chain $y_\beta^r \in L_0^r(K_\beta)$ satisfying the following conditions:

a) $$x_\beta^r - y_\beta^r \in H_0^r(K_\beta), \qquad \Delta_\beta x_\beta^r = \Delta_\beta y_\beta^r;$$

b) $$q(y_\beta^r) < q(x_\beta^r).$$

Proof. According to the definition of the number $q > r$, the chain x_β^r is on K_β^q but not on K_β^{q-1}. The pseudomanifolds $U_{\beta i}^q$ are open subcomplexes of K_β^q, so that

$$\Delta_{iq} U_{\beta i}^q x_\beta^r = U_{\beta i}^q \Delta_\beta x_\beta^r = 0.$$

Hence $U_{\beta i}^q x_\beta^r$ is a cycle on $U_{\beta i}^q$. Since $r < q$ and $U_{\beta i}^q$ is a simple pseudomanifold, $U_{\beta i}^q x_\beta^r \sim 0$ on $U_{\beta i}^q$.

Suppose that $x_{\beta i}^{r+1} \in L^{r+1}(U_{\beta i}^q)$ satisfies the condition

$$\Delta_{iq} x_{\beta i}^{r+1} = U_{\beta i}^q x_\beta^r.$$

Then

$$\Delta_\beta x_{\beta i}^{r+1} = U_{\beta i}^q x_\beta^r \quad \text{on} \quad U_{\beta i}^q,$$
$$\Delta_\beta x_{\beta i}^{r+1} = 0 \quad \text{on} \quad K_\beta \setminus (U_{\beta i}^q \cup B_{\beta i}^{q-1}).$$

Setting

$$\begin{cases} y_{\beta i}^r = 0 \quad \text{on} \quad U_{\beta i}^q \quad \text{and on} \quad K_\beta \setminus (U_{\beta i}^q \cup B_{\beta i}^{q-1}), \\ y_{\beta i}^r = \Delta_\beta x_{\beta i}^{r+1} \quad \text{on} \quad B_{\beta i}^{q-1}, \end{cases}$$

we have

$$\Delta_\beta x_{\beta i}^{r+1} = U_{\beta i}^q x_\beta^r + y_{\beta i}^r$$

and

$$x_\beta^r - K_\beta^{q-1} x_\beta^r = \textstyle\sum_i U_{\beta i}^q x_\beta^r = \Delta_\beta \sum_i x_{\beta i}^{r+1} - \sum_i y_{\beta i}^r,$$

or

(2.341) $$x_\beta^r - (K_\beta^{q-1} x_\beta^r - \textstyle\sum_i y_{\beta i}^r) = \Delta_\beta \sum_i x_{\beta i}^{r+1}.$$

Since $\sum_i y_{\beta i}^r$ is on K_β^{q-1}, the chain

$$K_\beta^{q-1} x_\beta^r - \textstyle\sum_i y_{\beta i}^r = y_\beta^r$$

is on K_β^{q-1}. Hence by (2.341)

$$x_\beta^r - y_\beta^r = \Delta_\beta \textstyle\sum x_{\beta i}^{r+1} \in H_0^r(K_\beta).$$

Since $x_\beta^r - y_\beta^r \in H_0^r(K_\beta) \subset Z_0^r(K_\beta)$,

$$\Delta_\beta(x_\beta^r - y_\beta^r) = \Delta_\beta x_\beta^r - \Delta_\beta y_\beta^r = 0,$$

that is,

(2.342) $$\Delta_\beta x_\beta^r = \Delta_\beta y_\beta^r.$$

This completes the proof of Lemma 2.34.

Proof of Condition 3°. Let $z_\beta^r \in Z_0^r(K_\beta)$. Applying Lemma 2.34 ν times, $\nu = q(x_\beta^r) - r$, with $x_\beta^r = z_\beta^r$, $\Delta_\beta x_\beta^r = 0$, we obtain a cycle $z_{\beta\nu}^r$ on K_β^r

which is homologous to the cycle z_β^r on K_β. By Lemma 2.33, $z_{\beta\nu}^r$ is of the form $\sum c_i u_{\beta i}^r$, which was to be proved.

Proof of Condition 4°. Let

$$z_\beta^r = \sum c_i u_{\beta i}^r,$$

$$z_\beta^r = \Delta_\beta x_\beta^{r+1},$$

$$x_\beta^{r+1} \in L_0^{r+1}(K_\beta),$$

and suppose that

$$q(x_\beta^{r+1}) > r + 1.$$

The boundary of x_β^{r+1} is on K_β^r. Hence application of Lemma 2.34 to x_β^{r+1} yields a chain y_β^{r+1} satisfying the conditions

$$q(y_\beta^{r+1}) < q(x_\beta^{r+1}),$$

$$\Delta_\beta y_\beta^{r+1} = \Delta_\beta x_\beta^{r+1} = z_\beta^r.$$

Applying Lemma 2.34 as many times as necessary we finally obtain a chain v_β^{r+1} such that

$$q(v_\beta^{r+1}) = r + 1,$$

$$\Delta v_\beta^{r+1} = z_\beta^r.$$

Hence, according to Lemma 2.33, the chain v_β^{r+1} is a linear combination of the chains $u_{\beta i}^{r+1}$. This is what we wished to prove.

From 2.26 and 2.32 we obtain

2.350. *If K_β is an elementary subdivision of a complex K_α, the chains $s_\beta^\alpha t_{\alpha i}^r$, where the $t_{\alpha i}^r$ are the oriented simplexes of K_α, form a fundamental system of chains of K_β.*

We shall say (as in III, 2.4) that a subdivision K_β of a complex K_α is a *regular subdivision* of K_α if it is the result of a finite number of successive elementary subdivisions of K_α.

2.35. (This theorem is true for an arbitrary subdivision K_β of K_α and follows from Theorem 2.261.) *If K_β is a regular subdivision of K_α and $t_{\alpha i}^r$ are the oriented simplexes of K_α, then the chains $s_\beta^\alpha t_{\alpha i}^r$ form a fundamental system of chains of K_β.*

The assertion is valid, in particular, if K_β is a barycentric subdivision of K_α.

Proof. We shall say that a regular subdivision K_β of K_α has rank p if K_β can be obtained from K_α by means of p elementary subdivisions. We shall suppose that Theorem 2.35 is true for all regular subdivisions of rank $\leq p - 1$ and show that it holds for a regular subdivision $K_\beta = K_{\alpha p}$ of rank p. Let

$$K_\alpha = K_{\alpha 0}, K_{\alpha 1}, \cdots, K_{\alpha p}$$

be a sequence of complexes each of which, beginning with the second, is an elementary subdivision of its predecessor.

It is required to prove that the chains $s_{\alpha p}{}^{\alpha}t_{\alpha i}{}^{r}$ satisfy Conditions 1°–4° of Def. 2.31. This is obvious for 1° and 2°. Passing to Condition 3°, let $\mathfrak{z}_{\alpha p}{}^{r} \in \Delta_0{}^{r}(K_{\alpha p})$. It is required to find in $\mathfrak{z}_{\alpha p}{}^{r}$ a cycle of the form $\sum_i a_i s_{\alpha p}{}^{\alpha} t_{\alpha i}{}^{r}$. Since $K_{\alpha p}$ is an elementary subdivision of $K_{\alpha(p-1)}$, $\mathfrak{z}_{\alpha p}{}^{r}$ contains a cycle of the form

$$z_{\alpha p}{}^{r} = \sum_i c_i s_{\alpha p}{}^{\alpha(p-1)} t_{\alpha(p-1)i}{}^{r}.$$

Since $K_{\alpha(p-1)}$ is a subdivision of rank $p - 1$ of K_{α}, the cycle

$$z_{\alpha(p-1)}{}^{r} = \sum_i c_i t_{\alpha(p-1)i}{}^{r}$$

is homologous to a cycle of the form $\sum_i a_i s_{\alpha(p-1)}{}^{\alpha} t_{\alpha i}{}^{r}$ on $K_{\alpha(p-1)}$. Hence

$$z_{\alpha p}{}^{r} = s_{\alpha p}{}^{\alpha(p-1)} z_{\alpha(p-1)}{}^{r} \sim \sum_i a_i s_{\alpha p}{}^{\alpha} t_{\alpha i}{}^{r} \quad \text{on} \quad K_{\alpha p}$$

and consequently the cycle $\sum_i a_i s_{\alpha p}{}^{\alpha} t_{\alpha i}{}^{r}$ is contained in $\mathfrak{z}_{\alpha p}{}^{r}$.

To prove Condition 4° let $z_{\alpha p}{}^{r} \in H_0{}^{r}(K_{\alpha p})$ be of the form

$$z_{\alpha p}{}^{r} = \sum_i a_i s_{\alpha p}{}^{\alpha} t_{\alpha i}{}^{r}.$$

It is required to show that $z_{\alpha p}{}^{r}$ is the boundary of a chain $s_{\alpha p}{}^{\alpha} x_{\alpha}{}^{r+1}$, where $x_{\alpha}{}^{r+1} \in L_0{}^{r+1}(K_{\alpha})$. Since $K_{\alpha p}$ is an elementary subdivision of $K_{\alpha(p-1)}$, the cycle $z_{\alpha p}{}^{r} = s_{\alpha p}{}^{\alpha(p-1)} \sum_i a_i s_{\alpha(p-1)}{}^{\alpha} t_{\alpha i}{}^{r}$ is the boundary of a chain

$$s_{\alpha p}{}^{\alpha(p-1)} x_{\alpha(p-1)}{}^{r+1}, \qquad x_{\alpha(p-1)}{}^{r+1} \in L_0{}^{r+1}(K_{\alpha(p-1)}):$$

$$z_{\alpha p}{}^{r} = s_{\alpha p}{}^{\alpha(p-1)} \sum_i a_i s_{\alpha(p-1)}{}^{\alpha} t_{\alpha i}{}^{r} = \Delta s_{\alpha p}{}^{\alpha(p-1)} x_{\alpha(p-1)}{}^{r+1} = s_{\alpha p}{}^{\alpha(p-1)} \Delta x_{\alpha(p-1)}{}^{r+1}$$

and

$$s_{\alpha(p-1)}{}^{\alpha} \sum_i a_i t_{\alpha i}{}^{r} = \sum_i a_i s_{\alpha(p-1)}{}^{\alpha} t_{\alpha i}{}^{r} = \Delta x_{\alpha(p-1)}{}^{r+1}.$$

Since $K_{\alpha(p-1)}$ is a subdivision of rank $p - 1$, the last equation implies that

$$s_{\alpha(p-1)}{}^{\alpha} \sum_i a_i t_{\alpha i}{}^{r} = \Delta s_{\alpha(p-1)}{}^{\alpha} x_{\alpha}{}^{r+1}, \qquad x_{\alpha}{}^{r+1} \in L_0{}^{r+1}(K_{\alpha}).$$

Hence

$$z_{\alpha p}{}^{r} = s_{\alpha p}{}^{\alpha} \sum_i a_i t_{\alpha i}{}^{r} = s_{\alpha p}{}^{\alpha(p-1)} s_{\alpha(p-1)}{}^{\alpha} \sum_i a_i t_{\alpha i}{}^{r} = s_{\alpha p}{}^{\alpha(p-1)} \Delta s_{\alpha(p-1)}{}^{\alpha} x_{\alpha}{}^{r+1}$$
$$= \Delta s_{\alpha p}{}^{\alpha} x_{\alpha}{}^{r+1}.$$

This completes the proof.

§2.4. The a-complex defined by a given fundamental system of chains. Let $u_{\beta i}{}^{r}$, $i = 1, 2, \cdots, \rho^{r}$; $r = 0, 1, \cdots, m$, be a fundamental system of chains of a complex K_{β}.

We shall construct a cell complex $\mathfrak{K}$ as follows:

a) The elements of $\mathfrak{K}$ are cells

$$u_1{}^{r}, \cdots, u_{\rho(r)}{}^{r}; \qquad -u_1{}^{r}, \cdots, -u_{\rho(r)}{}^{r}$$

corresponding (1-1) to the chains $\pm u_{\beta i}^r$, where the element u_i^r corresponds to the chain $u_{\beta i}^r$ and the element $-u_i^r$ to the chain $-u_{\beta i}^r$. The resulting (1-1) mapping of the set of all u_i^r onto the set of all chains $\pm u_{\beta i}^r$ will be denoted by β:

$$\beta u_i^r = u_{\beta i}^r, \qquad \beta(-u_i^r) = -u_{\beta i}^r.$$

b) The incidence numbers $(\pm u_i^r : \pm u_j^{r-1})$ are defined in the following way. By Def. 2.31, Condition 1° we have

$$\Delta u_{\beta i}^r = \textstyle\sum_j \eta_{ij}^{r-1} u_{\beta j}^{r-1}, \tag{2.41}$$

where the η_{ij}^{r-1} are integers.

We shall put

$$(u_i^r : u_j^{r-1}) = \eta_{ij}^{r-1},$$

and

$$(u_i^r : -u_j^{r-1}) = -\eta_{ij}^{r-1}; \qquad (-u_i^r : u_j^{r-1}) = -\eta_{ij}^{r-1};$$
$$(-u_i^r : -u_j^{r-1}) = \eta_{ij}^{r-1}.$$

Hence we may take

$$\begin{array}{c|lll}
 & u_1^{r-1} & \cdots u_j^{r-1} & \cdots u_{\rho(r-1)}^{r-1} \\ \hline
u_1^r & \eta_{11}^{r-1} & \cdots \eta_{1j}^{r-1} & \cdots \eta_{1\rho(r-1)}^{r-1} \\
\vdots & \vdots & & \\
u_i^r & \eta_{i1}^{r-1} & \cdots \eta_{ij}^{r-1} & \cdots \eta_{i\rho(r-1)}^{r-1} \\
\vdots & \vdots & & \\
u_{\rho(r)}^r & \eta_{\rho(r)1}^{r-1} & \cdots \eta_{\rho(r)j}^{r-1} & \cdots \eta_{\rho(r)\rho(r-1)}^{r-1}
\end{array}$$

as the incidence matrix of the cell complex $\mathfrak{K}$. Accordingly, we have

$$\Delta u_i^r = \textstyle\sum_j \eta_{ij}^{r-1} u_j^{r-1}; \tag{2.42}$$

$$\Delta u_j^{r-1} = \textstyle\sum_k \eta_{jk}^{r-2} u_k^{r-2}. \tag{2.43}$$

2.4. *The fundamental condition*

$$\Delta\Delta u_i^r = 0$$

is satisfied, so that $\mathfrak{K}$ is an a-complex.

Indeed, in consequence of (2.42), (2.43), and (2.41),

$$\begin{aligned}
\Delta\Delta u_i^r &= \Delta \textstyle\sum_j \eta_{ij}^{r-1} u_j^{r-1} = \sum_j \eta_{ij}^{r-1} \Delta u_j^{r-1} \\
&= \textstyle\sum_j \eta_{ij}^{r-1} \sum_k \eta_{jk}^{r-2} u_k^{r-2} = \sum_k (\sum_j \eta_{ij}^{r-1} \eta_{jk}^{r-2}) u_k^{r-2}, \\
\Delta\Delta u_{\beta i}^r &= \textstyle\sum_k (\sum_j \eta_{ij}^{r-1} \eta_{jk}^{r-2}) u_{\beta k}^{r-2}.
\end{aligned}$$

Since $\Delta\Delta u_{\beta i}^r = 0$, $\sum_k(\sum_j \eta_{ij}^{r-1} \eta_{jk}^{r-2}) u_{\beta k}^{r-2} = 0$ and therefore the linear independence of the chains $u_{\beta k}^{r-2}$ implies that $\sum_j \eta_{ij}^{r-1}\eta_{jk}^{r-2} = 0$ for arbitrary i, k. Hence $\Delta\Delta u_i^r = 0$.

Therefore, the cell complex $\mathfrak{K}$ is an a-complex. We shall refer to it as the *a-complex defined by the given fundamental system of chains.*

REMARK 1. Let K_β be an elementary [the following proposition is true for an arbitrary subdivision K (see Theorems 2.35 and 2.261)] subdivision of a complex K_α. Consider the fundamental system consisting of the chains $s_\beta^\alpha t_{\alpha i}^r$, where the $t_{\alpha i}^r$ are the oriented simplexes of K_α, and denote by $\mathfrak{K}$ the a-complex defined by this fundamental system. Then $\mathfrak{K}$ is isomorphic to the a-complex $\mathfrak{K}_\alpha$ consisting of the oriented simplexes of K_α.

This proposition is an immediate consequence of Theorem 2.12 and the definition of $\mathfrak{K}$.

REMARK 2. Let us return to Examples 1 and 2 of 2.2 and consider the fundamental systems of subcomplexes of the complex K constructed there. In Example 1, K is a triangulation of the torus, and in Example 2 it is a triangulation of the projective plane. The oriented elements of these fundamental systems of subcomplexes constitute, by 2.32, the fundamental system of chains

$$u_{\beta 1}^2, \quad u_{\beta 1}^1, \quad u_{\beta 2}^1, \quad u_{\beta 1}^0,$$

respectively,

$$u_{\beta 1}^2, \quad u_{\beta 1}^1, \quad u_{\beta 1}^0.$$

The corresponding a-complexes

$$\{\pm u_1^2, \pm u_1^1, \pm u_2^1, \pm u_1^0\} \quad \text{and} \quad \{\pm u_1^2, \pm u_1^1, \pm u_1^0\}$$

are none other than the a-complexes considered in VII, 4.2, Examples 2 and 3.

§2.5. The isomorphism β of $L^r(\mathfrak{K})$ into $L^r(K_\beta)$. Let us assign to every chain

$$x^r = \sum a_i u_i^r \in L^r(\mathfrak{K})$$

the chain

$$\beta x^r = \sum a_i u_{\beta i}^r \in L^r(K_\beta).$$

The resulting mapping β of $L^r(\mathfrak{K})$ into $L^r(K_\beta)$ is obviously a homomorphism. But β is also an isomorphism, since $\beta x^r = 0$ implies, because of the linear independence of the chains $u_{\beta i}^r$, that all the $a_i = 0$, i.e., $x^r = 0$.

Moreover,

$$\Delta\beta u_i^r = \Delta u_{\beta i}^r = \sum \eta_{ij}^{r-1} u_{\beta j}^{r-1} = \sum \eta_{ij}^{r-1}\beta u_j^{r-1} = \beta \sum \eta_{ij}^{r-1} u_j^{r-1} = \beta\Delta u_i^r$$

and

(2.51) $$\Delta\beta x^r = \beta\Delta x^r$$

for an arbitrary chain $x^r \in L^r(\mathfrak{K})$. Hence β maps $Z_\Delta{}^r(\mathfrak{K})$ into $Z_\Delta{}^r(K_\beta)$ and $H_\Delta{}^r(\mathfrak{K})$ into $H_\Delta{}^r(K_\beta)$. It follows that β induces a homomorphism (denoted by the same letter) of $\Delta^r(\mathfrak{K})$ into $\Delta^r(K_\beta)$.

It follows further from Def. 2.31, Condition 3° that the homomorphism β of $\Delta^r(\mathfrak{K})$ into $\Delta^r(K_\beta)$ is a homomorphism *onto* and hence by Condition 4°, β is an isomorphism of $\Delta^r(\mathfrak{K})$.

Consequently

2.51. *The isomorphism β of $L^r(\mathfrak{K})$ into $L^r(K_\beta)$ induces an isomorphism (denoted by the same letter β) of $\Delta^r(\mathfrak{K})$ onto $\Delta^r(K_\beta)$.*

Because of this isomorphism of $\Delta^r(\mathfrak{K})$ and $\Delta^r(K_\beta)$ and the results of IX, 3, 4, we have:

The groups $\nabla^r(\mathfrak{K})$ and $\nabla^r(K_\beta)$ are also isomorphic.

Hence

2.5. *The groups $\Delta^r(\mathfrak{K}, \mathfrak{A})$ $[\nabla^r(\mathfrak{K}, \mathfrak{A})]$ are isomorphic to the groups $\Delta^r(K_\beta, \mathfrak{A})$ $[\nabla^r(K_\beta, \mathfrak{A})]$ for every coefficient domain $\mathfrak{A}$.*

Remark 1. Let us recall the triangulations of the torus and the projective plane of 2.2, Examples 1 and 2 and the fundamental systems of chains of these triangulations considered in 2.4, Remark 2. We saw there that the a-complexes defined by these fundamental systems are none other than the a-complexes which occurred in VII, 4.2, Examples 2 and 3 and in VIII, 1.4, Exercise 4; the Betti groups of these complexes (noted in VIII, 1.4, Exercise 4) are isomorphic, by Theorem 2.5, to the Betti groups of the triangulation K of the torus and the projective plane, respectively.

Remark 2. The isomorphic mapping β of $L^r(\mathfrak{K})$ into $L^r(K_\beta)$ does not, in general, commute with ∇. In other words, $\nabla\beta x^r$ may not be equal to $\beta\nabla x^r$.

To see this, it is enough to consider the complex K_β consisting of the segments

$$|t_{\beta 1}^1| = (AB), \qquad |t_{\beta 2}^1| = (BC)$$

and their endpoints

$$A = t_{\beta 1}^0, \qquad B = t_{\beta 3}^0, \qquad C = t_{\beta 2}^0.$$

Setting

$t_{\beta 1}^1 = |AB|$, $t_{\beta 2}^1 = |BC|$, $u_\beta^1 = t_{\beta 1}^1 + t_{\beta 2}^1$ and $u_{\beta 1}^0 = t_{\beta 1}^0, u_{\beta 2}^0 = t_{\beta 2}^0$ and defining u_1^0, u_2^0, u^1 in accordance with 2.4, we have

$$\beta u_1^0 = t_{\beta 1}^0, \qquad \nabla\beta u_1^0 = t_{\beta 1}^1, \qquad \nabla u_1^0 = u^1,$$

$$\beta\nabla u_1^0 = \beta u^1 = u_\beta^1 = t_{\beta 1}^1 + t_{\beta 2}^1,$$

i.e.,

$$\nabla\beta u_1^0 \neq \beta\nabla u_1^0.$$

§2.6. Isomorphism of the Betti groups under elementary and barycentric subdivisions of a complex K. Let K_β be an elementary subdivision of a complex K_α relative to a simplex $T \in | K_\alpha |$. Then, by Theorem 2.350, the chains $u_{\beta i}^r = s_\beta{}^\alpha t_{\alpha i}^r$, where the $t_{\alpha i}^r$ are the oriented simplexes of K_α, form a fundamental system of chains of K_β.

But the a-complex $\mathfrak{K}_\alpha$ made up of all the oriented simplexes of K_α is isomorphic to the a-complex $\mathfrak{K}$ defined by the fundamental system of chains $u_{\beta i}^r$ (2.4, Remark 1). Hence, by Theorem 2.5, the Betti groups of K_β are isomorphic to the Betti groups of $\mathfrak{K}$ and therefore to the Betti groups of $\mathfrak{K}_\alpha$, that is, to the Betti groups of K_α.

Hence

2.60. *The Betti groups of a complex K_α are isomorphic to the Betti groups of an elementary subdivision of K_α relative to an arbitrary simplex $T \in | K_\alpha |$.*

Since a barycentric subdivision is the result of consecutive elementary subdivisions, we have the following fundamental theorem:

THEOREM 2.61. (The theorem is also valid for an arbitrary subdivision K_β of K_α. If K_β is a triangulation, this assertion follows, for example, from 2.261 and the parenthetical remark to Theorem 2.35.) *Let K_α be a triangulation or an open subcomplex of a triangulation and let K_β be a barycentric or an arbitrary regular subdivision of K_α. Then the group $\Delta^r(K_\alpha, \mathfrak{A})$ is isomorphic to the group $\Delta^r(K_\beta, \mathfrak{A})$ and the group $\nabla^r(K_\alpha, \mathfrak{A})$ to the group $\nabla^r(K_\beta, \mathfrak{A})$ for every r and every coefficient domain $\mathfrak{A}$.*

§3. Normal and canonical displacements in polyhedra

§3.1. Normal displacements of subdivisions of triangulations. Let K_β be a subdivision of a triangulation K_α. Every point of the polyhedron $P = \| K_\beta \| = \| K_\alpha \|$ is contained in a unique simplex of K_α, the carrier of the point in K_α. In particular, every vertex $e_{\beta j}$ of K_β has its carrier in K_α.

Let us assign to every vertex $e_{\beta j}$ of K_β a definite vertex $e_{\alpha j} = S_\alpha^\beta e_{\beta j}$ of K_α subject only to the condition that *$e_{\alpha j}$ is a vertex of the carrier of $e_{\beta j}$ in K_α.*

We shall prove that this vertex mapping defines a simplicial mapping of K_β into K_α. Suppose that the vertices $e_{\beta 0}, \cdots, e_{\beta r}$ form a skeleton in K_β and denote by T_α the carrier (in K_α) of the simplex $(e_{\beta 0} \cdots e_{\beta r}) = T_\beta$. Since every vertex $e_{\beta k}$ is a limit point of the simplex $(e_{\beta 0} \cdots e_{\beta r})$ and hence of the simplex $T_\alpha \supset (e_{\beta 0} \cdots e_{\beta r})$, all the vertices $e_{\beta 0}, \cdots, e_{\beta r}$ are contained in $\overline{T}_\alpha$. It follows that the carrier in K_α of each of the vertices $e_{\beta 0}, \cdots, e_{\beta r}$ is a face of T_α. Hence $S_\alpha^\beta e_{\beta 0}, \cdots, S_\alpha^\beta e_{\beta r}$ are vertices of T_α.

Consequently:

The mapping S_α^β of the set of all vertices of K_β into the set of vertices of K_α defines a simplicial mapping, also denoted by S_α^β, of K_β into K_α.

Every simplicial mapping S_α^β of a subdivision K_β of a complex K_α into K_α so constructed is called a *normal mapping or a normal displacement.*

As is the case with every simplicial mapping (see VII, 8.2), a normal displacement S_α^β induces a homomorphism S_α^β of $L_\beta^r = L^r(K_\beta, \mathfrak{A})$ into $L_\alpha^r = L^r(K_\alpha, \mathfrak{A})$, referred to in this case as a *normal homomorphism* of L_β^r into L_α^r.

It follows from the general results of VII, 8.31 and VIII, 1.8 that a normal homomorphism S_α^β induces a homomorphism (also referred to as a normal homomorphism) S_α^β of $\Delta_\beta^r = \Delta^r(K_\beta, \mathfrak{A})$ into $\Delta_\alpha^r = \Delta^r(K_\alpha, \mathfrak{A})$.

3.11. *A normal homomorphism S_α^β of L_β^r into L_α^r is a mapping onto L_α^r.*

This proposition obviously follows from the following fundamental identity which is valid for an arbitrary chain $x_\alpha^r \in L_\alpha^r$ and an arbitrary normal homomorphism S_α^β:

$$S_\alpha^\beta s_\beta^\alpha x_\alpha^r = x_\alpha^r, \tag{3.11}$$

where $s_\beta^\alpha x_\alpha^r$ is the subdivision of the chain x_α^r in K_β.

It suffices to prove (3.11) for a monomial chain $x_\alpha^r = t_\alpha^r$. The proof is based on the following lemma:

ALEXANDER'S LEMMA. *Let S be a simplicial mapping of a complex K into the complex $| T^n |$ consisting of a simplex T^n and all its faces (we assume merely that K is a simplicial complex whose oriented simplexes form an a-complex). Let t^n be any orientation of the simplex T^n. If*

$$S\Delta x^n = \Delta t^n$$

for an integral chain $x^n \in L^n(K)$, then

$$Sx^n = t^n.$$

Proof. Sx^n is an integral n-chain of $| T^n |$ and since $| T^n |$ contains exactly one n-simplex, T^n, Sx^n is necessarily a monomial chain. Hence it is of the form ct^n, with c and integer and $| t^n | = T^n$.

We shall prove that $c = 1$. $Sx^n = ct^n$ implies that

$$S\Delta x^n = \Delta Sx^n = c\Delta t^n.$$

Since $S\Delta x^n = \Delta t^n$ by hypothesis, $c = 1$ and this proves the lemma.

REMARK 1. Alexander's lemma and its proof remain valid if the coefficient domain is the ring J_m.

We shall now use Alexander's lemma to prove the identity

$$S_\alpha^\beta s_\beta^\alpha t_\alpha^r = t_\alpha^r \tag{3.111}$$

which immediately implies (3.11).

(3.111) is valid for $r = 0$ since $s_\beta^\alpha t_\alpha^0 = t_\alpha^0$, and S_α^β reduces in this case to the identity mapping of the vertex $| t_\alpha^0 |$ onto itself.

Let us now assume that (3.111) is true for $r = n - 1$ and prove it for $r = n$.

Choosing the orientations $t_{\alpha 0}{}^{n-1}, \cdots, t_{\alpha n}{}^{n-1}$ of all the $(n - 1)$-faces of $T_\alpha{}^n$ so that

$$\Delta t_\alpha{}^n = t_{\alpha 0}{}^{n-1} + \cdots + t_{\alpha n}{}^{n-1},$$

we have

$$s_\beta{}^\alpha \Delta t_\alpha{}^n = s_\beta{}^\alpha t_{\alpha 0}{}^{n-1} + \cdots + s_\beta{}^\alpha t_{\alpha n}{}^{n-1}.$$

Since

$$S_\alpha{}^\beta s_\beta{}^\alpha t_{\alpha 0}{}^{n-1} = t_{\alpha 0}{}^{n-1}, \cdots, S_\alpha{}^\beta s_\beta{}^\alpha t_{\alpha n}{}^{n-1} = t_{\alpha n}{}^{n-1}$$

by assumption, it follows that

$$S_\alpha{}^\beta \Delta s_\beta{}^\alpha t_\alpha{}^n = S_\alpha{}^\beta s_\beta{}^\alpha \Delta t_\alpha{}^n = t_{\alpha 0}{}^{n-1} + \cdots + t_{\alpha n}{}^{n-1} = \Delta t_\alpha{}^n.$$

Hence, by Alexander's lemma (putting $t^n = t_\alpha{}^n$, $x^n = s_\beta{}^\alpha t_\alpha{}^n$, $S = S_\alpha{}^\beta$),

$$S_\alpha{}^\beta s_\beta{}^\alpha t_\alpha{}^n = t_\alpha{}^n.$$

This completes the proof.

REMARK 2. If the coefficient domain is J_2, it is easy to see that (3.111) becomes Sperner's lemma (see V, 2.1).

The identity (3.11) immediately leads to a series of important corollaries:

3.12. *A normal mapping $S_\alpha{}^\beta$ is a mapping of the complex K_β onto the complex K_α.*

Indeed, let $T_{\alpha i}{}^r$ be an arbitrary simplex of K_α, let $t_{\alpha i}{}^r$ be its orientation, and let $T_{\beta 1}{}^r, \cdots, T_{\beta q}{}^r$ be the r-simplexes of K_β of which $T_{\alpha i}{}^r$ is the carrier. $S_\alpha{}^\beta$ maps at least one of the simplexes $T_{\beta j}{}^r$, $j = 1, 2, \cdots, q$, onto $T_{\alpha i}{}^r$, since in the contrary case we would have $S_\alpha{}^\beta s_\beta{}^\alpha t_{\alpha i}{}^r = 0$ and not

$$S_\alpha{}^\beta s_\beta{}^\alpha t_{\alpha i}{}^r = t_{\alpha i}{}^r.$$

We shall now assume until the end of this section that K_α is a triangulation.

3.13. *A homomorphism $S_\alpha{}^\beta$ of $L_\beta{}^r$ onto $L_\alpha{}^r$ maps $Z_\beta{}^r$ onto $Z_\alpha{}^r$ and $H_\beta{}^r$ onto $H_\alpha{}^r$.*

Proposition 3.13 follows immediately from (3.11).

In consequence of 3.13, the homomorphism $S_\alpha{}^\beta$ of $\Delta_\beta{}^r = \Delta^r(K_\beta, \mathfrak{A})$ into $\Delta_\alpha{}^r(K_\alpha, \mathfrak{A})$ induced by a homomorphism $S_\alpha{}^\beta$ of $L_\beta{}^r$ onto $L_\alpha{}^r$ is a homomorphism onto all of $\Delta_\alpha{}^r$.

We shall prove that *$S_\alpha{}^\beta$ is an isomorphism of $\Delta_\beta{}^r$ onto $\Delta_\alpha{}^r$*. The reader may carry out the proof of this proposition given below under the assumption of various degrees of generality for the subdivision K_β: he may restrict himself to the case of K_β an elementary subdivision of the triangulation K_α; or he may assume that K_β is a regular subdivision of K_α (see 2.350,

2.35); this case includes, in particular, the case of K_β a barycentric subdivision of K_α (even the case of K_β a higher order barycentric subdivision); finally, he may assume that K_β is an arbitrary subdivision of K_α. In accordance with the chosen degree of generality it will be necessary to refer to 2.350 or to 2.35 or finally to the parenthetical remark to Theorem 2.35. In the sequel we shall use only regular subdivisions (and almost exclusively first order or higher order barycentric subdivisions).

We shall therefore prove

3.14. *A normal mapping S_α^β of a subdivision K_β of a triangulation K_α onto K_α induces an isomorphism S_α^β of Δ_β^r onto Δ_α^r (inverse to the isomorphism s_β^α).*

It is sufficient to prove the following lemma:

3.15. *If $z_\beta^r \in Z_\beta^r$ and $S_\alpha^\beta z_\beta^r \in H_\alpha^r$, then $z_\beta^r \in H_\beta^r$.*

To prove 3.15 we note that by 2.350 (or 2.35 or the parenthetical remark to 2.35), there are chains $x_\beta^{r+1} \in L_\beta^{r+1}$ and $z_\alpha^r \in Z_\alpha^r$ such that

$$\Delta x_\beta^{r+1} = z_\beta^r - s_\beta^\alpha z_\alpha^r. \tag{3.15}$$

Hence

$$\begin{aligned}\Delta S_\alpha^\beta x_\beta^{r+1} &= S_\alpha^\beta \Delta x_\beta^{r+1} = S_\alpha^\beta z_\beta^r - S_\alpha^\beta s_\beta^\alpha z_\alpha^r \\ &= S_\alpha^\beta z_\beta^r - z_\alpha^r,\end{aligned}$$

that is, $S_\alpha^\beta z_\beta^r - z_\alpha^r \in H_\alpha^r$. Therefore, if $S_\alpha^\beta z_\beta^r \in H_\alpha^r$, then $z_\alpha^r \in H_\alpha^r$. Consequently $s_\beta^\alpha z_\alpha^r \in H_\beta^r$, i.e., by (3.15), $z_\beta^r \in H_\beta^r$. This completes the proof.

REMARK. The normal homomorphism $\bar{S}_\beta^\alpha$ of L_α^r into L_β^r dual to (see IX, 6.1) a normal homomorphism S_α^β induces a homomorphism $\bar{S}_\beta^\alpha$ of $\nabla^r(K_\alpha, \mathfrak{A})$ into $\nabla^r(K_\beta, \mathfrak{A})$. The latter homomorphism is an isomorphism of $\nabla^r(K_\alpha, \mathfrak{A})$ onto $\nabla^r(K_\beta, \mathfrak{A})$; we shall prove this proposition in the following section.

§3.2. Examples of normal homomorphisms S_α^β and $\bar{S}_\beta^\alpha$.

1°. Let $K_\alpha = |T_\alpha^2|$ be the complex consisting of a triangle, its sides, and vertices, and let K_β be a barycentric subdivision of K_α.

Let S_α^β be a simplicial mapping of K_β into K_α obtained by assigning to every vertex of K_β a vertex of its carrier in K_α. In Fig. 130, the letter in parentheses beside each vertex of K_β denotes the image of that vertex under the mapping S_α^β.

Denote by t_α^2 the triangle *abc* oriented counterclockwise (as well as the chain assuming the value 1 on t_α^2).

For brevity, we put $x_\alpha^1 = \Delta t_\alpha^2$ and denote by x_β^2 the chain whose value is 1 on all the triangles, oriented counterclockwise, of K_β. (We could obviously write $x_\beta^2 = s_\beta^\alpha t_\alpha^2$, with $\Delta x_\beta^2 = s_\beta^\alpha \Delta t_\alpha$. We shall not use the notation s_β^α here in order not to make difficulties for the reader who has come to 3.2 directly from IX, 6.2.) We shall denote the hatched triangle of K_β, oriented counterclockwise, by t_β^2. Finally let x_β^1 be the chain of K_β whose value is 1 on all the heavily drawn segments of K_β oriented by arrows as indicated in Fig. 130.

We may immediately verify that

$$S_\alpha^\beta x_\beta^2 = t_\alpha^2, \qquad S_\alpha^\beta \Delta x_\beta^2 = \Delta t_\alpha^2,$$

in accordance with the general theorems. Furthermore,

$$\nabla x_\alpha^1 = 3t_\alpha^2,$$

$$\bar{S}_\beta^\alpha x_\alpha^1 = x_\beta^1, \qquad \bar{S}_\beta^\alpha t_\alpha^2 = t_\beta^2,$$

$$\nabla x_\beta^1 = 3t_\beta^2,$$

so that

$$\nabla \bar{S}_\beta^\alpha x_\alpha^1 = \bar{S}_\beta^\alpha \nabla x_\alpha^1,$$

in agreement with IX, 6.2.

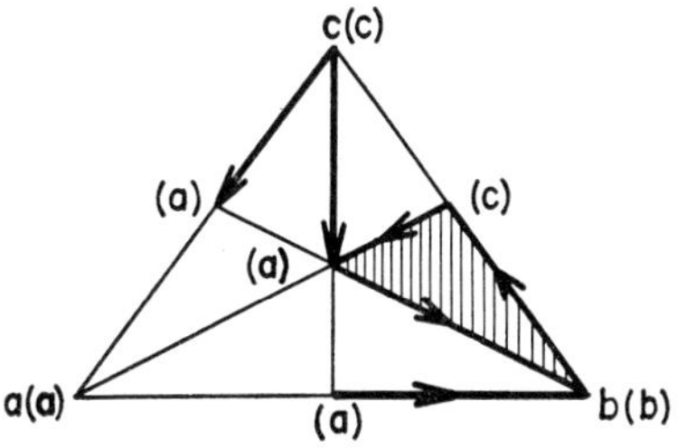

FIG. 130

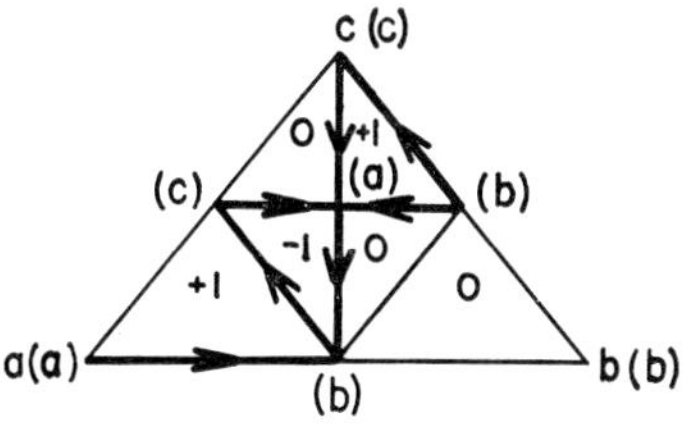

FIG. 131

2°. Again, let $K_\alpha = |T_\alpha^2|$ and let K_β be the subdivision of K_α indicated in Fig. 131. Here again, the letter in parentheses next to each vertex of K_β indicates its image under S_α^β. We adopt the following notation: t_α^2 is the triangle T_α^2 oriented counterclockwise; x_β^2 is the chain whose value is 1 on all the triangles, oriented counterclockwise, of K_β ; y_β^2 is the chain assuming on the counterclockwise oriented triangles of K_β the values indicated in Fig. 131; y_β^1 is the chain whose value is 1 on the heavily drawn segments of K_β (oriented by arrows as indicated) and 0 on all the remaining 1-simplexes of K_β .

It is immediately verified that

$$S_\alpha^\beta x_\beta^2 = t_\alpha^2, \qquad S_\alpha^\beta y_\beta^2 = 3t_\alpha^2,$$

$$\nabla \Delta t_\alpha^2 = 3t_\alpha^2,$$

$$\bar{S}_\beta^\alpha \Delta t_\alpha^2 = y_\beta^1, \qquad \bar{S}_\beta^\alpha t_\alpha^2 = y_\beta^2,$$

$$\Delta y_\beta^1 = 3y_\beta^2.$$

§4. Canonical systems of bases for subdivisions K_β of a triangulation K_α. The homomorphism $\bar{S}_\beta^\alpha$ dual to a normal homomorphism S_α^β

§4.1. Canonical systems of bases of K_β. In this section the reader may assume that K_β is either an elementary subdivision of a triangulation K_α or an arbitrary regular subdivision of a triangulation K_α or finally an arbitrary subdivision of K_α. Depending on the degree of generality chosen it will be necessary to use either Theorem 2.350 or Theorem 2.35, or finally the parenthetical remark to Theorem 2.35. For the applications which we have in mind, the case of an elementary subdivision is quite sufficient.

Let $\mathfrak{A} = J$. Instead of $L_0^r(K_\alpha)$, $Z_0^r(K_\alpha)$, $\cdots$, $Z_0^r(K_\beta)$, $\cdots$ we shall write $L_\alpha^r, Z_\alpha^r, \cdots, Z_\beta^r, \cdots$.

We shall consider the subgroups $L_{0\beta}^r$, $Z_{0\beta}^r$, $H_{0\beta}^r$ of L_β^r, Z_β^r, H_β^r, respectively, consisting of chains of the form $s_\beta^\alpha x_\alpha^r$, where x_α^r belongs to L_α^r, Z_α^r, H_α^r, respectively, and s_β^α, as usual, is the subdivision operator (see 2.1). Obviously

$$Z_{0\beta}^r = L_{0\beta}^r \cap Z_\beta^r,$$

$$H_{0\beta}^r = L_{0\beta}^r \cap H_\beta^r,$$

$$\hat{H}_{0\beta}^r = L_{0\beta}^r \cap \hat{H}_\beta^r.$$

The groups $Z_{0\beta}^r$, $L_{0\beta}^r$ are division closed in L_β^r (and of course $Z_{0\beta}^r$ is division closed in Z_β^r). Consequently $Z_\beta^r/Z_{0\beta}^r$ is a free Abelian group and every basis of $Z_{0\beta}^r$ can be completed to a basis of Z_β^r by elements *arbitrarily chosen from each one of the classes* $\mathfrak{h}_\beta^r \in Z_\beta^r/Z_{0\beta}^r$.

4.11. *Every class* $\mathfrak{h}_\beta^r \in Z_\beta^r/Z_{0\beta}^r$ *contains at least one element* $h_\beta^r \in H_\beta^r$.

Indeed, let z_β^r be an arbitrary element of the class $\mathfrak{h}_\beta^r \in Z_\beta^r/Z_{0\beta}^r$. By Theorem 2.350 (or 2.35 or the parenthetical remark to 2.35) there is a cycle $'z_{0\beta}^r = s_\beta^\alpha z_\alpha^r \in Z_{0\beta}^r$ homologous to z_β^r in K_β; the cycle

$$h_\beta^r = z_\beta^r - {'z_{0\beta}^r} \in H_\beta^r$$

is contained in the same class $\mathfrak{h}_\beta^r$ as z_β^r.

It follows therefore that:

4.12. *An arbitrary basis of* $Z_{0\beta}^r$ *can be completed to a basis of* Z_β^r *with cycles* $h_{\beta i}^r$ *chosen from each one of the classes* $\mathfrak{h}_\beta^r \in Z_\beta^r/Z_{0\beta}^r$ *and belonging to* H_β^r.

We now wish to construct a system of canonical bases for L_β^r, $r = 0, 1, \cdots, n$, satisfying certain supplementary conditions having to do with the fact that K_β is a subdivision of K_α.

We shall first construct, for each $r = 0, 1, \cdots, n$, a canonical basis

$$z_{\alpha 1}^r, \cdots, z_{\alpha\pi(r)}^r; \qquad u_{\alpha 1}^r, \cdots, u_{\alpha\tau(r)}^r; \qquad v_{\alpha 1}^r, \cdots, v_{\alpha\sigma(r)}^r$$

of Z_α^r. This construction defines a basis for $Z_{0\beta}^r$ consisting of the cycles

$$z_{0h}^r = s_\beta^\alpha z_{\alpha h}^r; \qquad u_{0k}^r = s_\beta^\alpha u_{\alpha k}^r; \qquad v_{0l}^r = s_\beta^\alpha v_{\alpha l}^r.$$

To this basis we adjoin cycles $v_{j'}^r \in H_\beta^r$ to obtain a *canonical basis*

$$z_{01}^r, \cdots, z_{0\pi(r)}^r; \qquad u_{01}^r, \cdots, u_{0\tau(r)}^r; \qquad v_{01}^r, \cdots, v_{0\sigma(r)}^r; \qquad v_1^r, \cdots, v_{\sigma'(r)}^r$$

of Z_β^r.

Because of the isomorphism Δ between L_β^r/Z_β^r and H_β^{r-1}, the basis

$$\theta_1^{r-1}u_{01}^{r-1}, \cdots, \theta_{\tau(r-1)}^{r-1}u_{0\tau(r-1)}^{r-1}; \qquad v_{01}^{r-1}, \cdots, v_{0\sigma(r-1)}^{r-1};$$
$$v_1^{r-1}, \cdots, v_{\sigma'(r-1)}^{r-1}$$

of H_β^{r-1} corresponds to the basis

$$\mathfrak{x}_{01}^r, \cdots, \mathfrak{x}_{0\tau(r-1)}^r; \qquad \mathfrak{y}_{01}^r, \cdots, \mathfrak{y}_{0\sigma(r-1)}^r; \qquad \mathfrak{y}_1^r, \cdots, \mathfrak{y}_{\sigma'(r-1)}^r$$

of L_β^r/Z_β^r and since $\theta_i^{r-1}u_{0i}^{r-1} = s_\beta^\alpha\theta_i^{r-1}u_{\alpha i}^{r-1}$ and $v_{0j}^{r-1} = s_\beta^\alpha v_{\alpha j}^{r-1}$, we may choose chains $x_{0i}^r = s_\beta^\alpha x_{\alpha i}^r$, $y_{0j}^r = s_\beta^\alpha y_{\alpha i}^r$ in $\mathfrak{x}_{0i}^r$, $\mathfrak{y}_{0j}^r$, respectively. Choosing such chains and also arbitrary chains $y_{j'}^r \in \mathfrak{y}_{j'}^r$ for all $i = 1, \cdots, \tau^{r-1}$, $j = 1, \cdots, \sigma^{r-1}$, $j' = 1, \cdots, \sigma'^{r-1}$, we obtain the required *system of canonical bases*

$$(4.13_\Delta) \quad r = 0, \cdots, n \begin{cases} x_{01}^r, \cdots, x_{0\tau(r-1)}^r; \\ \qquad y_{01}^r, \cdots, y_{0\sigma(r-1)}^r; \qquad y_1^r, \cdots, y_{\sigma'(r-1)}^r; \\ z_{01}^r, \cdots, z_{0\pi(r)}^r; \\ \qquad u_{01}^r, \cdots, u_{0\tau(r)}^r; \qquad v_{01}^r, \cdots, v_{0\sigma(r)}^r; \\ \qquad v_1^r, \cdots, v_{\sigma'(r)}^r \end{cases}$$

of the groups L_β^r with the relations

$$(4.14_\Delta) \quad \begin{cases} \Delta x_{0i}^r = \theta_i^{r-1}u_{0i}^{r-1}; \qquad \Delta y_{0j}^r = v_{0j}^{r-1}; \qquad \Delta y_{j'}^r = v_{j'}^{r-1}, \\ \Delta z_{0h}^r = \Delta u_{0k}^r = \Delta u_{0l}^r = \Delta v_{l'}^r = 0. \end{cases}$$

The bases dual to the canonical system of bases above will be written as

$$(4.13_\nabla) \quad \bar{u}_{0i}^r; \qquad \bar{v}_{0j}^r; \qquad \bar{v}_{j'}^r; \qquad \bar{z}_{0h}^r; \qquad \bar{x}_{0k}^r; \qquad \bar{y}_{0l}^r; \qquad \bar{y}_{l'}^r$$

and satisfy the conditions

$$(4.14_\nabla) \quad \begin{cases} \nabla\bar{u}_{0i}^r = \nabla\bar{v}_{0j}^r = \nabla\bar{v}_{j'}^r = \nabla\bar{z}_{0h}^r = 0, \\ \nabla\bar{x}_{0k}^r = \theta_k^r\bar{u}_{0k}^{r+1}; \qquad \nabla\bar{y}_{0l}^r = \bar{v}_{0l}^{r+1}; \qquad \nabla\bar{y}_{l'}^r = \bar{v}_{l'}^{r+1}. \end{cases}$$

Indeed, in Table (4.15) every row (regarded as a linear form in x_{0i}^{r-1}, y_{0j}^{r-1}, $y_{j'}^{r-1}$, z_{0h}^{r-1}, u_{0h}^{r-1}, v_{0l}^{r-1}, $v_{l'}^{r-1}$) represents the Δ-boundary of the r-element of the basis (4.13_Δ) heading the row; consequently, every column of Table (4.15) (regarded as a linear form in $\bar{u}_{0i}^r$, $\bar{v}_{0j}^r$, $\bar{v}_{j'}^r$, $\bar{z}_{0h}^r$, $\bar{x}_{0k}^r$, $\bar{y}_{0l}^r$, $\bar{y}_{l'}^r$) represents the ∇-boundary of the $(r-1)$-element of the basis (4.13_∇)

heading this column. We therefore have at once

$$\nabla \bar{u}_{0i}^{r-1} = \nabla \bar{v}_{0j}^{r-1} = \nabla \bar{v}_{j'}^{r-1} = \nabla \bar{z}_{0h}^{r-1} = 0,$$

$$\nabla \bar{x}_{0k}^{r-1} = \theta_k^{r-1} \bar{u}_{0k}^{r}, \qquad \nabla \bar{y}_{0l}^{r-1} = \bar{v}_{0l}^{r}, \qquad \nabla \bar{y}_{l'}^{r-1} = \bar{v}_{l'}^{r}.$$

which becomes (4.14_∇) if $r - 1$ is replaced by r.

As in IX, 3.5, we easily verify that the cycles

$$\bar{z}_{0h}^{r}, \qquad \bar{u}_{0k}^{r}, \qquad \bar{v}_{0l}^{r}, \qquad \bar{v}_{l'}^{r}$$

form a basis of $Z_{\beta\nabla}^{r}$ while the cycles

$$\theta_k^{r-1} \bar{u}_{0k}^{r}, \qquad \bar{v}_{0l}^{r}, \qquad \bar{v}_{l'}^{r}$$

are a basis of $H_{\beta\nabla}^{r}$.

§4.2. Normal homomorphisms relative to canonical bases. Let us now consider the form which the matrix of a normal homomorphism S_α^β of L_β^r into L_α^r takes on relative to a system of canonical bases of the complexes K_β and K_α.

(4.15)

		$\bar{u}_{0i}^{r-1}$	$\bar{v}_{0j}^{r-1}$	$\bar{v}_{j'}^{r-1}$	$\bar{z}_{0h}^{r-1}$	$\bar{x}_{0k}^{r-1}$	$\bar{y}_{0l}^{r-1}$	$\bar{y}_{l'}^{r-1}$
		x_{0i}^{r-1}	y_{0j}^{r-1}	$y_{j'}^{r-1}$	z_{0h}^{r-1}	u_{0k}^{r-1}	v_{0l}^{r-1}	$v_{l'}^{r-1}$
$\bar{u}_{0i}^{r}$	x_{0i}^{r}	0	0	0	0	θ_1^{r-1} . . . $\theta_{\tau(r-1)}^{r-1}$	0	0
$\bar{v}_{0j}^{r}$	y_{0j}^{r}	0	0	0	0	0	1 . . . 1	0
$\bar{v}_{j'}^{r}$	$y_{j'}^{r}$	0	0	0	0	0	0	1 . . . 1
$\bar{z}_{0h}^{r}$	z_{0h}^{r}	0	0	0	0	0	0	0
$\bar{x}_{0k}^{r}$	u_{0k}^{r}	0	0	0	0	0	0	0
$\bar{y}_{0l}^{r}$	v_{0l}^{r}	0	0	0	0	0	0	0
$\bar{y}_{l'}^{r}$	$v_{l'}^{r}$	0	0	0	0	0	0	0

We have

$$S_\alpha^\beta x_{0i}^r = x_{\alpha i}^r, \qquad S_\alpha^\beta y_{0j}^r = y_{\alpha j}^r, \qquad S_\alpha^\beta z_{0h}^r = z_{\alpha h}^r,$$
$$S_\alpha^\beta u_{0k}^r = u_{\alpha k}^r, \qquad S_\alpha^\beta v_{0l}^r = v_{\alpha l}^r.$$

Furthermore, let

$$S_\alpha^\beta y_{j'}^r = \sum_i a_{j'i} x_{\alpha i}^r + \sum_j b_{j'j} y_{\alpha j}^r + \sum_h c_{j'h} z_{\alpha h}^r + \sum_k d_{j'k} u_{\alpha k}^r + \sum_l e_{j'l} v_{\alpha l}^r,$$

where the coefficients on the right are integers.

Finally, since $v_{l'}^r \in H_\beta^r$, it follows that $S_\alpha^\beta v_{l'}^r \in H_\alpha^r$, that is, $S_\alpha^\beta v_{l'}^r = \sum_k p_{l'k} \theta_k^r u_{\alpha k}^r + \sum_l q_{l'l} v_{\alpha l}^r$, where $p_{l'k}$, $q_{l'l}$ are integers. This may all be written in the form of Table (4.21), where every row (regarded as a linear form in $x_{\alpha i}^r$, $y_{\alpha j}^r$, $z_{\alpha h}^r$, $u_{\alpha k}^r$, $v_{\alpha l}^r$) represents the image under the homomorphism S_α^β of the element of the basis (4.13_Δ) heading the row.

(4.21)

		$\bar{u}_{\alpha i}^r$	$\bar{v}_{\alpha j}^r$	$\bar{z}_{\alpha h}^r$	$\bar{x}_{\alpha k}^r$	$\bar{y}_{\alpha l}^r$
		$x_{\alpha i}^r$	$y_{\alpha j}^r$	$z_{\alpha h}^r$	$u_{\alpha k}^r$	$v_{\alpha l}^r$
$\bar{u}_{0i}^r$	x_{0i}^r	1	0	0	0	0
$\bar{v}_{0j}^r$	y_{0j}^r	0	1	0	0	0
$\bar{v}_{j'}^r$	$y_{j'}^r$	$a_{j'i}$	$b_{j'j}$	$c_{j'h}$	$d_{i'k}$	$e_{j'l}$
$\bar{z}_{0h}^r$	z_{0h}^r	0	0	1	0	0
$\bar{x}_{0k}^r$	u_{0k}^r	0	0	0	1	0
$\bar{y}_{0l}^r$	v_{0l}^r	0	0	0	0	1
$\bar{y}_{l'}^r$	$v_{l'}^r$	0	0	0	$p_{l'k}\theta_k^r$	$q_{l'l}$

§4.3. The homomorphism dual to a normal homomorphism. Passing to the dual bases and recalling that S_α^β and $\bar{S}_\beta^\alpha$ are dual homomorphisms, we remark that by Theorem 5.4, Appendix 2, every column of Table (4.21) (as a linear form in $\bar{u}_{0i}^r$, $\bar{v}_{0j}^r$, $\bar{v}_{j'}^r$, $\bar{z}_{0h}^r$, $\bar{x}_{0k}^r$, $\bar{y}_{0l}^r$, $\bar{y}_{l'}^r$) represents the image under $\bar{S}_\beta^\alpha$ of the element of the basis $\{\bar{u}_{\alpha i}^r, \bar{v}_{\alpha j}^r, \bar{z}_{\alpha h}^r, \bar{x}_{\alpha k}^r, \bar{y}_{\alpha l}^r\}$ heading this column.

In particular,

$$\bar{S}_\beta{}^\alpha \bar{u}_{\alpha i}{}^r = \bar{u}_{0i}{}^r + \sum_{j'} a_{j'i} \bar{v}_{j'}{}^r, \qquad i = 1, \cdots, \tau^{r-1},$$

$$\bar{S}_\beta{}^\alpha \bar{z}_{\alpha h}{}^r = \bar{z}_{0h}{}^r + \sum_{j'} c_{j'h} \bar{v}_{j'}{}^r, \qquad h = 1, \cdots, \pi^r.$$

But $\bar{v}_{j'}{}^r \in H_{\beta\nabla}{}^r$ and consequently

$$(4.3) \qquad \bar{S}_\beta{}^\alpha \bar{\mathfrak{u}}_{\alpha i}{}^r = \bar{\mathfrak{u}}_{0i}{}^r, \qquad \bar{S}_\beta{}^\alpha \bar{\mathfrak{z}}_{\alpha h}{}^r = \bar{\mathfrak{z}}_{0h}{}^r,$$

where $\bar{\mathfrak{u}}_{\alpha i}{}^r$, $\bar{\mathfrak{u}}_{0i}{}^r$, $\bar{\mathfrak{z}}_{\alpha h}{}^r$, $\bar{\mathfrak{z}}_{0h}{}^r$ are, respectively, the cohomology classes of the elements $\bar{u}_{\alpha i}{}^r$, $\bar{u}_{0i}{}^r$, $\bar{z}_{\alpha h}{}^r$, $\bar{z}_{0h}{}^r$; but the cohomology classes $\bar{\mathfrak{u}}_{\alpha i}{}^r$, $\bar{\mathfrak{z}}_{\alpha h}{}^r$ form a basis of $\nabla^r(K_\alpha)$ and the cohomology classes $\bar{\mathfrak{u}}_{0i}{}^r$, $\bar{\mathfrak{z}}_{0h}{}^r$ form a basis of $\nabla^r(K_\beta)$. It follows from (4.3), that $\bar{S}_\beta{}^\alpha$ establishes a $(1-1)$ correspondence between these two bases, that is, $\bar{S}_\beta{}^\alpha$ is an isomorphism of $\nabla^r(K_\alpha)$ onto $\nabla^r(K_\beta)$.

Summing up the results of this section, we have

4.3. *Let $S_\alpha{}^\beta$ be a normal mapping of a subdivision K_β of a triangulation K_α onto K_α and let $S_\alpha{}^\beta$ also denote the corresponding homomorphism of the group $L_\beta{}^r$ onto $L_\alpha{}^r$. If $\bar{S}_\beta{}^\alpha$ is the dual homomorphism of $L_\alpha{}^r$ into $L_\beta{}^r$, $\bar{S}_\beta{}^\alpha$ induces an isomorphism $\bar{S}_\beta{}^\alpha$ of the group $\nabla^r(K_\alpha)$ onto $\nabla^r(K_\beta)$.*

§5. The complexes $K(R, \epsilon)$. Small displacements in polyhedra and compacta. The Pflastersatz and the invariance of the Betti numbers

§5.1. The complex $K(R, \epsilon)$; ϵ-chains of a metric space R. Let R be a metric space and ϵ a positive number. We shall call every finite set of points of R whose diameter is $<\epsilon$ an ϵ-skeleton or an ϵ-simplex of R. It is clear that the ϵ-skeletons of a space R form an unrestricted skeleton complex; we denote this complex by $K(R, \epsilon)$; the chains of $K(R, \epsilon)$ will be referred to as ϵ-chains of the metric space R. We may define the ϵ-chains of R as follows: *an ϵ-chain of the metric space R is a finite linear form $x^r = \sum c_i t_i^r$, where the t_i^r are oriented ϵ-simplexes of R and the c_i are elements of a given coefficient domain $\mathfrak{A}$.*

The cycles of the complex $K(R, \epsilon)$ will be called ϵ-cycles of the metric space R. In other words, *an ϵ-chain $z^r = \sum c_i t_i^r$ is said to be an ϵ-cycle of R if $\Delta z^r = \sum c_i \Delta t_i^r = 0$.*

If $\epsilon < \epsilon'$, every ϵ-skeleton is at the same time an ϵ'-skeleton, so that $K(R, \epsilon) \subseteq K(R, \epsilon')$.

If an ϵ-cycle z^r is homologous to zero in $K(R, \epsilon')$, it is said to be *ϵ'-homologous to zero in R.*

Otherwise stated, we say that *an ϵ-cycle z^r of R is ϵ'-homologous to zero in R and write*

$$z^r(\epsilon'\text{-}\sim)\, 0 \quad \textit{in } R$$

if there exists an ϵ'-chain x^{r+1} of R with z^r as its ϵ-boundary: $\Delta x^{r+1} = z^r$.

Furthermore, the ϵ-chains of R [since they are, by definition, chains of the complex $K(R, \epsilon)$] form a group $L^r[K(R, \epsilon), \mathfrak{A}]$; hence addition and subtraction of ϵ-chains and, in particular, of ϵ-cycles requires no further explanation. In connection with this, we note that two ϵ-cycles z_1^r and z_2^r are said to be ϵ'-homologous in R:

$$z_1^r(\epsilon'\text{-}\sim)z_2^r \quad \text{in } R$$

if their difference is ϵ'-homologous to zero in R.

§5.2. ϵ-displacements. Let M be a subset of a metric space R. Suppose that every point $x \in M$ is assigned a point $y = S(x) \in R$ subject to the single condition that the distance between x and $S(x)$ is less than a prescribed $\epsilon > 0$. A mapping S of the set M in R satisfying this condition is called an ϵ-displacement of M in R. We make no assumptions about the continuity of the mapping S.

Now suppose that $\epsilon > 0$, $\epsilon' > 0$ and that S is an ϵ'-displacement of M in R. Then if $p, p' \in M$ and $q = S(p)$, $q' = S(p')$, we have

$$\rho(q, q') \leq \rho(q, p) + \rho(p, p') + \rho(p', q') < \rho(p, p') + 2\epsilon.$$

Hence S maps every ϵ-skeleton of the subspace $M \subseteq R$ into an $(\epsilon + 2\epsilon')$-skeleton of R, that is, S induces a simplicial mapping of the complex $K(R, \epsilon)$ into the complex $K(R, \epsilon + 2\epsilon')$. This mapping is called an ϵ'-displacement of the complex $K(R, \epsilon)$ and is also denoted by S.

The simplicial mapping S maps every ϵ-chain

$$x^r = \sum c_i t_i^r \in L^r[K(R, \epsilon)]$$

into an $(\epsilon + 2\epsilon')$-chain

$$Sx^r = \sum c_i S t_i^r \in L^r[K(R, \epsilon + 2\epsilon')];$$

if $t_i^r = | e_0 \cdots e_r |$, then $St_i^r = | Se_0 \cdots Se_r |$. The chain Sx^r is called an ϵ'-displacement of the chain x^r.

§5.3. Canonical displacements. Let Φ be a compactum (a case which occurs particularly often is that of Φ a closed bounded subset of the Euclidean space R^n).

Suppose that M is a subset of Φ, that

$$\alpha = \{A_1, \cdots, A_s\}, \qquad A_1 \cup \cdots \cup A_s = \Phi \tag{5.31}$$

is a closed ϵ-covering of Φ, and that K_α is the nerve of α realized in Φ; or if $\Phi \subset R^n$, we shall assume that K_α is realized in a prescribed neighborhood of Φ. Then the vertex e_i of the nerve K_α corresponding to the element A_i of the covering α is a point of A_i, or a point of R^n, whose distance from A_i is less than some prescribed $\delta > 0$, which we will in all cases take to be so small that it is a Lebesgue number of α, so that $\delta(A_i) + \delta < \epsilon$ for every i [$\delta(A_i)$ is the diameter of A_i] (see IV, 2.1, Remark 4 and I, Def. 8.33).

We shall now define an ϵ-displacement S_α^Φ of a set $M \subseteq \Phi$ as follows. Let p be an arbitrary point of M and let A_i be an element of α which contains p (if there are several sets A_i with this property, choose a definite one). If e_i is the vertex of the nerve corresponding to A_i, set

$$q = S_\alpha^\Phi(p) = e_i .$$

Since both p and $e_i = S_\alpha^\Phi(p)$ are contained in the same set of diameter $< \epsilon$, S_α^Φ is an ϵ-displacement of M. The ϵ-displacement obtained in this way is called a *canonical displacement of* $M \subseteq \Phi$ *relative to the covering* α of the compactum Φ.

REMARK 1. In order to completely define a canonical displacement of $M \subseteq \Phi$ relative to a given covering α of Φ, it is necessary first to choose a definite realization of the nerve of the covering α and secondly to choose for each point $p \in M$ (which is contained in more than one set A_i) a definite set A_i containing p.

THEOREM AND DEFINITION 5.31. *Let* S_α^Φ *be a canonical displacement of a compactum* Φ *relative to a closed* ϵ*-covering* α *of* Φ *and let* δ *be a Lebesgue number of* α. *Then* S_α^Φ *induces a simplicial mapping* S_α^Φ *of the complex* $K(\Phi, \delta)$ *into the nerve* K_α *of* α.

The simplicial mapping S_α^Φ, *as well as the homomorphism of the group* $L^r[K(\Phi, \delta)]$ *into the group* $L^r(K_\alpha)$ *induced by it, will be referred to as a canonical displacement relative to the covering* α.

To prove Theorem 5.31 it is enough to note that if $| t^r | = (p_0 \cdots p_r)$ is a δ skeleton in Φ and $p_k \in A_{i(k)}$, then by the definition of the number δ, $A_{i(0)} \cap \cdots \cap A_{i(r)} \neq 0$. Consequently, the vertices $e_{i(0)} = S_\alpha^\Phi p_0, \cdots, e_{i(r)} = S_\alpha^\Phi p_r$ determine a simplex $S_\alpha^\Phi | t^r |$ of K_α.

REMARK 2. Since

$$S_\alpha^\Phi\{Z^r[K(\Phi, \delta)]\} \subseteq Z^r(K_\alpha), \qquad S_\alpha^\Phi\{H^r[K(\Phi, \delta)]\} \subseteq H^r(K_\alpha)$$

(see VII, Theorem 5.31), a canonical displacement maps every δ-cycle of a compactum into a cycle of the complex K_α and every δ-cycle, δ-homologous to zero, into a cycle homologous to zero in the complex K_α.

§5.4. The numbers $\eta(K)$. Canonical displacements in polyhedra. Let Φ be a polyhedron and let K_α be a triangulation of Φ:

$$\Phi = \| K_\alpha \|.$$

We denote by α a barycentric covering dual to the triangulation K_α (see IV, 5.3).

Whenever it is convenient we shall call a canonical displacement of $M \subseteq \| K_\alpha \|$ relative to a covering α a *canonical displacement relative to the complex* K_α. In the same way, Lebesgue numbers of α will be called Lebesgue numbers of K_α.

We recall a property of the barycentric covering α:

Let $T_{\alpha i} \in K_\alpha$; the closed simplex $\overline{T}_{\alpha i}$ intersects only those elements of α whose centers are vertices of $T_{\alpha i}$. Denote by $\epsilon(K_\alpha)$ (as in IV, Theorem 5.43) the minimum of the distances between any simplex $T_{\alpha i} \in K_\alpha$ and the union of all the elements of α whose centers are not vertices of $T_{\alpha i}$.

Finally, we denote by $\eta(K)$ any Lebesgue number of α which is less than $\epsilon(K_\alpha)$. The nerve of α is the complex K_α; consequently, a canonical displacement of Φ relative to α induces a simplicial mapping of the complex $K(\Phi, \delta)$ into K_α, where δ is a Lebesgue number of α.

If M, in particular, is the set of all vertices of a subdivision K_β of a triangulation K_α, a canonical displacement of M relative to K_α maps each vertex e_β of K_β into the center of a closed barycentric star of K_α containing this vertex, that is (see IV, Theorem 5.42), into a vertex of the carrier of e_β in K_α. Hence every canonical displacement of the set of vertices of K_β relative to K_α induces a normal displacement of K_β into K_α.

We state this result briefly as:

5.41. *Every canonical displacement of a subdivision K_β of a triangulation K_α relative to K_α is a normal displacement.*

We shall now state a stronger theorem which will find repeated application in the sequel.

FUNDAMENTAL THEOREM 5.42. *Let K_β be a subdivision of a triangulation K_α of a polyhedron $\Phi = \| K_\alpha \| = \| K_\beta \|$. Let $\eta = \eta(K_\alpha)$ and let $d_\Phi{}^\beta$ be an η-displacement of the set M of all vertices of K_β in Φ. If $S_\alpha{}^\Phi$ is a canonical displacement of Φ relative to K_α, then $S_\alpha{}^\Phi d_\Phi{}^\beta$ is a normal mapping of K_β onto K_α, so that*

$$S_\alpha{}^\Phi d_\Phi{}^\beta s_\beta{}^\alpha x_\alpha{}^r = x_\alpha{}^r \tag{5.42}$$

for an arbitrary chain $x_\alpha{}^r \in L^r(K_\alpha)$ and, in particular,

$$S_\alpha{}^\Phi d_\Phi{}^\beta s_\beta{}^\alpha t_\alpha = t_\alpha . \tag{5.420}$$

for an arbitrary oriented simplex t_α of K_α.

Proof. If $e_{\beta j}$ is a vertex of K_β, then according to the definition of $\eta(K_\alpha)$, the point $d_\Phi{}^\beta e_{\beta j}$ can be contained in only those elements of the barycentric covering α whose centers are vertices of the carrier of $e_{\beta j}$ in K_α; it follows at once that $S_\alpha{}^\Phi d_\Phi{}^\beta e_{\beta j}$ is a vertex of the carrier of $e_{\beta j}$ in K_α. This completes the proof.

§5.5. The Pflastersatz. Invariance of the Betti numbers. We shall now show that both the Pflastersatz (see V, Theorems II′ and II at the beginning of the chapter) and the equality

$$b^r(\| K_\alpha \|) = \pi^r(K_\alpha) \tag{5.51}$$

follow immediately from Theorem 5.42.

Proof of the Pflastersatz. Let K_α be a triangulation of an n-dimensional polyhedron $\Phi : \Phi = \| K_\alpha \|$. Let us set $\epsilon = \eta(K_\alpha)$ and prove that every closed ϵ-covering

$$\varphi = \{A_1, \cdots, A_\nu\}$$

of Φ has order $\geq n + 1$. We shall denote the nerve of the covering φ by K_φ.

Let δ be a Lebesgue number of the covering φ and consider a subdivision K_β of K_α of mesh $<\delta$ (the mesh of a triangulation is the maximum of the diameters of its simplexes). Denote by d_φ^β a canonical displacement of K_β relative to φ. d_φ^β is an ϵ-displacement; hence if S_α^Φ is a canonical displacement relative to K_α, the hypotheses of Theorem 5.42 are satisfied and $S_\alpha^\Phi d_\varphi^\beta$ is a normal simplicial mapping of K_β onto K_α.

Hence K_α is the image of the subcomplex $d_\varphi^\beta K_\beta$ of K_φ under S_α^Φ and it follows that the dimension number of $d_\varphi^\beta K_\beta$, and consequently of K_φ, is $\geq n$. This completes the proof.

Proof of (5.51). There exists a closed covering γ of $\Phi = \| K_\alpha \|$ of arbitrarily small mesh whose nerve K_γ satisfies the equality $\pi^r(K_\gamma) = \pi^r(K_\alpha)$. To obtain such a covering it is enough to define γ as the barycentric covering (see IV, 5.3) dual to a barycentric subdivision K_γ of K_α of sufficiently high order. Consequently

$$b^r(\|K_\alpha\|) \leq \pi^r(K_\alpha)$$

and it merely remains to be proved that

$$\pi^r(K_\varphi) \geq \pi^r(K_\alpha) \tag{5.511}$$

for every closed covering φ of $\Phi = \| K_\alpha \|$ of sufficiently small mesh (where K_φ denotes the nerve of φ). To prove this inequality, let us again set

$$\epsilon = \eta(K_\alpha)$$

and show that (5.511) holds for an arbitrary ϵ-covering.

Let K_β be a barycentric subdivision of K_α of order such that the mesh of $K_\beta < \delta$, where δ is a Lebesgue number of the prescribed ϵ-covering. As in the proof of the Pflastersatz, denote by d_φ^β a canonical displacement of K_β relative to φ and by S_α^Φ a canonical displacement relative to α. Then $S_\alpha^\Phi d_\varphi^\beta$ is a normal simplicial mapping of K_β onto K_α. Let

$$\mathfrak{z}_{\alpha 1}^r, \cdots, \mathfrak{z}_{\alpha \pi(r)}^r$$

be a set of linearly independent elements of $\Delta_0^r(K_\alpha)$ and choose arbitrary cycles $z_{\alpha i}^r \in \mathfrak{z}_{\alpha i}^r$. These cycles are *lirh*, i.e.,

$$\textstyle\sum_i c_i z_{\alpha i}^r \sim 0 \quad \text{in} \quad K_\alpha,$$

where the c_i are integers, if, and only if, all the c_i are 0.

Since $S_\alpha^\Phi d_\varphi^\beta$ is a normal simplicial mapping,

$$S_\alpha^\Phi d_\varphi^\beta s_\beta^\alpha z_{\alpha i}^r = z_{\alpha i}^r.$$

This implies that the cycles $d_\varphi^\beta s_\beta^\alpha z_{\alpha i}^r$ are *lirh* in K_φ. For, if there were a chain $x_\varphi^{r+1} \in L_0^{r+1}(K_\varphi)$ bounded by a linear combination of the cycles $d_\varphi^\beta s_\beta^\alpha z_{\alpha i}^r$:

$$\Delta x_\varphi^{r+1} = \sum_i c_i d_\varphi^\beta s_\beta^\alpha z_{\alpha i}^r,$$

it would follow that

$$\Delta x_\alpha^{r+1} = \Delta S_\alpha^\Phi x_\varphi^{r+1} = S_\alpha^\Phi \Delta x_\varphi^{r+1} = \sum_i c_i S_\alpha^\Phi d_\varphi^\beta s_\beta^\alpha z_{\alpha i}^r = \sum_i c_i z_{\alpha i}^r,$$

where $x_\alpha^{r+1} = S_\alpha^\Phi x_\varphi^{r+1}$. However, since the cycles $z_{\alpha i}^r$ are *lirh*, the relation

$$\Delta x_\alpha^{r+1} = \sum_i c_i z_{\alpha i}^r$$

implies that all the c_i are equal to zero.

Hence there are in K_φ at least $\pi = \pi^r(K_\alpha)$ cycles $d_\varphi^\beta s_\beta^\alpha z_{\alpha i}^r$ *lirh*. This proves (5.511) and therefore (5.51).

The invariance of the Euler characteristic follows from the invariance of the Betti numbers and the Euler-Poincaré formula:

5.52. *If the polyhedra* $\Phi = \| K \|$ *and* $\Phi' = \| K' \|$ *are homeomorphic, the triangulations* K *and* K' *have the same Euler characteristic.*

This may also be stated as

5.520. *All topological triangulations of a topological polyhedron or of two homeomorphic topological polyhedra have the same Euler characteristic.*

REMARK. The proof of the equality

$$b_m^r(\| K_\alpha \|) = \pi_m^r(K_\alpha), \qquad m \text{ a prime},$$

can be carried out in exactly the same way as the proof of (5.51) and is left to the reader.

§6. Invariance of the Betti groups

§6.1. In this section we shall give a proof of Theorem 1.32. This will also prove Theorems 1.33–1.35, as well as the generalization of these theorems to the case of polyhedral complexes. It follows from the formulation of Theorem 1.32 and the definition of the groups $\mathfrak{B}^r(\Phi)$ that to prove Theorem 1.32 it is sufficient to prove the following two propositions:

6.11. *If* $\Phi = \| K_\alpha \|$ *is a polyhedron, the system of groups* $\mathfrak{G}^r(\Phi)$ *is non-empty.*

6.12. *The group* $\Delta_0^r(K_\alpha)$ *is isomorphic to a subgroup of* $\mathfrak{B}$ *for every* $\mathfrak{B} \in \mathfrak{G}^r(\Phi)$.

To prove 6.11 it is enough to note that for every $\epsilon > 0$ there is a closed ϵ-covering γ with nerve K_γ such that $\Delta_0^r(K_\gamma)$ is isomorphic to $\Delta_0^r(K_\alpha)$. To construct the covering γ it suffices to choose a natural number h such that the closed barycentric stars of the barycentric subdivision $K_{\alpha h}$ of order h of K_α are of diameter $<\epsilon$. These stars form a closed ϵ-covering of Φ with nerve $K_{\alpha h}$; the group $\Delta_0^r(K_{\alpha h})$ is isomorphic to the group $\Delta_0^r(K_\alpha)$.

We shall now prove 6.12. Let $\mathfrak{B} \in \mathfrak{B}^r(\Phi)$ and put $\eta = \frac{1}{2}\eta(K_\alpha)$. Choose a closed η-covering λ of Φ such that $\mathfrak{B}$ is isomorphic to $\Delta_0^r(K_\lambda)$, with K_λ the nerve of λ.

Let K_β be a barycentric subdivision of K_α of sufficiently high order so that the mesh of K_β is less than a Lebesgue number of λ. If S_λ^β is a canonical displacement of the set of all vertices of K_β relative to λ and S_α^λ is a canonical displacement of the resulting set relative to K_α, $S_\alpha^\lambda S_\lambda^\beta$ is a normal mapping of K_β onto K_α.

Let us assign to each chain x_α^r of K_α the chain

(6.11) $$S_\lambda^\beta s_\beta^\alpha x_\alpha^r \in L_0^r(K_\lambda)$$

(where s_β^α, as usual, is the subdivision operator). The result is a homomorphism of $L_0^r(K_\alpha)$ into $L_0^r(K_\lambda)$ and an induced homomorphism (also denoted by $S_\lambda^\beta s_\beta^\alpha$) of $\Delta_0^r(K_\alpha)$ into $\Delta_0^r(K_\lambda)$. We shall show that the latter homomorphism is an isomorphism. To do this, it is enough to show that if $z_\alpha^r \in Z_0^r(K_\alpha)$ and

(6.12) $$S_\lambda^\beta s_\beta^\alpha z_\alpha^r \sim 0 \quad \text{in} \quad K_\lambda,$$

then $z_\alpha^r \sim 0$ in K_α. But (6.12) implies that there is a chain $y_\lambda^{r+1} \in L_0^r(K_\lambda)$ whose boundary is $S_\lambda^\beta s_\beta^\alpha z_\alpha^r$:

$$S_\lambda^\beta s_\beta^\alpha z_\alpha^r = \Delta y_\lambda^{r+1}.$$

Hence

$$S_\alpha^\lambda S_\lambda^\beta s_\beta^\alpha z_\alpha^r = S_\alpha^\lambda \Delta y_\lambda^{r+1}.$$

Since $S_\alpha^\lambda S_\lambda^\beta$ is a normal mapping of K_β onto K_α, the left side of the last inequality is z_α^r, so that

$$z_\alpha^r = S_\alpha^\lambda \Delta y_\lambda^{r+1} = \Delta S_\alpha^\lambda y_\lambda^{r+1},$$

where $S_\alpha^\lambda y_\lambda^{r+1} \in L_0^{r+1}(K_\alpha)$. This completes the proof.

§6.2. Invariance of the Betti groups of polyhedral complexes. Let K be a polyhedral complex (see IV, 1.2). According to the results of VII, Addendum, the oriented elements of K form an a-complex; the Betti groups of this a-complex are, by definition, the Betti groups of the polyhedral complex K.

THEOREM 6.2. *The Betti groups of a polyhedral complex K are isomorphic to the Betti groups of an arbitrary triangulation K' of the polyhedron $\| K \|$.*

It is sufficient to prove that the Betti groups of a polyhedral complex K

are isomorphic to the Betti groups of a barycentric subdivision K_1 of K. To this end it is enough to show that if $\{T_i^r\}$ are the elements of K and if $\{U_i^r\}$ are their barycentric subdivisions, then the set $\{U_i^r\}$ is a fundamental system of subcomplexes of K_1. The last assertion will obviously follow if we show that each U_i^r is a simple combinatorial pseudomanifold.

The proof of the fact that U_i^r is a pseudomanifold can be carried out very simply (e.g., by induction over the dimension number) and may be left to the reader. The orientability of the pseudomanifold U_i^r is immediate: it suffices to choose any orientation of the convex polyhedral domain T_i^r; the corresponding orientations of all the r-simplexes of U_i^r define an orientation of U_i^r. It remains to be proved that every p-cycle $z^p \in Z_0^p(U_i^r)$ is homologous to zero in U_i^r for $0 < p < r$. Denoting by Δ_1 the boundary operator in K_1, we note that $\Delta_1 z^p$ is a $(p - 1)$-cycle z^{p-1} of the complex

$$B_i^{r-1} = | U_i^r | \setminus U_i^r$$

and that the polyhedron $\| B_i^{r-1} \|$ is homeomorphic to an $(r - 1)$-sphere. Hence

$$z^{p-1} \sim 0 \quad \text{in} \quad B_i^{r-1}.$$

Choosing a chain $x^p \in L_0^p(B_i^{r-1})$ which bounds z^{p-1}, we see that $z^p - x^p \in Z_0^p(| U_i^r |)$. Since U_i^r is a convex polyhedral domain and $| U_i^r |$ is a triangulation of this domain, $z^p - x^p \sim 0$ in $| U_i^r |$, so that there is a chain

$$x^{p+1} \in L_0^{p+1}(| U_i^r |)$$

which bounds $z^p - x^p$. Hence

$$\Delta_i x^{p+1} = z^p,$$

where Δ_i is the boundary operator in U_i^r. This completes the proof.

§7. Invariance of pseudomanifolds

§7.1. Formulation of the theorems. In this section we shall prove

7.11. *If a triangulation K of a polyhedron Φ is a combinatorial pseudomanifold, every triangulation of every polyhedron homeomorphic to Φ is also a combinatorial pseudomanifold.*

This theorem leads naturally to the following definition:

7.12. *A polyhedron Φ is said to be a pseudomanifold if some (and hence an arbitrary) triangulation of Φ is a combinatorial pseudomanifold.*

To prove Theorem 7.11, we make the following definition:

7.13. A point of an n-dimensional polyhedron Φ is called a *regular* point of Φ if it has a neighborhood homeomorphic to R^n; in the contrary case, the point is said to be *singular*.

This definition implies that the set of all regular points of Φ is open in Φ; hence the set of singular points is closed.

Theorem 7.11 is obviously contained in the following proposition:

7.14. *If at least one triangulation of an n-dimensional polyhedron Φ is a combinatorial pseudomanifold, then*

1°) *Φ is a strongly connected compactum* (VI, 5.2);

2°) *the dimension of the set of singular points of Φ is $\leq n - 2$.*

Conversely, if a polyhedron Φ satisfies Conditions 1°) *and* 2°), *then every triangulation of Φ is a combinatorial pseudomanifold.*

On the basis of Theorem 7.14 we may give Def. 7.12 of an n-dimensional pseudomanifold the following so called invariant form:

7.120. *A polyhedron is called an n-dimensional pseudomanifold if it is a strongly connected compactum of dimension n and its set of singular points has dimension $\leq n - 2$.*

We note finally that the invariance of the Betti groups (Theorem 1.33) and the definition of the orientability of a combinatorial pseudomanifold (VIII, 3.2) imply

7.15. *If a triangulation of a polyhedron Φ is an orientable (a nonorientable) combinatorial pseudomanifold, then every triangulation of Φ is an orientable (a nonorientable) combinatorial pseudomanifold.*

Hence we have only to prove Theorem 7.14.

§7.2. Proof of Theorem 7.14. Suppose that the triangulation K of the polyhedron Φ is an n-dimensional combinatorial pseudomanifold. Then K is strongly connected and consequently Φ is also strongly connected (VI, Theorem 5.251). Moreover, all points of Φ whose carriers in K are of dimension $\geq n - 1$ are obviously regular points of Φ; hence the singular points of Φ form a closed set of dimension $\leq n - 2$. This proves the first half of Theorem 7.14.

Now suppose that an n-dimensional polyhedron Φ satisfies Conditions 1° and 2° of Theorem 7.14; we must then show that an arbitrary triangulation K of Φ is an n-dimensional combinatorial pseudomanifold.

Since Φ is a strongly connected polyhedron, K is a strongly connected complex (VI, Theorem 5.252). It remains to be proved that every $(n - 1)$-simplex T^{n-1} of K is a face of exactly two n-simplexes of K. This is a consequence of

7.21. *If an $(n - 1)$-simplex T^{n-1} of an n-dimensional triangulation K is a face of only one or of at least three n-simplexes of K, then every point $p \in T^{n-1}$ is a singular point of $\| K \|$.*

Proof. We shall consider two cases.

1°. Suppose that T^{n-1} is a face of exactly one n-simplex $T^n \in K$; let V^n be the carrying n-plane of T^n. If $p \in T^{n-1}$ had a neighborhood $U^n \subset V^n$ relative to K, homeomorphic to R^n, then U^n would be open in V^n (V,

Theorem 3.14) and hence p would be an interior point relative to V^n of the simplex $\overline{T}^n$. This is obviously impossible.

2°. Suppose T^{n-1} is a face of T_1^n, T_2^n, $T_3^n \in K$ (and perhaps of still other n-simplexes of K).

The set $\Gamma^n = T_1^n \cup T^{n-1} \cup T_2^n$ is obviously homeomorphic to R^n. Let V^n be a neighborhood of $p \in T^{n-1}$ relative to $\| K \|$ and homeomorphic to R^n; then, since $\rho(p, \Phi \setminus V^n)$ is positive, there exists a similitude with center p of the set Γ^n into a subset $U^n \subseteq \Gamma^n \cap V^n$. Then U^n is open in V^n (V, Theorem 3.140) and hence also in $\| K \|$. This is a contradiction, since p is a limit point of T_3^n and T_3^n and U^n are disjoint.

Chapter XI

THE Δ GROUPS OF COMPACTA

§1. Definition of the groups $\Delta^r(\Phi, \mathfrak{A})$

§1.1. Proper cycles. Let Φ be a compactum. A sequence

$$z_1^r, z_2^r, \cdots, z_h^r, \cdots,$$

where z_h^r is an r-dimensional δ_h-cycle (see X, 5.1, 5.2) of Φ over $\mathfrak{A}$ is called a *proper* (or *true*) r-cycle of Φ over $\mathfrak{A}$ if the following conditions are satisfied:

1°. $$\lim_{h\to\infty} \delta_h = 0.$$

2°. For every $\epsilon > 0$ there exists a natural number $h(\epsilon)$ such that

$$z_{h'}^r \ (\epsilon\text{-}\sim)\ z_{h''}^r \quad \text{for} \quad h' > h(\epsilon), \qquad h'' > h(\epsilon).$$

Proper cycles will be designated by

$$(1.11) \qquad \mathfrak{z}^r = (z_1^r, z_2^r, \cdots, z_h^r, \cdots).$$

The sum of two proper cycles $\mathfrak{z}_1^r = (z_{11}^r, z_{12}^r, \cdots, z_{1h}^r, \cdots)$, $\mathfrak{z}_2^r = (z_{21}^r, z_{22}^r, \cdots, z_{2h}^r, \cdots)$ is the proper cycle $\mathfrak{z}^r = (z_{11}^r + z_{21}^r, z_{12}^r + z_{22}^r, \cdots, z_{1h}^r + z_{2h}^r, \cdots)$.

This definition of addition converts the set of all proper r-cycles of Φ over the coefficient domain $\mathfrak{A}$ into a group $Z^r(\Phi, \mathfrak{A})$. Indeed, the addition defined above is associative; the zero or null proper cycle (the identity) is obtained by putting $z_i^r = 0$ in (1.11); and the cycle inverse to (1.11) is

$$-\mathfrak{z}^r = (-z_1^r, -z_2^r, \cdots, -z_h^r, \cdots),$$

which therefore satisfies the condition

$$\mathfrak{z}^r + (-\mathfrak{z}^r) = 0.$$

A proper cycle (1.11) is said to be *homologus to zero* ($\mathfrak{z}^r \sim 0$) *in* Φ if for every $\epsilon > 0$ there is an $h(\epsilon)$ such that

$$z_h^r(\epsilon\text{-}\sim)0 \text{ in } \Phi$$

for all $h > h(\epsilon)$.

Two proper cycles $\mathfrak{z}_1^r$ and $\mathfrak{z}_2^r$ are said to be homologous if their difference is homologous to zero.

If the proper cycles

$$\mathfrak{z}_1^r = (z_{11}^r, z_{12}^r, \cdots, z_{1h}^r, \cdots), \qquad \mathfrak{z}_2^r = (z_{21}^r, z_{22}^r, \cdots, z_{2h}^r, \cdots)$$

are each homologous to zero in Φ, then their sum is also homologous to zero in Φ.

Indeed, for arbitrary $\epsilon > 0$ and $h > h_1(\epsilon)$, $h > h_2(\epsilon)$ there are ϵ-chains x_{1h}^{r+1} and x_{2h}^{r+1} which are bounded by the cycles z_{1h}^r and z_{2h}^r, respectively. Hence, if $h > h_1(\epsilon)$, $h_2(\epsilon)$,

$$\Delta(x_{1h}^{r+1} + x_{2h}^{r+1}) = z_{1h}^r + z_{2h}^r,$$

and this proves the assertion.

Moreover, since the null proper cycle is homologous to zero and since $\mathfrak{z}^r \sim 0$ implies that $(-\mathfrak{z}^r) \sim 0$, it follows that the proper r-cycles homologous to zero in Φ form a subgroup $H^r(\Phi, \mathfrak{A})$ of $Z^r(\Phi, \mathfrak{A})$.

Definition 1.1. The factor group

$$\Delta^r(\Phi, \mathfrak{A}) = Z^r(\Phi, \mathfrak{A})/H^r(\Phi, \mathfrak{A})$$

is called the *r-dimensional* Δ*-group* (rth homology group) or simply the Δ^r-group of the compactum Φ over the coefficient domain $\mathfrak{A}$; the elements of the group $\Delta^r(\Phi, \mathfrak{A})$ are the homology classes of Φ, that is, classes of homologous proper r-cycles.

Remark 1. Since a proper cycle is defined as a sequence of δ_h-cycles z_h^r, the term "subsequence of a proper cycle" needs no explanation. It is also quite natural to refer to the cycles z_h^r as the "members" or elements of the proper cycle (1.11).

It is clear that every subsequence of a proper cycle is itself a proper cycle. Condition 2° of the definition of a proper cycle immediately implies that every subsequence of a proper cycle $\mathfrak{z}^r$ is a proper cycle homologous to $\mathfrak{z}^r$.

Remark 2. If $\Phi' \subseteq \Phi$, every proper cycle of Φ' is at the same time a proper cycle of Φ; but a proper cycle (1.11) of Φ need not be a proper cycle of Φ', even if $\Phi' \subset \Phi$ contains the vertices of all the cycles z_h^r. [The vertices of a δ-chain $x^r = \sum c_i t_i^r$ are the vertices of the simplexes t_i^r appearing in the linear form $x^r = \sum c_i t_i^r$ with nonvanishing coefficients (see VII, 5.2, Remark 3).]

The following definition therefore makes sense:

Let (1.11) be a proper cycle of a compactum Φ; every closed subset $\Phi' \subseteq \Phi$ such that (1.11) is also a proper cycle of Φ' is called a *carrier* of (1.11) in Φ. Examples illustrating the concept of proper cycle and, in particular, the meaning of Remark 2 will be given in the following section.

§2. Lemmas on ϵ-displacements and ϵ-homologies

[See X, 5.1 and 5.2.]

§2.1. Prisms and ϵ-displacements. Let R be a metric space and let x^r be an ϵ-chain of R. Suppose that S is an ϵ'-displacement of the set of vertices of the chain x^r transforming x^r into the chain $x'^r = Sx^r$. Having enumerated the vertices of x^r in a definite order, let us consider the prism

$K_{[01]}$ over the skeleton complex $K = |x^r|$ (see IV, 2.4 and VII, 5.2, Remark 3). We shall define a mapping S' of the set of all vertices of $K_{[01]}$ into R as follows: if e_i is a vertex of the lower base of $K_{[01]}$ (that is, a vertex of the chain x^r), we shall put $S'e_i = e_i$; but if e_i' is a vertex of the upper base of $K_{[01]}$ and e_i is the corresponding vertex of the lower base, then we shall set $S'e_i' = Se_i$.

If $(e_0' \cdots e_k' e_k \cdots e_r)$ is any skeleton of the prism $K_{[01]}$, its image under the mapping S' is obviously the set $\{Se_0, \cdots, Se_k, e_k, \cdots, e_r\}$ whose diameter is equal to the maximum of the numbers $\rho(e_i, e_j)$, $\rho(Se_i, Se_j)$, $\rho(e_i, Se_j)$, with e_i, e_j any two vertices of the same skeleton of K.

However,

$$\rho(e_i, e_j) < \epsilon,$$

$$(2.11) \quad \begin{cases} \rho(Se_i, Se_j) \leq \rho(Se_i, e_i) + \rho(e_i, e_j) + \rho(e_j, Se_j) \\ \qquad\qquad < \epsilon' + \epsilon + \epsilon' = \epsilon + 2\epsilon', \\ \rho(e_i, Se_j) \leq \rho(e_i, e_j) + \rho(e_j, Se_j) < \epsilon + \epsilon'. \end{cases}$$

Hence S' maps every skeleton of $K_{[01]}$ into an $(\epsilon + 2\epsilon')$-skeleton of R, i. e., S' is a simplicial mapping of $K_{[01]}$ into $K(R, \epsilon + 2\epsilon')$.

The simplicial mapping S' of $K_{[01]}$ into $K(R, \epsilon + 2\epsilon')$ maps the prism $x_{[01]}^{r+1}$ over the chain x^r (see VII, 9.3) into an $(\epsilon + 2\epsilon')$-chain which we denote by $\Pi_S x^r$ and call the prism spanned by the chain x^r and its ϵ'-displacement $x'^r = Sx^r$. In the same way S' maps the prism $z_{[01]}^r$ over the chain $z^{r-1} = \Delta x^r$ (in the complex $K_{[01]}$) into the prism $\Pi_S \Delta x^r$ spanned by Δx^r and $S\Delta x^r$.

From VII, (9.34), we now have

$$(2.12) \qquad \Delta \Pi_S x^r = x^r - x'^r - \Pi_S \Delta x^r.$$

In particular, if x^r is an ϵ-cycle, then $\Pi_S \Delta x^r = 0$, so that (2.12) reduces to

$$(2.120) \qquad \Delta \Pi_S x^r = x^r - x'^r,$$

which yields the following important proposition:

2.12. *Every ϵ-cycle of a metric space R is $(\epsilon + 2\epsilon')$-homologous in R to every one of its ϵ'-displacements.*

REMARK 1. Let Φ be a compactum and let α be a closed ϵ-covering of Φ. Suppose that K_α is the nerve of the covering $\alpha = \{A_1, \cdots, A_s\}$, realized in Φ (or in a neighborhood of Φ if $\Phi \subset R^n$) in such a way that $\delta(a_i \cup A_i) < \epsilon$, with a_i the vertex of the nerve corresponding to $A_i \in \alpha$; and let δ be a Lebesgue number of α satisfying the condition

$$2\delta + \delta(a_i \cup A_i) < \epsilon$$

for every i. Finally, denote by S_α^Φ a canonical displacement of Φ relative to α (see X, 5.3).

We saw above that every prism spanned by a δ-chain x^r and a canonical ϵ-displacement $S_\alpha^\Phi x^r$ of x^r is a $(\delta + 2\epsilon)$-chain. However, in our case, instead of (2.11), we have the more precise estimates

$$\rho(e_i, e_j) < \delta < \epsilon,$$

$$\rho(e_i, S_\alpha^\Phi e_j) \leq \rho(e_i, e_j) + \rho(e_j, S_\alpha^\Phi e_j) < \delta + \epsilon' < \epsilon,$$

$$\rho(S_\alpha^\Phi e_i, S_\alpha^\Phi e_j) \leq \rho(S_\alpha^\Phi e_i, e_i) + \rho(e_i, e_j) + \rho(e_j, S_\alpha^\Phi e_j)$$
$$< \epsilon' + \delta + \epsilon' < \epsilon + \epsilon' < 2\epsilon,$$

where ϵ' is the maximum of the numbers $\delta(a_i \cup A_i)$.

It follows that the prism spanned by a δ-chain and a canonical displacement of the chain is a 2ϵ-chain and, therefore, in particular:

2.13. *If δ is a Lebesgue number of a closed ϵ-covering α of a compactum Φ, every δ-cycle of Φ is 2ϵ-homologous to every one of its canonical displacements relative to α.*

§2.2. The case of a polyhedron $\Phi = \| K_\alpha \|$. If K_α is a triangulation of a polyhedron Φ and K_β is a subdivision of K_α, every normal mapping (see X, 3.1) S_α^β of K_β onto K_α is obviously an ϵ-displacement, where $\epsilon = \max \{\delta(T_{\alpha i}); T_{\alpha i} \in K_\alpha\}$.

In this case then all the vertices $e_{\beta j(0)}, \cdots, e_{\beta j(r)}$ of an arbitrary simplex $T_{\beta j}$ of K_β, as well as their images $e_{\alpha i(0)} = S_\alpha^\beta e_{\beta j(0)}, \cdots, e_{\alpha i(r)} = S_\alpha^\beta e_{\beta j(r)}$, are contained in $T_{\alpha i}$, where $T_{\alpha i}$ is the carrier of $T_{\beta j}$; hence all simplexes of the form $(e_{\alpha i(0)} \cdots e_{\alpha i(k)} e_{\beta j(k)} \cdots e_{\beta j(r)})$ are ϵ-simplexes and every prism spanned by a chain $x_\beta^r \in L^r(K_\beta)$ and its normal image $S_\alpha^\beta x_\beta^r \in L^r(K_\alpha)$ is an ϵ-chain.

Therefore

2.21. *Let K_α be a triangulation of the polyhedron $\Phi = \| K_\alpha \|$ and let ϵ be the mesh of K_α (the maximum of the diameters of the simplexes of K_α). If K_β is a subdivision of K_α and S_α^β is a normal mapping of K_β onto K_α,*

$$S_\alpha^\beta z_\beta^r \ (\epsilon\text{-}\sim) \ z_\beta^r$$

in Φ for every cycle z_β^r of K_β.

Corollary. Let

$$K_{\alpha 0} = K_\alpha, K_{\alpha 1}, \cdots, K_{\alpha h}, \cdots$$

be the consecutive barycentric subdivisions of a triangulation K_α of a polyhedron Φ; the mesh of $K_{\alpha h}$ tends to zero with increasing h. If z_α^r is a cycle of K_α and $z_{\alpha 1}^r, z_{\alpha 2}^r, \cdots, z_{\alpha h}^r, \cdots$ are the consecutive subdivisions of z_α^r in $K_{\alpha 1}, K_{\alpha 2}, \cdots, K_{\alpha h}, \cdots$, then

$$\mathfrak{z}_{(\alpha)}^r = (z_\alpha^r, z_{\alpha 1}^r, z_{\alpha 2}^r, \cdots, z_{\alpha h}^r, \cdots) \tag{2.21}$$

is a proper cycle of Φ.

Indeed, if ϵ_h is the mesh of $K_{\alpha h}$, $\lim \epsilon_h = 0$ by hypothesis; and, because of 2.21,

$$z_{\alpha(h+k)}^{\ r} \ (\epsilon_h\text{-}\sim) \ z_{\alpha h}^{\ r}.$$

REMARK. If $z_\alpha^{\ r} \sim 0$ in K_α, then $\mathfrak{z}_{(\alpha)}^{\ r} \sim 0$ in Φ. For suppose that $x_\alpha^{\ r+1} \in L^{r+1}(K_\alpha)$ and $\Delta x_\alpha^{\ r+1} = z_\alpha^{\ r}$. If $x_{\alpha h}^{\ r+1}$ is the subdivision of the chain $x_\alpha^{\ r+1}$ in $K_{\alpha h}$, then $x_{\alpha h}^{\ r+1}$ is an ϵ_h-chain and moreover, according to X, Theorem 2.12,

$$\Delta x_{\alpha h}^{\ r+1} = z_{\alpha h}^{\ r}, \qquad \text{q.e.d.}$$

The corollary to Theorem 2.21 yields an unlimited supply of proper cycles of polyhedra; moreover, the study of the homology theory of polyhedra can be confined to the investigation of proper cycles of the type (2.21). This follows from the fact (proved in §4) that every proper polyhedral cycle is homologous to a proper cycle whose elements are the consecutive barycentric subdivisions of a cycle of an arbitrary triangulation K_α of the polyhedron.

FIG. 132

To conclude this section, we shall give some further examples of proper cycles.

1°. Let us write all the rational numbers in the interval $0 < \theta < 1$ in a sequence $r_1, r_2, \cdots, r_h, \cdots$.

Denote by $\bar{z}^1$ an oriented circumference and by M^2 the torus obtained by revolving $\bar{z}^1$ about an axis pp' in its plane but not intersecting it. Now consider the oriented circumference $\bar{z}_h^{\ 1}$ which is the result of turning $\bar{z}^1$ about the axis pp' through the angle $2r_h\pi$ measured from the original position of $\bar{z}^1$. Divide the circumference $\bar{z}_h^{\ 1}$ into 2^{h+2} $(h = 1, 2, 3, \cdots)$ equal parts oriented in accordance with the chosen orientation z^1 of the circumference $\bar{z}^1$. Denote the resulting 1-cycle by $z_h^{\ 1}$. Then $\mathfrak{z}^1 = (z_1^{\ 1}, z_2^{\ 1}, \cdots, z_h^{\ 1}, \cdots)$ is a proper cycle on the torus M^2 which is not homologous to zero on M^2.

2°. Consider the closed curve Φ^1 of Fig. 132 consisting of the segment $-1 \leq y \leq 1$ of the ordinate axis, the portion of the curve $y = \sin 1/x$, $0 < x \leq 1/\pi$, and any smooth arc connecting the points $(0, -1)$ and $(1/\pi, 0)$ and having no points except these two in common with the ordinate axis and the curve $y = \sin 1/x$, $0 < x \leq 1/\pi$.

Let

$$e_0, \cdots, e_s, e_{s+1} = e_0 = (1/\pi, 0)$$

be a finite sequence of points of Φ^1 cyclicly ordered counterclockwise and with

$$\rho(e_i, e_{i+1}) < \delta_h .$$

This sequence defines a δ_h-cycle of the compactum Φ^1. Constructing sequences of this sort (always starting from the same point $e_0 = (1/\pi, 0)$ and proceeding counterclockwise) for a sequence of values of δ_h which approaches zero, we obtain a proper cycle

$$\mathfrak{z}^1 = (z_1^1, \cdots, z_h^1, \cdots)$$

which is not homologous to zero in Φ^1.

The groups $\Delta^0(\Phi^1)$ and $\Delta^1(\Phi^1)$ are infinite cyclic; $\Delta^r(\Phi^1)$ is the null group for $r > 1$; the proof is left to the reader as an exercise. HINT: For every ϵ, there is an ϵ-displacement of Φ^1 which maps it into a closed polygon without multiple points.

If Φ^1 is revolved about an axis pp' which does not intersect it, the result is a ring-like surface Φ^2. The proper cycle $\mathfrak{z}^1$ constructed on $\Phi^1 \subset \Phi^2$ is not homologous to zero on Φ^2 either. We may construct on Φ^2 a proper cycle analogous to that of Example 1°. The group $\Delta^1(\Phi^2)$ is the free Abelian group of rank 2, the groups $\Delta^2(\Phi^2)$ and $\Delta^0(\Phi^2)$ are infinite cyclic; $\Delta^r(\Phi^2)$ is the null group for $r > 2$. Hence the groups $\Delta^r(\Phi^2)$ are isomorphic to the groups Δ^r of the torus (for each dimension r).

EXERCISE. Prove that $\Delta^0(\Phi)$ is infinite cyclic if Φ is a connected compactum (continuum). If Φ has n components, $\Delta^0(\Phi, \mathfrak{A})$ is the direct sum of n copies of $\mathfrak{A}$.

§3. The homomorphism of the groups $\Delta^r(\Phi)$ induced by a continuous mapping of a compactum

§3.1. The continuous image of a proper cycle. If C is a continuous mapping of a compactum X into a compactum Y, for each $\epsilon > 0$ let

$$\epsilon_C = \sup \{\rho[C(x), C(x')]; \rho(x, x') < \epsilon\}$$

(the least upper bound is to be taken over the set of all pairs of points $x, x' \in X$ for which $\rho(x, x') < \epsilon$).

The continuous mapping C maps every ϵ-skeleton of X into an ϵ_C-skeleton of Y; hence C induces a simplicial mapping (denoted by the same letter) of the complex $K(X, \epsilon)$ into the complex $K(Y, \epsilon_C)$ and this in turn induces a homomorphism of the group $L^r[K(X, \epsilon)]$ into $L^r[K(Y, \epsilon_C)]$.

This homomorphism (again denoted by C) assigns to each chain

$$x^r = \sum a_i t_i^r \in L^r[K(X, \epsilon)]$$

the chain

$$Cx^r = \sum a_i C t_i^r \in L^r[K(Y, \epsilon_C)].$$

Since the homomorphism C is induced by a simplicial mapping (see VII, 8.3), it commutes with the boundary operator Δ:

$$\Delta C x^r = C \Delta x^r. \tag{3.1}$$

It follows immediately that C maps every ϵ-cycle of X into an ϵ_C-cycle of Y; every ϵ-cycle of X, ϵ-homologous to zero in X, into an ϵ_C-cycle of Y, ϵ_C-homologous to zero in Y; and every pair of ϵ-homologous ϵ-cycles of X into a pair of ϵ_C-homologous ϵ_C-cycles of Y. Since C, as a continuous mapping of a compactum, is uniformly continuous, as ϵ approaches zero so does ϵ_C. Consequently,

If

$$\mathfrak{z}^r = (z_1^r, z_2^r, \cdots, z_h^r, \cdots) \tag{3.11}$$

is a proper cycle of X, *then*

$$(Cz_1^r, Cz_2^r, \cdots, Cz_h^r, \cdots) \tag{3.12}$$

is a proper cycle of Y; *we shall denote the proper cycle* (3.12) *by* $C\mathfrak{z}^r$ *and refer to it as the continuous image of* $\mathfrak{z}^r$ *under the continuous mapping* C.

The mapping C of the group $Z^r(X)$ into the group $Z^r(Y)$ defined in this way is a homomorphism. Furthermore, it follows from the above that if $\mathfrak{z}^r \sim 0$ in X, then $C\mathfrak{z}^r \sim 0$ in Y. Hence the homomorphism C maps

$$H^r(X) \subseteq Z^r(X)$$

into $H^r(Y)$; so that (Appendix 2, 1.1) the homomorphism C of $Z^r(X)$ into $Z^r(Y)$ induces a homomorphism (also denoted by C) of $\Delta^r(X)$ into $\Delta^r(Y)$.

If C is a topological mapping of X onto Y, the homomorphism C is an isomorphism of $Z^r(X)$ $[\Delta^r(X)]$ onto $Z^r(Y)$ $[\Delta^r(Y)]$.

Hence

3.1. *Homeomorphic compacta have isomorphic homology groups.*

§**3.2.** An immediate consequence of the definition of the homomorphism C is the proposition

3.2. *If* C_2^1 *is a continuous mapping of a compactum* Φ_1 *onto a compactum* Φ_2 *and* C_3^2 *is a continuous mapping of* Φ_2 *onto a compactum* Φ_3, *then* (see I, 1.2)

$$C_3^1 = C_3^2 C_2^1 \tag{3.2}$$

is a continuous mapping of Φ_1 *onto* Φ_3 *and* (3.2) *remains valid if* C_2^1, C_3^2, C_3^1 *are interpreted as the corresponding induced homomorphisms of* $\Delta^r(\Phi_1)$ *into* $\Delta^r(\Phi_2)$, $\Delta^r(\Phi_2)$ *into* $\Delta^r(\Phi_3)$, *and* $\Delta^r(\Phi_1)$ *into* $\Delta^r(\Phi_3)$, *respectively.*

§**3.3. Homology classification of mappings.** Two continuous mappings C_1 and C_2 of a compactum X into a compactum Y are said to be $(r, \mathfrak{A})$-*homologous* if they induce the same homomorphism $C = C_1 = C_2$ of $\Delta^r(X, \mathfrak{A})$ into $\Delta^r(Y, \mathfrak{A})$. If C_1 and C_2 are $(r, \mathfrak{A})$-homologous for a given r and all $\mathfrak{A}$

(for a given $\mathfrak{A}$ and all r), they are said to be *r-homologous* (*$\mathfrak{A}$-homologous*). Finally, if C_1 and C_2 are $(r, \mathfrak{A})$-homologous for all r and all $\mathfrak{A}$, we shall say that they are *completely homologous*. It is clear that each of these relations is reflexive, symmetric, and transitive; hence it is an equivalence relation, and partitions the set $\mathfrak{L}$ of all continuous mappings of X into Y into $(r, \mathfrak{A})$-, r-, and $\mathfrak{A}$-classes, respectively. However, an even more subtle and much more significant classification is achieved by partitioning $\mathfrak{L}$ into homology classes, i.e., into classes of completely homologous mappings.

THEOREM 3.3. *Two homotopic mappings* (see I, 7.4) *are completely homologous.*

Let C_0 and C_1 be two homotopic continuous mappings of a compactum Φ into a compactum Φ'. To prove that C_0 and C_1 are completely homologous it is enough to show that if

$$\mathfrak{z}^r = (z_1^r, z_2^r, \cdots, z_k^r, \cdots)$$

is a proper cycle of Φ, then $C_0\mathfrak{z}^r \sim C_1\mathfrak{z}^r$ in Φ'. Hence it is sufficient to prove that for every $\epsilon > 0$ there exists a $k(\epsilon)$ such that

$$C_0 z_k^r (\epsilon\text{-}\sim) C_1 z_k^r$$

for $k > k(\epsilon)$. We shall prove the last assertion. Let C_θ, $0 \le \theta \le 1$, be a deformation of C_0 into C_1. Choose a $\delta > 0$ such that $\rho(p', p'') < \delta$ in Φ and $| \theta' - \theta'' | < \delta$ imply

$$\rho(C_{\theta'}p', C_{\theta''}p'') < \epsilon/3 \qquad \text{in } \Phi', \tag{3.31}$$

and let $k(\delta)$ be sufficiently large so that z_k^r is a δ-cycle for $k > k(\delta)$.

We shall assume that $k > k(\delta)$ and write the vertices of the complex $| z_k^r |$ (see VII, 5.2, Remark 3) in a definite but arbitrary order, say, $a_1, \cdots, a_\mu$. The prism over the skeleton complex $| z_k^r |$ (see IV, 2.1 and 2.4) will be denoted by Π_k. Now divide the interval $0 \le \theta \le 1$ into equal segments

$$(0, \theta_1), (\theta_1, \theta_2), \cdots, (\theta_{s-1}, 1)$$

of length $<\delta$. Setting $\theta_0 = 0$, $\theta_s = 1$, we shall prove the relation

$$C_{\theta(i+1)} z_k^r - C_{\theta(i)} z_k^r \sim 0 \qquad \text{in } K(\Phi', \epsilon) \tag{3.33}$$

for arbitrary $i = 0, 1, \cdots, s - 1$.

Let S be the mapping of the set of all vertices a_ν and b_ν [where a_ν (b_ν) is the set of vertices of the lower (upper) base of the prism Π_k (see IV, 2.4)] defined as:

$$S(a_\nu) = C_{\theta(i)}(a_\nu),$$
$$S(b_\nu) = C_{\theta(i+1)}(a_\nu), \qquad \nu = 1, 2, \cdots, \mu.$$

Because of the choice of δ (Condition (3.31)), it is easy to see that S maps every skeleton of Π_k into an ϵ-skeleton of Φ', that is, into a skeleton of the complex $K(\Phi', \epsilon)$. Hence S is a simplicial mapping of Π_k into $K(\Phi', \epsilon)$; if we let Πz_k^r denote the prism over z_k^r in the complex Π_k (see VII, 9.3), then S maps the chain Πz_k^r into the ϵ-chain $S\Pi z_k^r$ and

$$\Delta S\Pi z_k^r = C_{\theta(i+1)} z_k^r - C_{\theta(i)} z_k^r.$$

This proves (3.33). Summing (3.33) for $i = 0, 1, \cdots, s-1$, we obtain the relation

$$C_1 z_k^r - C_0 z_k^r \sim 0 \qquad \text{in } K(\Phi', \epsilon),$$

which proves Theorem 3.3.

§3.4. Deformation of a continuous image of a proper cycle. Deformation of a proper cycle. We have just proved the following theorem:

3.4. *If C_0 and C_1 are homotopic mappings of a compactum Φ into a compactum Φ' and $\mathfrak{z}^r$ is a proper cycle of Φ, then the proper cycles $C_0\mathfrak{z}^r$ and $C_1\mathfrak{z}^r$ are homologous in Φ'.*

Related to this result is the following concept:

DEFINITION 3.41. Assume as given a compactum Φ, a closed set $\Psi \subseteq \Phi$, a proper cycle $\mathfrak{z}$ of Ψ, a continuous mapping C_0 of Ψ into a compactum Φ', and a deformation C_θ, $0 \leq \theta \leq 1$, of C_0. The family of proper cycles $C_\theta(\mathfrak{z})$ of Φ', indexed by the parameter θ, is called a *deformation of the continuous mapping C_0 of the proper cycle $\mathfrak{z}$*. If $\Psi \subseteq \Phi'$ and C_0 is the identity mapping of Ψ into Φ', this will be referred to simply as a *deformation of $\mathfrak{z}$ in Φ'*.

Theorem 3.4 then implies

3.42. *A deformation in a compactum Φ maps every proper cycle $\mathfrak{z}$ of Φ into a cycle homologous to $\mathfrak{z}$ in Φ, that is, a deformation in Φ does not take a proper cycle $\mathfrak{z}$ of Φ out of its homology class.*

§4. The fundamental theorem on the Δ^r-groups of polyhedra (The second proof of the invariance of the homology and cohomology groups)

§4.1. Fundamental Theorem 4.1. *Let K_α be a triangulation of a polyhedron Φ. Then $\Delta^r(\Phi, \mathfrak{A})$ is isomorphic to $\Delta^r(K_\alpha, \mathfrak{A})$.*

REMARK. Theorem 1.35 of Chapter X follows from Theorem 4.1.

We shall now proceed to prove Theorem 4.1.

We shall write Z_Φ^r, H_Φ^r, Δ_Φ^r, Z_α^r, H_α^r, Δ_α^r instead of $Z^r(\Phi, \mathfrak{A})$, $H^r(\Phi, \mathfrak{A})$, $\Delta^r(\Phi, \mathfrak{A})$, $Z^r(K_\alpha, \mathfrak{A})$, $H^r(K_\alpha, \mathfrak{A})$, $\Delta^r(K_\alpha, \mathfrak{A})$; and in place of $L^r[K(\Phi, \delta), \mathfrak{A}]$ we shall write $L_{\Phi,\delta}^r$, etc.

§4.2. Construction of the homomorphism S_α^Φ of Δ_Φ^r into Δ_α^r. Let α be the barycentric covering of Φ dual to the triangulation K_α and let δ be a

Lebesgue number of α. We shall use S_α^Φ to denote a canonical displacement (see X, 5.3) of Φ relative to α, as well as the induced simplicial mapping of $K(\Phi, \delta)$ into K_α and the induced homomorphism of $L_{\Phi,\delta}^r$ into L_α^r which commutes with Δ. The homomorphism S_α^Φ assigns to every chain

(4.21) $$x^r = \sum c_i t_i^r$$

of Φ the chain

(4.22) $$S_\alpha^\Phi x^r = \sum c_i S_\alpha^\Phi t_i^r;$$

it maps every δ-cycle z^r into a cycle $S_\alpha^\Phi z^r \in Z_\alpha^r$, every δ-homologous to zero δ-cycle z^r into a cycle $S_\alpha^\Phi z^r$ homologous to zero in K_α, and every pair of δ-homologous δ-cycles into a pair of homologous cycles of K_α.

If

(4.23) $$\mathfrak{z}^r = (z_1^r, z_2^r, \cdots, z_h^r, \cdots)$$

is a proper cycle of Φ, starting with some h all the z_h^r are δ-cycles which are δ-homologous to each other. Consequently, for sufficiently large h all the cycles $S_\alpha^\Phi z_h^r$ are homologous, that is, they are all contained in the same homology class $\mathfrak{z}_\alpha^r \in \Delta_\alpha^r$. We shall denote this homology class by $S_\alpha^\Phi \mathfrak{z}^r$, and use the same symbol S_α^Φ to designate the homomorphism of Z_Φ^r into Δ_α^r which assigns to every proper cycle $\mathfrak{z}^r$ of Φ the homology class $S_\alpha^\Phi \mathfrak{z}^r \in \Delta_\alpha^r$.

Since a canonical displacement S_α^Φ maps every δ-cycle, δ-homologous to zero in Φ, into a cycle of K_α, homologous to zero in K_α, the homomorphism S_α^Φ maps a proper cycle homologous to zero into the identity of Δ_α^r. Hence S_α^Φ maps every pair of homologous proper cycles of Φ into the same element of Δ_α^r. Therefore, a canonical displacement S_α^Φ induces a homomorphism (also denoted by S_α^Φ) of Δ_Φ^r into Δ_α^r.

§4.3. S_α^Φ is a mapping onto Δ_α^r. Let $\mathfrak{z}_\alpha^r$ be an element of Δ_α^r and let z_α^r be a cycle of the homology class $\mathfrak{z}_\alpha^r$. Denote by $K_{\alpha h}$ the barycentric subdivision of order h of K_α, and by $z_{\alpha h}^r$ the subdivision of the cycle z_α^r in $K_{\alpha h}$. A canonical displacement S_α^Φ applied to the vertices of $K_{\alpha h}$ induces a normal simplicial mapping of $K_{\alpha h}$ onto K_α, with $S_\alpha^\Phi z_{\alpha h}^r = z_\alpha^r$; that is, if $\mathfrak{z}^r = (z_\alpha^r, z_{\alpha 1}^r, \cdots, z_{\alpha h}^r, \cdots)$ is a proper cycle whose homology class is $\mathfrak{z}_\Phi^r \in \Delta_\Phi^r$, then

$$S_\alpha^\Phi \mathfrak{z}^r = z_\alpha^r, \qquad S_\alpha^\Phi \mathfrak{z}_\Phi^r = \mathfrak{z}_\alpha^r.$$

Hence S_α^Φ is a homomorphism *onto* Δ_α^r.

We are now ready for the last step of the proof of the invariance theorem.

§4.4. The homomorphism S_α^Φ of Δ_Φ^r onto Δ_α^r is an isomorphism. This assertion follows from the following proposition:

4.41. *If*

$$\mathfrak{z}^r = (z_1^r, z_2^r, \cdots, z_k^r, \cdots)$$

is a proper cycle of Φ *and*

$$S_\alpha^\Phi \mathfrak{z}^r \sim 0 \qquad \text{in } K_\alpha ,$$

then $\mathfrak{z}^r \sim 0$ *in* Φ, *i. e., for every* $\epsilon > 0$ *there is a* $k(\epsilon)$ *such that*

$$z_k^r(\epsilon\text{-}\sim)0 \qquad \text{in } \Phi$$

for $k \geq k(\epsilon)$.

Proof. Suppose that $2\delta_\alpha < \epsilon$ is a Lebesgue number of K_α (that is, of the closed barycentric covering dual to K_α) and choose h so that $K_{\alpha h}$ is a δ_α-complex and the corresponding barycentric covering is a δ_α-covering.

If $\delta_{\alpha h} < \delta_\alpha$ is a Lebesgue number of $K_{\alpha h}$, choose $k(\epsilon)$ so that all the z_k^r are $\delta_{\alpha h}$-cycles for $k \geq k(\epsilon)$. Assume that $k \geq k(\epsilon)$.

Then (by 2.13)

$$z_k^r(2\delta_\alpha\text{-}\sim)S_{\alpha h}^\Phi z_k^r. \tag{4.410}$$

Since $K_{\alpha h}$ is a regular subdivision of K_α, according to X, Theorem 2.35, there exists a cycle z_α^r of K_α such that

$$S_{\alpha h}^\Phi z_k^r \sim s_{\alpha h}^\alpha z_\alpha^r \qquad \text{in } K_{\alpha h} ;$$

hence

$$z_k^r(2\delta_\alpha\text{-}\sim)s_{\alpha h}^\alpha z_\alpha^r. \tag{4.411}$$

The canonical displacement S_α^Φ maps the left side of the homology (4.411) into $S_\alpha^\Phi z_k^r$ and the right side into z_α^r, so that

$$S_\alpha^\Phi z_k^r \sim z_\alpha^r \qquad \text{in } K_\alpha .$$

By assumption, $S_\alpha^\Phi z_k^r \sim 0$ in K_α ; hence $s_{\alpha h}^\alpha z_\alpha^r \sim 0$ in $K_{\alpha h}$. Since $K_{\alpha h}$ is a δ_α-complex,

$$s_{\alpha h}^\alpha z_\alpha^r(\epsilon\text{-}\sim)0 \qquad \text{in } \Phi.$$

Therefore, by (4.411),

$$z_k^r(\epsilon\text{-}\sim)0 \qquad \text{in } \Phi.$$

This completes the proof of the invariance theorem.

§4.5. Rules for finding the images of the isomorphisms S_α^Φ and $(S_\alpha^\Phi)^{-1}$. We shall formulate once more the concrete realizations of the isomorphism S_α^Φ of Δ_Φ^r onto Δ_α^r and its inverse $(S_\alpha^\Phi)^{-1}$.

First Rule. Given an element $\mathfrak{z}_\Phi^r$ of Δ_Φ^r it is required to find the corresponding element $\mathfrak{z}_\alpha^r = S_\alpha^\Phi \mathfrak{z}_\Phi^r$ of Δ_α^r. To this end, choose an arbitrary proper cycle $\mathfrak{z}^r$ of the homology class $\mathfrak{z}_\Phi^r$ and an h sufficiently large so that all the cycles z_k^r, $k > h$, are δ-homologous δ-cycles (where δ is a Lebesgue number of the covering α). The canonical displacement S_α^Φ relative to α of an arbitrary cycle z_k^r, $k > h$, is a cycle of K_α, and its homology class is the required element of Δ_α^r.

Second Rule. If $\mathfrak{z}_\alpha^r \in \Delta_\alpha^r$, it is required to find the corresponding element $\mathfrak{z}_\Phi^r = (S_\alpha^\Phi)^{-1}\mathfrak{z}_\alpha^r$ of Δ_Φ^r.

To this end, choose a cycle $z_\alpha^r \in \mathfrak{z}_\alpha^r$. Then z_α^r and its consecutive barycentric subdivisions

$$z_{\alpha 1}^r, z_{\alpha 2}^r, \cdots, z_{\alpha h}^r, \cdots$$

form a proper cycle $\mathfrak{z}^r$ whose homology class is the desired element of Δ_Φ^r.

REMARK. The preceding considerations contain as a special case the proof of an assertion made in 2.2:

Every proper cycle

$$\mathfrak{z}^r = (z_1^r, z_2^r, \cdots, z_k^r, \cdots)$$

of a polyhedron $\Phi = \| K_\alpha \|$ *is homologous to a proper cycle of the form*

$$\mathfrak{z}_{(\alpha)}^r = (z_\alpha^r, z_{\alpha 1}^r, \cdots, z_{\alpha h}^r, \cdots),$$

where

$$z_\alpha^r \in Z_\alpha^r \quad \textit{and} \quad z_{\alpha 1}^r, z_{\alpha 2}^r, \cdots, z_{\alpha h}^r, \cdots$$

are the consecutive barycentric subdivisions of z_α^r.

Indeed, suppose that a canonical displacement S_α^Φ relative to α maps $\mathfrak{z}^r$ into z_α^r. Then according to 4.3, $S_\alpha^\Phi \mathfrak{z}_{(\alpha)}^r = z_\alpha^r$. Therefore,

$$S_\alpha^\Phi(\mathfrak{z}^r - \mathfrak{z}_{(\alpha)}^r) = 0,$$

so that, by 4.4, $\mathfrak{z}^r \sim \mathfrak{z}_{(\alpha)}^r$ in Φ.

§4.6. Cycles $z^r \in Z_\alpha^r$ and homologies in $\Phi = \| K_\alpha \|$. Since the second rule of 4.5 gives a perfectly definite realization of the isomorphism between Z_α^r and Z_Φ^r, we are justified, whenever convenient, in identifying an arbitrary cycle $z_\alpha^r \in Z_\alpha^r$ with the proper cycle

$$\mathfrak{z}_{(\alpha)}^r = (z_\alpha^r, z_{\alpha 1}^r, \cdots, z_{\alpha h}^r, \cdots) \tag{4.61}$$

consisting of z_α^r and its consecutive barycentric subdivisions. In particular, it is natural to say that a cycle $z_\alpha^r \in Z_\alpha^r$ is homologous to zero in Φ if $\mathfrak{z}_{(\alpha)}^r$ is homologous to zero in Φ. In the same way we say that the cycles z_α^r and $z_\alpha'^r \in Z_\alpha^r$ are homologous in Φ, etc.

4.6. *If* $z_\alpha^r \in Z_\alpha^r$ *is homologous to zero in a polyhedron* $\Phi = \| K_\alpha \|$, *then* $z_\alpha^r \in H_\alpha^r$.

Proof. Let ϵ be a Lebesgue number of K_α. Then there exist a natural number h and an ϵ-chain x^{r+1} in Φ such that $\Delta x^{r+1} = z_{\alpha h}^r$. A canonical displacement relative to K_α maps x^{r+1} into a chain $x_\alpha^{r+1} \in L_\alpha^{r+1}$ and $z_{\alpha h}^r$ into z_α^r, with $\Delta x_\alpha^{r+1} = z_\alpha^r$.

COROLLARY 4.61. *If the cycles z_α^r, $z_\alpha'^r \in Z_\alpha^r$ are homologous in Φ, they are homologous in K_α.*

§4.7. The image of a cycle $z_\alpha^r \in Z_\alpha^r$ under a continuous mapping C of a polyhedron $\Phi = \| K_\alpha \|$ into a compactum Φ'. Parametric representation and deformation of singular cycles. Let K_α be a triangulation of a polyhedron Φ and let C be a continuous mapping of Φ into a compactum Φ'. The considerations of 4.6 lead to the following definitions:

4.71. We shall call the proper cycle

$$\mathfrak{z}^r = C(z_\alpha^r) = (Cz_\alpha^r,\ Cz_{\alpha 1}^r,\ \cdots,\ Cz_{\alpha k}^r,\ \cdots)$$

(see 3.1), where $z_{\alpha 1}^r, \cdots, z_{\alpha k}^r, \cdots$ are the consecutive barycentric subdivisions of z_α^r, the *image of the cycle* $z_\alpha^r \in Z_\alpha^r$ under the continuous mapping C of the polyhedron Φ into the compactum Φ'.

4.72. A proper cycle $\mathfrak{z}^r$ of a compactum Φ' is said to be a *singular cycle* of Φ' if it can be represented as a continuous image of a cycle $z_\alpha^r \in Z_\alpha^r$:

$$\mathfrak{z}^r = (Cz_\alpha^r,\ Cz_{\alpha 1}^r,\ \cdots,\ Cz_{\alpha k}^r,\ \cdots), \tag{4.7}$$

where K_α is a triangulation of Φ, the cycles $z_{\alpha 1}^r, z_{\alpha 2}^r, \cdots, z_{\alpha k}^r, \cdots$ are the consecutive barycentric subdivisions of z_α^r, and C is a continuous mapping of $\| K_\alpha \|$ into Φ'. The representation (4.7) is called a *parametric representation* (relative to K_α and C) of the singular cycle $\mathfrak{z}^r$.

We shall call a deformation of a continuous mapping C of a proper cycle $(z_\alpha^r, z_{\alpha 1}^r, \cdots, z_{\alpha k}^r, \cdots)$ of a polyhedron $\| K_\alpha \|$ a *parametric deformation of the singular cycle* $\mathfrak{z}^r$ (represented by the proper cycle).

Two parametric representations

$$C_0(z_\alpha^r) = (C_0 z_\alpha^r,\ C_0 z_{\alpha 1}^r,\ \cdots,\ C_0 z_{\alpha k}^r,\ \cdots),$$

$$C_1(z_\alpha^r) = (C_1 z_\alpha^r,\ C_1 z_{\alpha 1}^r,\ \cdots,\ C_1 z_{\alpha k}^r,\ \cdots)$$

are said to be homotopic if there is a parametric deformation which takes one into the other.

REMARK. In all the applications of these definitions in Chapter XVI the compactum Φ' is assumed to be a polyhedron.

§4.8. Orientability and orientation of closed pseudomanifolds. The following proposition is an immediate consequence of the isomorphism of the groups $\Delta^n(\Phi)$ and $\Delta^n(K_\alpha)$ for every polyhedron Φ and every triangulation K_α of Φ:

4.81. If Φ is an n-dimensional closed pseudomanifold, only two cases are possible: either $\Delta_0^n(\Phi) = \Delta^n(\Phi, J)$ is *infinite cyclic* or $\Delta_0^n(\Phi)$ is the *null group*; in the first case every triangulation K_α of Φ is an orientable combinatorial pseudomanifold and Φ is said to be an *orientable pseudomanifold*; in the second case every K_α is a nonorientable combinatorial pseudomanifold and Φ is called a *nonorientable pseudomanifold.*

If Φ is an orientable n-dimensional closed pseudomanifold, each one of the two generators $\pm\mathfrak{z}_\Phi^n$ of $\Delta_0^n(\Phi)$ is called an *orientation* of Φ.

If K_α^n is any triangulation of an orientable pseudomanifold Φ and $\pm\mathfrak{z}_\alpha^n$ are the two orientations of the combinatorial pseudomanifold K_α^n, the orientation $\mathfrak{z}_\alpha^n$ corresponds to that orientation of $\Phi = \| K_\alpha^n \|$ which contains the proper cycle

$$(z_\alpha^n, z_{\alpha 1}^n, \cdots, z_{\alpha k}^n, \cdots); \tag{4.81}$$

in the same way, the orientation $\mathfrak{z}_\Phi^n$ of $\Phi = \| K_\alpha^n \|$ corresponds to the orientation $\mathfrak{z}_\alpha^n$ of K_α^n which satisfies the condition

$$\mathfrak{Z}_{(\alpha)}^n = (z_\alpha^n, z_{\alpha 1}^n, \cdots, z_{\alpha k}^n, \cdots) \in \mathfrak{z}_\Phi^n.$$

Hence there is a $(1-1)$ *correspondence between the orientations of a pseudomanifold* Φ *and those of an arbitrary triangulation* K_α *of* Φ.

Every proper cycle $\mathfrak{Z}^n$ contained in a given orientation $\mathfrak{z}_\Phi^n$ of a pseudomanifold Φ is called an *orienting cycle* of Φ (more precisely, an orienting cycle defined by the given orientation of Φ). If K_α^n is any triangulation of Φ, each of the two orientations $\pm\mathfrak{z}_\alpha^n$ of K_α^n is also often referred to as an orienting cycle of Φ.

REMARK 1. Let C be a continuous mapping of an orientable closed n-dimensional pseudomanifold Φ; the image of Φ is a compactum Φ'. C maps each orientation $\mathfrak{z}_\Phi^n$ of Φ into an element $C\mathfrak{z}_\Phi^n$ of $\Delta_0^n(\Phi')$, the *image of the orientation under* C.

REMARK 2. Sometimes, the term *oriented pseudomanifold* is used instead of the expressions *orientation* and *orienting cycle* of a pseudomanifold; the former refers of course to the pair: a pseudomanifold and an orientation (or an orienting cycle) of the pseudomanifold (or most often to a pseudomanifold and an orientation of some triangulation of the pseudomanifold). If the pseudomanifold is an n-sphere, the expression *oriented sphere* is commonly used in the above sense.

§4.9. The homomorphism C_σ^α of $\Delta_\alpha^r = \Delta^r(K_\alpha, \mathfrak{A})$ into $\Delta_\sigma^r = \Delta^r(M_\sigma, \mathfrak{A})$ induced by a continuous mapping C_Ψ^Φ of a polyhedron $\Phi = \| K_\alpha \|$ into a polyhedron $\Psi = \| M_\sigma \|$. Because of the isomorphism established above between the groups Δ_Φ^r and Δ_α^r (and the groups Δ_Ψ^r, Δ_σ^r), the homomorphism C_Ψ^Φ of Δ_Φ^r into Δ_Ψ^r induced by a continuous mapping C_Ψ^Φ of Φ into Ψ defines a homomorphism of Δ_α^r into Δ_σ^r which we shall denote by C_σ^α. The mapping is easily constructed by 4.5. We shall state the rule explicitly:

Given an element $\mathfrak{z}_\alpha^r$ of the group Δ_α^r it is required to find the element

$$C_\sigma^\alpha \mathfrak{z}_\alpha^r = S_\sigma^\Psi C_\Psi^\Phi (S_\alpha^\Phi)^{-1} \mathfrak{z}_\alpha^r \tag{4.90}$$

of the group Δ_σ^r. To this end, recalling the definition of S_σ^Ψ and $(S_\alpha^\Phi)^{-1}$, letting $K_{\alpha h}$ stand for the barycentric subdivision of K_α of order h, and $s_{\alpha h}^\alpha$ for the isomorphism of Δ_α^r onto $\Delta_{\alpha h}^r$ induced by the subdivision (see X, 2.1), we may rewrite (4.90) in its final form

$$C_\sigma^\alpha \mathfrak{z}_\alpha^r = S_\sigma^\Psi C_\Psi^\Phi s_{\alpha h}^\alpha \mathfrak{z}_\alpha^r, \tag{4.9}$$

where h is a sufficiently large natural number (see below). Equation (4.9) may be stated as the following proposition:

4.9. *A continuous mapping* C_Ψ^Φ *of a polyhedron* $\Phi = \| K_\alpha \|$ *into a polyhedron* $\Psi = \| M_\sigma \|$ *induces a homomorphism* C_σ^α *of* Δ_α^r *into* Δ_σ^r *defined as follows*: *Set* (see X, 5.4) $\eta = \eta(M_\sigma)$ *and choose* $\delta > 0$ *sufficiently small so that* $\delta_C < \eta$ (see 3.1). *Now choose h sufficiently large so that the mesh of* $K_{\alpha h}$ is *less than* δ.

If $\mathfrak{z}_\alpha^r \in \Delta_\alpha^r$ *and* z_α^r *is an arbitrary cycle of* $\mathfrak{z}_\alpha^r$, *consider the subdivision* $z_{\alpha h}^r = s_{\alpha h}^\alpha z_\alpha^r$ *of* $\mathfrak{z}_\alpha^r$ *in* $K_{\alpha h}$. *The cycle* $C_\Psi^\Phi z_{\alpha h}^r$ *is an* η-*cycle of* Ψ; *a canonical displacement* S_σ^Ψ *of* $C_\Psi^\Phi z_{\alpha h}^r$ *relative to* M_σ *is a cycle* z_σ^r *of* M_σ *and its homology class is the desired element* $C_\sigma^\alpha \mathfrak{z}_\alpha^r$ *of* Δ_σ^r.

If C_Ψ^Φ is a topological mapping of a polyhedron $\Phi = \| K_\alpha \|$ onto a polyhedron $\Psi = \| M_\sigma \|$, C_σ^α is an *isomorphism* of Δ_α^r *onto* Δ_σ^r; finally, if $\Phi = \Psi$ and C_Ψ^Φ is the identity mapping, the homomorphism C_σ^α, which in this case takes the form

$$C_\sigma^\alpha \mathfrak{z}_\alpha^r = S_\sigma^\Phi s_{\alpha h}^\alpha \mathfrak{z}_\alpha^r, \tag{4.9'}$$

becomes an isomorphism of Δ_α^r onto Δ_σ^r, where K_α and M_σ are two (in general, topological) triangulations of Φ.

§5. Simplicial approximations to continuous mappings of a polyhedron into a polyhedron

§5.1. Definition of a simplicial approximation to a continuous mapping C_Ψ^Φ of a polyhedron $\Phi = \| K_\alpha \|$ into a polyhedron $\Psi = \| M_\sigma \|$. Let $\eta = \eta(M_\sigma)$ and choose a $\delta > 0$ such that $\delta_C < \eta$ (see 3.1) and a subdivision $K_{\alpha h}$ of K_α of mesh $< \delta$. The mapping C_Ψ^Φ takes every δ-skeleton of Φ into an η-skeleton of Ψ; hence, if S_σ^Ψ is a canonical displacement of Ψ relative to the covering σ, the mapping $S_\sigma^\Psi C_\Psi^\Phi$ defined on the set of vertices of $K_{\alpha h}$ induces a simplicial mapping

$$S_\sigma^\Psi C_\Psi^\Phi = S_\sigma^{\alpha h}$$

of $K_{\alpha h}$ into M_σ.

The mapping $S_\sigma^{\alpha h}$ of $K_{\alpha h}$ into M_σ, and also the simplicial mapping $\tilde{S}_\sigma^{\alpha h}$ of $\Phi = \| K_{\alpha h} \|$ into $\Psi = \| M_\sigma \|$ induced by it, is called a *simplicial* $(\alpha h, \sigma)$-*approximation* to the continuous mapping C_Ψ^Φ.

The composition of the homomorphism $s_{\alpha h}^\alpha$ (the subdivision operator) of L_α^r into $L_{\alpha h}^r$ and of the homomorphism $S_\sigma^{\alpha h}$ induced by the simplicial mapping $S_\sigma^{\alpha h}$ of $K_{\alpha h}$ into M_σ yields a homomorphism $S_\sigma^{\alpha h} s_{\alpha h}^{\alpha}$ of L_α^r into L_σ^r:

$$(5.11) \qquad S_\sigma^{\alpha h} s_{\alpha h}^{\alpha} z_\alpha^r = S_\sigma^{\Psi} C_\Psi^{\Phi} s_{\alpha h}^{\alpha} z_\alpha^r, \qquad z_\alpha^r \in L_\alpha^r.$$

Since $S_\sigma^{\alpha h}$ commutes with Δ, it induces a homomorphism $S_\sigma^{\alpha h} s_{\alpha h}^{\alpha}$ of Δ_α^r into Δ_σ^r:

$$(5.12) \qquad S_\sigma^{\alpha h} s_{\alpha h}^{\alpha} \mathfrak{z}_\alpha^r = S_\sigma^{\Psi} C_\Psi^{\Phi} s_{\alpha h}^{\alpha} \mathfrak{z}_\alpha^r.$$

Comparing (5.12) with (4.9), we conclude that

$$(5.1) \qquad C_\sigma^{\alpha} \mathfrak{z}_\alpha^r = S_\sigma^{\alpha h} s_{\alpha h}^{\alpha} \mathfrak{z}_\alpha^r,$$

that is,

5.1. *The homomorphism of Δ_α^r into Δ_σ^r induced by a continuous mapping of a polyhedron $\Phi = \| K_\alpha \|$ into a polyhedron $\Psi = \| M_\sigma \|$ is realized by taking a subdivision $K_{\alpha h}$ of K_α of sufficiently small mesh and applying to the elements of Δ_α^r first the subdivision operator $s_{\alpha h}^{\alpha}$ and then any simplicial $(\alpha h, \sigma)$-approximation to the continuous mapping.*

§5.2. Fundamental property of $\tilde{S}_\sigma^{\alpha h}$.

THEOREM 5.21. *A simplicial approximation $\tilde{S}_\sigma^{\alpha h}$ (of $\Phi = \| K_\alpha \| = \| K_{\alpha h} \|$ into $\Psi = \| M_\sigma \|$) to a continuous mapping C_Ψ^{Φ} is homotopic to C_Ψ^{Φ}.*

LEMMA 5.210. *Suppose that C_0 and C_1 are two continuous mappings of a compactum Φ into a polyhedron $\Psi = \| M_\sigma \|$ with the following property: if $x \in \Phi$, both points C_0x and C_1x are contained in the closure of a simplex of M_σ. Then C_0 is homotopic to C_1.*

Proof of Lemma 5.210. The polyhedron Ψ is contained in some R^n. Every closed simplex is a convex set; hence, by hypothesis, for every $x \in \Phi$, the straight line segment $[C_0x, C_1x]$ joining the points C_0x and C_1x in R^n is contained in Ψ. Let us denote by $C_\theta x$, $0 < \theta < 1$, the point of $[C_0x, C_1x]$ which divides this segment in the ratio $\theta : (1 - \theta)$ in the direction from C_0x to C_1x. The resulting deformation $C_\theta x$ maps C_0 into C_1.

Proof of Theorem 5.21. According to Lemma 5.210, it is enough to show that if $p \in \Phi$, the carrier of the point $\tilde{S}_\sigma^{\alpha h} p$ in M_σ is a face of the carrier of the point $C_\Psi^{\Phi} p$.

Suppose $p \in T_{\alpha h} \in K_{\alpha h}$, $T_{\alpha h} = (a_0 \cdots a_r)$, and $C_\Psi^{\Phi} p \in T_\sigma \in M_\sigma$. Since the diameter of the set $C_\Psi^{\Phi} \overline{T}_{\alpha h}$ is less than η and $C_\Psi^{\Phi} \overline{T}_{\alpha h}$ is known to have the point $C_\Psi^{\Phi} p$ in common with T_σ, the set $C_\Psi^{\Phi} \overline{T}_{\alpha h}$ is contained in the union of the elements of the covering σ (the barycentric covering dual to M_σ) whose centers are the vertices of T_σ and which do not intersect the remaining elements of σ. Therefore the points $S_\sigma^{\Psi} C_\Psi^{\Phi} a_0 = a_0'$, $\cdots$, $S_\sigma^{\Psi} C_\Psi^{\Phi} a_r = a_r'$ are vertices of the simplex T_σ so that the simplex

$$S_\sigma^{\alpha h} T_{\alpha h} = (a_0' \cdots a_r')$$

is a face of T_σ. This is what we wished to prove. Hence

5.2. *A continuous mapping of a polyhedron* Φ *into a polyhedron* Ψ *is homotopic, and consequently homologous, to every simplicial approximation to the mapping.*

A consequence of the fact that the carrier of the point $\tilde{S}_\sigma^{\alpha h} p$ in M_σ is a face of the carrier of the point $C_\Psi^{\ \Phi} p$ in the same complex is the following proposition:

5.22. *If the mesh of* M_σ *is less than a prescribed* $\epsilon > 0$, *then*

$$\rho(C_\Psi^{\ \Phi} p, \tilde{S}_\sigma^{\alpha h} p) < \epsilon$$

for arbitrary $p \in \Phi$.

For a triangulation M_σ of sufficiently small mesh, we conclude from 5.22 that:

5.23. *Let* C *be a continuous mapping of a polyhedron* $\Phi = \| K_\alpha \|$ *into a polyhedron* Ψ. *For every* $\epsilon > 0$ *there exists a simplicial approximation* $\tilde{S}_\sigma^{\alpha h}$ *to* C *(which is a simplicial mapping of some subdivision* $K_{\alpha h}$ *of* K_α*) of* K_α *into a triangulation* M_σ *of* Ψ *of sufficiently small mesh such that*

$$\rho(Cp, \tilde{S}_\sigma^{\alpha h} p) < \epsilon$$

for every $p \in \Phi$.

§6. Degree of a continuous mapping of closed pseudomanifolds

In this section Φ and Ψ are closed n-dimensional pseudomanifolds, assumed to be orientable. C denotes a continuous mapping of Φ into Ψ; K_α and M_σ are arbitrary triangulations of Φ, Ψ, respectively.

§6.1. Definition of the degree. Choose definite orientations $\mathfrak{z}_\Phi$ and $\mathfrak{z}_\Psi$ of Φ and Ψ. Then $\Delta_\Phi^{\ n}$ consists of the elements $m\mathfrak{z}_\Phi$ and $\Delta_\Psi^{\ n}$ of the elements $m\mathfrak{z}_\Psi$, where m is an arbitrary integer. The homomorphism C of $\Delta_\Phi^{\ n}$ into $\Delta_\Psi^{\ n}$ induced by the mapping C yields

$$C(\mathfrak{z}_\Phi) = \gamma \mathfrak{z}_\Psi ,$$

with γ an integer, known as the *degree* of the continuous mapping C.

Remark. The number γ is completely determined by the mapping C and the orientations of Φ and Ψ; replacing one of these orientations by its opposite changes the sign of γ.

§6.2. Definition of the degree of a continuous mapping of an n-cycle into an n-dimensional orientable pseudomanifold. The following notion is an immediate generalization of Def. 6.1.

Suppose that C is a continuous mapping of a compactum Φ into an n-dimensional orientable pseudomanifold Ψ. Let us choose a definite orientation $\mathfrak{z}_\Psi$ of Ψ, that is, a definite generator of the infinite cyclic group $\Delta_\Psi^{\ n}$. Suppose that z^n is a proper n-cycle of Φ mapped by C into the proper cycle Cz^n of Ψ and that Cz^n is contained in the homology class $\gamma\, \mathfrak{z}_\Psi^{\ n} \in \Delta_\Psi^{\ n}$;

the integer γ is called the *degree of the mapping of the proper cycle* $\mathfrak{z}^n$ *into* Ψ

In the majority of applications of this definition, the compactum Φ will be a polyhedron and the proper cycle $\mathfrak{z}^n$ will be a cycle of a triangulation of Φ in the sense of 4.6; consequently, $C\mathfrak{z}^n$ is a singular cycle.

§6.3. Calculation of the degree of a mapping. The isomorphism S_α^Φ maps a generator $\mathfrak{z}_\Phi$ of Δ_Φ^n into a generator $\mathfrak{z}_\alpha = z_\alpha^n$ of the cyclic group $\Delta_\alpha^n = Z_\alpha^n$ and it is always possible to choose the orientations $t_{\alpha i}$ of the simplexes of K_α so that

$$z_\alpha^n = \sum_i t_{\alpha i}^n.$$

In the same way one may choose the orientations $t_{\sigma i}^n$ of the n-simplexes of M_σ so as to satisfy the relations

$$S_\sigma^\Psi \mathfrak{z}_\Psi = \mathfrak{z}_\sigma = z_\sigma^n = \sum_i t_{\sigma i}^n$$

for the chosen orientation $\mathfrak{z}_\Psi$ of Ψ.

Then the cycles

$$\mathfrak{z}_{(\alpha)} = (z_\alpha^n, z_{\alpha 1}^n, \cdots, z_{\alpha k}^n, \cdots)$$

$$\mathfrak{z}_{(\sigma)} = (z_\sigma^n, z_{\sigma 1}^n, \cdots, z_{\sigma k}^n, \cdots)$$

are contained in the homology classes $\mathfrak{z}_\Phi$ and $\mathfrak{z}_\Psi$, respectively, and

$$C_\Psi^\Phi \mathfrak{z}_{(\alpha)} \sim \gamma \mathfrak{z}_{(\sigma)} = (\gamma z_\sigma^n, \gamma z_{\sigma 1}^n, \cdots, \gamma z_{\sigma k}^n, \cdots).$$

Furthermore, $(S_\alpha^\Phi)^{-1}\mathfrak{z}_\alpha = \mathfrak{z}_\Phi$; but $\mathfrak{z}_\Phi$ is the homology class of the proper cycle $\mathfrak{z}_{(\alpha)} = (z_\alpha^n, z_{\alpha 1}^n, \cdots, z_{\alpha k}^n, \cdots)$, mapped by C_Ψ^Φ into the homology class of the proper cycle

$$\gamma \mathfrak{z}_{(\sigma)} = (\gamma z_\sigma^n, \gamma z_{\sigma 1}^n, \cdots, \gamma z_{\sigma k}^n, \cdots).$$

A canonical displacement S_σ^Ψ maps the latter homology class into the class containing $\gamma\mathfrak{z}_\sigma = \gamma z_\sigma^n$. Hence $S_\sigma^\Psi C_\Psi^\Phi (S_\alpha^\Phi)^{-1} z_\alpha^n = \gamma z_\sigma^n$, that is, by (4.90),

$$C_\sigma^\alpha z_\alpha^n = \gamma z_\sigma^n.$$

Therefore,

6.31. *If* C_Ψ^Φ *is a continuous mapping of an oriented pseudomanifold* Φ *into an oriented pseudomanifold* Ψ, *the homomorphism* C_Ψ^Φ *of* Δ_Φ^n *into* Δ_Ψ^n *and the homomorphism* C_σ^α *of* Δ_α^n *into* Δ_σ^n *induced by* C_Ψ^Φ *are defined in accordance with the formulas*

$$C_\Psi^\Phi \mathfrak{z}_\Phi = \gamma \mathfrak{z}_\Psi, \tag{6.31$_1$}$$

$$C_\sigma^\alpha z_\alpha^n = \gamma z_\sigma^n; \tag{6.31$_2$}$$

where

$$z_\alpha^n = \sum_i t_{\alpha i}^n, \qquad z_\sigma^n = \sum_i t_{\sigma i}^n$$

are the orientations of K_α *and* M_σ *corresponding to the orientations* $\mathfrak{z}_\Phi$ *and* $\mathfrak{z}_\Psi$ *and* γ *is the degree of the mapping* $C_\Psi{}^\Phi$.

REMARK 1. We adopted $(6.31)_1$ as the definition of the degree γ of the mapping C; we see now that it would be equally valid to take $(6.31)_2$ as the definition of γ.

Hence

6.32. *Suppose that* K_α, M_σ *are arbitrary triangulations of* Φ, Ψ; *that* $K_{\alpha h}$ *is a sufficiently fine subdivision of* K_α; *and that* $S_\sigma{}^{\alpha h}$ *is an* $(\alpha h, \sigma)$-*simplicial approximation to the continuous mapping* $C_\Psi{}^\Phi$.

If the orientations of K_α *and* Φ, *as well as those of* M_σ *and* Ψ, *correspond to each other, the degree of the mapping* $C_\Psi{}^\Phi$ *is equal to the degree of the simplicial approximation* $S_\sigma{}^{\alpha h}$.

REMARK 2. Theorem 6.32 is a special case of Theorem 5.2. I have preferred, however, to prove Theorem 6.32 independently in order to show once more how simplicial approximations are used to study continuous mappings.

§**6.4. Fundamental properties of the degree of a mapping.** An immediate consequence of the definition of degree is

6.41. *Two* (n, J)-*homologous, and therefore two completely homologous or two homotopic, mappings of a closed orientable pseudomanifold into another have the same degree.*

REMARK: *The case* $n = 1$. Using Remark 3 of VIII, 5.2, it is easily shown that a "normal mapping of degree γ" of one oriented circumference into another defined in II, 2.5 is indeed of degree γ. Since both definitions of degree (that of II, 2.5 and XI, 6.1) yield the same degree for two mappings in the same homotopy class, and since every mapping of one circumference into another is homotopic to a normal mapping, *the degree of an arbitrary continuous mapping of one circumference into another in the sense of* Def. 6.1 *is equal to its degree in the sense of* II, 2.5.

Let us now return to the general case of two n-dimensional orientable pseudomanifolds Φ and Ψ. There exist precisely two isomorphic mappings of the infinite cyclic group $\Delta_\Phi{}^n$ onto the infinite cyclic group $\Delta_\Psi{}^n$: the isomorphism which maps $\mathfrak{z}_\Phi{}^n$ into $\mathfrak{z}_\Psi{}^n$ and that which maps $\mathfrak{z}_\Phi{}^n$ into $-\mathfrak{z}_\Psi{}^n$ (here $\mathfrak{z}_\Phi{}^n$ and $\mathfrak{z}_\Psi{}^n$ are definite orientations of Φ and Ψ).

Consequently

6.42. *The degree of a topological mapping of a pseudomanifold* Φ *onto a pseudomanifold* Ψ *is either* 1 *or* -1.

6.43. If $C_2{}^1$ $(C_3{}^2)$ is a continuous mapping of a closed orientable pseudomanifold Φ_1 (Φ_2) into a closed orientable pseudomanifold Φ_2 (Φ_3), then the degree of the mapping $C_3{}^1 = C_3{}^2C_2{}^1$ of Φ_1 into Φ_3 is equal to the product of the degrees of $C_2{}^1$ and $C_3{}^2$.

For,

$$C_2^1(\mathfrak{z}_1) = \gamma_1\mathfrak{z}_2 ,$$

$$C_3^2(\mathfrak{z}_2) = \gamma_2\mathfrak{z}_3 ,$$

$$C_3^2(C_2^1(\mathfrak{z}_1)) = C_3^2(\gamma_1\mathfrak{z}_2) = \gamma_1 C_3^2(\mathfrak{z}_2) = \gamma_1\gamma_2\mathfrak{z}_3 ,$$

where $\mathfrak{z}_1$, $\mathfrak{z}_2$, $\mathfrak{z}_3$ are orientations of Φ_1, Φ_2, Φ_3, respectively, and γ_1, γ_2 are the degrees of C_2^1 and C_3^2.

6.44. *If $C(\Phi) \subset \Psi$, that is, if $C(\Phi) \neq \Psi$, then the degree of C is zero.*

Proof. Suppose $p' \in \Psi$, $p' \notin C(\Phi)$ and let $\epsilon = \rho(p', C(\Phi))$. Take a triangulation M_σ of Ψ of mesh $<\epsilon$ and denote by M_σ^* the combinatorial closure of the complex consisting of all the simplexes of M_σ containing at least one point of $C(\Phi)$. M_σ^* is obviously a proper subcomplex of M_σ so that $\Delta^n(M_\sigma^*)$, and hence $\Delta^n(\| M_\sigma^* \|)$, is the null group. The mapping C takes Φ into the polyhedron $\| M_\sigma^* \|$ so that every cycle $C(\mathfrak{z}^n)$, with $\mathfrak{z}^n \in Z_\Phi{}^n$ any proper cycle of Φ, is homologous to zero in $\| M_\sigma^* \|$ and therefore in Ψ; in other words, $C(\mathfrak{z})$ is the identity of $\Delta_\Psi{}^n$ for every $\mathfrak{z} \in \Delta_\Phi{}^n$. This completes the proof.

Examples of continuous mappings of various degrees are given in XII, 5.4, which may be read at this point.

Chapter XII

RELATIVE CYCLES AND THEIR APPLICATIONS

In this chapter Φ denotes a compactum, Γ an open set in Φ, and $\Psi = \Phi \setminus \Gamma$.

§1. The complex $K(\Gamma, \epsilon)$

§1.1. Definition of $K(\Gamma, \epsilon)$ and basic notation. By $K(\Gamma, \epsilon)$ or, more precisely, $K(\Phi, \Gamma, \epsilon)$ we shall mean the complex $K(\Phi, \epsilon) \setminus K(\Psi, \epsilon)$. Hence the simplexes (skeletons) of the complex $K(\Gamma, \epsilon)$ are precisely those ϵ-simplexes of Φ which have at least one vertex in Γ.

Since $K(\Gamma, \epsilon) = K(\Phi, \epsilon) \setminus K(\Psi, \epsilon)$ and $K(\Psi, \epsilon)$ is an unrestricted, and hence closed, subcomplex of $K(\Phi, \epsilon)$, *$K(\Gamma, \epsilon)$ is an open subcomplex of the unrestricted simplicial complex $K(\Phi, \epsilon)$.*

Remark. We shall use the notation $K(\Phi, \Gamma, \epsilon)$ very infrequently and only when Γ is also an open subset of a second compactum (besides Φ) which enters into the discussion. The most important case occurs when, of two given compacta, one is a subset of the other. Thus, if $\Gamma \subset \Phi_0 \subset \Phi$, where Φ_0 is closed and Γ is open in Φ, then every skeleton of $K(\Phi_0, \Gamma, \epsilon)$, as is easily seen, is a skeleton of $K(\Phi, \Gamma, \epsilon)$; that is,

$$K(\Phi_0, \Gamma, \epsilon) \subseteq K(\Phi, \Gamma, \epsilon).$$

The converse inclusion does not hold in general: $K(\Phi, \Gamma, \epsilon)$ contains skeletons not in $K(\Phi_0, \Gamma, \epsilon)$, namely, all the skeletons of $K(\Phi, \epsilon)$ which have at least one vertex in Γ and at least one vertex in $\Phi \setminus \Phi_0$.

We shall use the following simplified notation. $L_{\Phi,\epsilon}^r$, $Z_{\Phi,\epsilon}^r$, etc. will stand for $L^r[K(\Phi, \epsilon, \mathfrak{A})]$, $Z^r[K(\Phi, \epsilon, \mathfrak{A})]$, etc. Instead of $L^r[K(\Gamma, \epsilon, \mathfrak{A})]$, $Z^r[K(\Gamma, \epsilon, \mathfrak{A})]$, etc. we shall write $L_{\Gamma,\epsilon}^r$, $Z_{\Gamma,\epsilon}^r$, etc. Finally, Δx^r will mean the boundary of a chain $x^r \in L_{\Phi,\epsilon}^r$ in $K(\Phi, \epsilon)$, and $\Delta_\Gamma x^r$ will denote the boundary of a chain $x^r \in L_{\Gamma,\epsilon}^r$ in $K(\Gamma, \epsilon)$.

§1.2. Cycles and homologies in $K(\Gamma, \epsilon)$. Since $K(\Gamma, \epsilon)$ is an open subcomplex of $K(\Phi, \epsilon)$, Theorem 6.51 of VII yields

$$\Delta_\Gamma K(\Gamma, \epsilon) x^r = K(\Gamma, \epsilon) \Delta x^r \tag{1.20}$$

for every $x^r \in L_{\Phi,\epsilon}^r$; in particular

$$\Delta_\Gamma x^r = K(\Gamma, \epsilon)\Delta x^r, \qquad x^r \in L_{\Gamma,\epsilon}^r. \tag{1.200}$$

It follows easily that:

1.21. If $x^r \in L_{\Phi,\epsilon}^r$, then $K(\Gamma, \epsilon)x^r$ is a cycle of $K(\Gamma, \epsilon)$ if, and only if,

$\Delta x^r \in Z_{\Psi,\epsilon}^{r-1}$. As a special case, if $x^r \in L_{\Gamma,\epsilon}^r$, then $x^r \in Z_{\Gamma,\epsilon}^r$ if, and only if, $\Delta x^r \in Z_{\Psi,\epsilon}^{r-1}$.

Another consequence of (1.20) is

1.22. If $x^r \in L_{\Phi,\epsilon}^r$, then

$$K(\Gamma,\ \epsilon)x^r \sim 0 \quad \text{in } K(\Gamma,\ \epsilon) \tag{1.220}$$

if, and only if, there is a chain $x^{r+1} \in L_{\Phi,\epsilon}^{r+1}$ such that

$$\Delta x^{r+1} - x^r \in L_{\Psi,\epsilon}^r. \tag{1.221}$$

For, if (1.220) is satisfied, there exists a chain $x^{r+1} \in L_{\Gamma,\epsilon}^{r+1}$ such that

$$\Delta_\Gamma x^{r+1} = K(\Gamma,\ \epsilon)x^r;$$

but $x^{r+1} = K(\Gamma,\ \epsilon)x^{r+1}$, so that

$$\Delta_\Gamma x^{r+1} = \Delta_\Gamma K(\Gamma,\ \epsilon)x^{r+1} = K(\Gamma,\ \epsilon)\Delta x^{r+1};$$

that is, $K(\Gamma,\ \epsilon)x^r = K(\Gamma,\ \epsilon)\Delta x^{r+1}$ or

$$K(\Gamma,\ \epsilon)(\Delta x^{r+1} - x^r) = 0.$$

Hence

$$\Delta x^{r+1} - x^r \in L_{\Psi,\epsilon}^r. \tag{1.221}$$

On the other hand, if (1.221) is satisfied, then

$$K(\Gamma,\ \epsilon)(\Delta x^{r+1} - x^r) = 0,$$

that is,

$$K(\Gamma,\ \epsilon)x^r = K(\Gamma,\ \epsilon)\Delta x^{r+1} = \Delta_\Gamma K(\Gamma,\ \epsilon)x^{r+1};$$

so that

$$K(\Gamma,\ \epsilon)x^r \sim 0 \quad \text{in } K(\Gamma,\ \epsilon).$$

As a special case, if $x^r \in Z_{\Gamma,\epsilon}^r$, we obtain

1.221. A cycle $z^r \in Z_{\Gamma,\epsilon}^r$ is homologous to zero in $K(\Gamma,\ \epsilon)$ if, and only if, there is a chain $x^{r+1} \in L_{\Phi,\epsilon}^r$ satisfying the condition

$$\Delta x^{r+1} - z^r \in L_{\Psi,\epsilon}^r.$$

Example. Let Φ be a square with interior Γ and boundary Ψ. Let K be a triangulation (Fig. 133) of Φ of mesh less than a prescribed ϵ. Denote the triangles of K oriented counterclockwise by $t_1^2,\ \cdots,\ t_{32}^2$ and assume that $| t_1^2 |, \cdots, | t_{16}^2 |$ are the triangles of the top half of the square. The oriented 1-simplexes $t_1^1, \cdots, t_{12}^1$ are shown in the figure. If

$$z^2 = \sum_{i=1}^{32} t_i^2, \qquad x^2 = \sum_{i=1}^{16} t_i^2, \qquad x^1 = t_1^1 + t_2^1 + t_3^1 + t_4^1,$$
$$y^1 = \sum_{i=5}^{12} t_i^1,$$

then $z^2 \in Z_{\Gamma,\epsilon}^2$. Further, $x^1 \in H_{\Gamma,\epsilon}^1$. This is because

$$\Delta x^2 = x^1 + y^1,$$

where

$$y^1 \in L_{\Psi,\epsilon}^1.$$

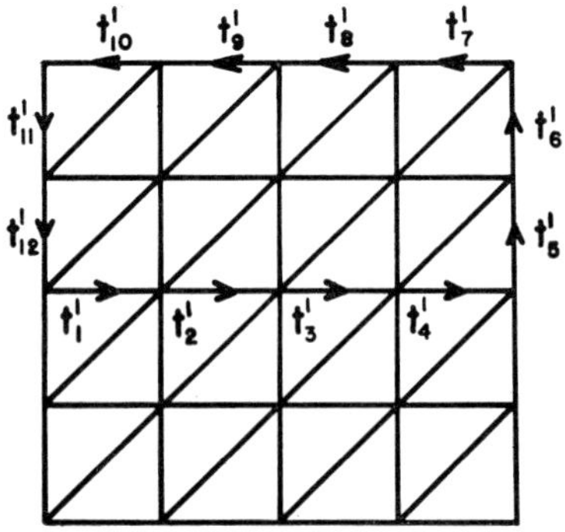

FIG. 133

§1.3. (ϵ, Ψ)-displacements.

DEFINITION 1.31. An (ϵ, Ψ)-*displacement* of a compactum Φ is a mapping S^Φ of Φ into itself with the following two properties:

1°. $\rho(x, S^\Phi x) < \epsilon$ for every $x \in \Phi$.

2°. $S^\Phi x \in \Psi$ for every $x \in \Psi = \Phi \setminus \Gamma$.

Let δ and ϵ be arbitrary positive numbers. Every (ϵ, Ψ)-displacement induces a simplicial mapping of the complex $K(\Phi, \delta)[K(\Psi, \delta)]$ into $K(\Phi, \delta + 2\epsilon)$ $[K(\Psi, \delta + 2\epsilon)]$ (see X, 5.2) and hence homomorphisms S^Φ, S^Ψ, and $S^\Gamma = K(\Gamma, \delta + 2\epsilon)S^\Phi$ of the groups $L_{\Phi,\delta}^r$ (see VII, 8.2 and 8.4), $L_{\Psi,\delta}^r$, and $L_{\Gamma,\delta}^r$ into $L_{\Phi,\delta+2\epsilon}^r$, $L_{\Psi,\delta+2\epsilon}^r$, $L_{\Gamma,\delta+2\epsilon}^r$, respectively. We are interested here in the third of these homomorphisms, S^Γ. If x^r is an arbitrary chain of $K(\Gamma, \delta)$:

$$x^r = \sum_j a_j t_{j,\delta}^r \in L_{\Gamma,\delta}^r,$$

then

$$S^\Gamma x^r = \sum_j a_j S^\Phi t_{j,\delta}^r,$$

where only those terms $a_j S^\Phi t_{j,\delta}^r$ of the linear form $\sum a_j S^\Phi t_{j,\delta}^r$ are to be retained for which $S^{\Phi}t_{j,\delta}^{r}$ are oriented r-simplexes of $K(\Gamma, \delta + 2\epsilon)$. The homomorphism S^{Γ} commutes with Δ, that is,

$$S^{\Gamma}\Delta_{\Gamma}x^{r} = \Delta_{\Gamma}S^{\Gamma}x^{r}$$

for an arbitrary $x^{r} \in L_{\Gamma,\delta}^{r}$ (see VII, 8.4).

THEOREM 1.32. *If z^{r} is an arbitrary cycle of $K(\Gamma, \delta)$ and S^{Φ} is an (ϵ, Ψ)-displacement of Φ, then*

$$z^{r} \sim S^{\Gamma}z^{r} \qquad \text{in } K(\Gamma, \delta + 2\epsilon).$$

The proof is similar to that of Theorem 2.12 of XI. Let Πz^{r} be the prism spanned by z^{r} and $S^{\Phi}z^{r}$ (see, XI, 2.1); its simplexes (that is, the simplexes on which the value of Πz^{r} is different from zero) have diameter $<\delta + 2\epsilon$ (XI, 2.1); hence $\Pi z^{r} \in L_{\Phi,\delta+2\epsilon}^{r}$. By XI, (2.12),

$$\Delta\Pi z^{r} = z^{r} - S^{\Phi}z^{r} - \Pi\Delta z^{r},$$

where $\Pi\Delta z^{r}$ is the prism spanned by Δz^{r} and $S^{\Phi}\Delta z^{r} = S^{\Psi}\Delta z^{r}$. Since $\Pi\Delta z^{r}$ is on $K(\Psi, \delta + 2\epsilon)$, according to 1.22,

$$z^{r} - S^{\Gamma}z^{r} \sim 0 \qquad \text{in } K(\Gamma, \delta + 2\epsilon).$$

This completes the proof.

§1.4. Canonical displacements. If $\alpha = \{A_1, \cdots, A_s\}$ is a closed ϵ-covering of a compactum Φ and $A_1, \cdots, A_u$ are the elements of α which intersect Ψ, then the sets $\Psi \cap A_1, \cdots, \Psi \cap A_u$ form a covering $\Psi\alpha$ of $\Psi \subseteq \Phi$. The nerve of α will be denoted by K_α and the vertices of K_α by $a_1, \cdots, a_s$, with a_i and A_i corresponding to each other. In this notation, $a_1, \cdots, a_u$ are the vertices of the nerve $K_{\Psi\alpha}$ of the covering $\Psi\alpha$. The complex $K_{\Psi\alpha}$ is a subcomplex of K_α; a set of vertices $a_{i(0)}, a_{i(1)}, \cdots, a_{i(r)}$ of $K_{\Psi\alpha}$ defines a simplex of $K_{\Psi\alpha}$ if, and only if,

$$\Psi \cap A_{i(0)} \cap \cdots \cap A_{i(r)} \neq 0.$$

We shall denote the groups $L^{r}(K_\alpha, \mathfrak{A})$, $L^{r}(K_{\Psi\alpha}, \mathfrak{A})$, $L^{r}(K_\alpha \setminus K_{\Psi\alpha}, \mathfrak{A})$, etc., by L_α^{r}, $L_{\Psi\alpha}^{r}$, $L_{\Gamma\alpha}^{r}$, etc.

A canonical displacement S_α^{Φ} of Φ relative to α assigns to each point $p \in \Phi$ a vertex a_i of K_α, and to each point $p \in \Psi$ a vertex $a_{i(j)}$ of $K_{\Psi\alpha}$.

Now let δ be sufficiently small so that it is a Lebesgue number of both coverings α and $\Psi\alpha$. As we know S_α^{Φ} maps every δ-skeleton of Φ into a skeleton of K_α; in addition, it maps every δ-skeleton $(e_0 \cdots e_r)$ of Ψ into a skeleton of $K_{\Psi\alpha}$. In fact, for each e_j, $j = 0, \cdots, r$, the point $S_\alpha^{\Phi}e_j$ is a vertex $a_{i(j)}$ subject to the condition that $e_j \in A_{i(j)}$; since all the e_j are contained in Ψ, the sets $\Psi \cap A_{i(0)}, \cdots, \Psi \cap A_{i(r)}$ all intersect the set consisting of the points $e_0, \cdots, e_r$ and having diameter $<\delta$. Since δ is a Lebesgue number of $\Psi\alpha$, the sets $\Psi \cap A_{i(0)}, \cdots, \Psi \cap A_{i(r)}$ have a nonempty intersection; hence the vertices $a_{i(0)}, \cdots, a_{i(r)}$ form a skeleton of $K_{\Psi\alpha}$.

Consequently

1.41. If $\alpha = \{A_1, \cdots, A_s\}$ is a closed ϵ-covering of a compactum Φ and $A_1, \cdots, A_u$ are the elements of α which intersect $\Psi \subset \Phi$, then the sets $\Psi \cap A_1, \cdots, \Psi \cap A_u$ are a covering $\Psi\alpha$ of Ψ whose nerve $K_{\Psi\alpha}$ is a subcomplex of the nerve K_α of α. If δ is a Lebesgue number of both α and $\Psi\alpha$, a canonical displacement S_α^Φ of Φ relative to α is also a canonical displacement of Ψ relative to $\Psi\alpha$. Hence it induces homomorphisms S_α^Φ, $S_{\Psi\alpha}^\Psi$, and $S_{\Gamma\alpha}^\Gamma = (K_\alpha \setminus K_{\Psi\alpha})S_\alpha^\Phi$ of the groups $L_{\Phi,\delta}^r$, $L_{\Psi,\delta}^r$, $L_{\Gamma,\delta}^r$ into L_α^r, $L_{\Psi\alpha}^r$, and $L_{\Gamma\alpha}^r$, respectively. (In reference to the homomorphism $(K_\alpha \setminus K_{\Psi\alpha})S_\alpha^\Phi$ see VII, 8.4; the analogous homomorphism was there denoted by $G_\alpha S_\alpha^\beta$.)

These homomorphisms commute with the corresponding boundary operators:

If $x^r \in L_{\Phi,\delta}^r$, or $x^r \in L_{\Psi,\delta}^r$, or $x^r \in L_{\Gamma,\delta}^r$, then

$$\Delta_1 S_\alpha^\Phi x^r = S_\alpha^\Phi \Delta x^r,$$
$$\Delta_2 S_{\Psi\alpha}^\Psi x^r = S_{\Psi\alpha}^\Psi \Delta x^r, \tag{1.41}$$
$$\Delta_3 S_{\Gamma\alpha}^\Gamma x^r = S_{\Gamma\alpha}^\Gamma \Delta_\Gamma x^r,$$

respectively; where Δ_1, Δ_2, Δ_3 are the boundary operators in K_α, $K_{\Psi\alpha}$, $K_\alpha \setminus K_{\Psi\alpha}$.

§2. Γ-cycles (relative cycles) and Γ-homologies in Φ; the groups $Z_\Phi^r(\Gamma, \mathfrak{A})$, $H_\Phi^r(\Gamma, \mathfrak{A})$, $\Delta_\Phi^r(\Gamma, \mathfrak{A})$

§2.1. Definitions.

DEFINITION 2.11. A sequence

$$\mathfrak{z}^r = (z_1^r, z_2^r, \cdots, z_k^r, \cdots), \tag{2.11}$$

with z_k^r a δ_k-cycle of $K(\Gamma, \delta_k)$ and $\lim_{k\to\infty} \delta_k = 0$, is called a Γ-*cycle* of a compactum Φ if for every $\epsilon > 0$ there is a $k(\epsilon)$ such that

$$z_p^r \sim z_q^r \quad \text{in } K(\Gamma, \epsilon)$$

for every two natural numbers $p, q > k(\epsilon)$.

EXAMPLE. Let Φ be a square, Γ its interior, and Ψ its boundary. If K is any triangulation of Φ, let K_n stand for the barycentric subdivision of K of order n. Suppose $t_{n1}^2, \cdots, t_{np(n)}^2$ are the triangles of K_n, all oriented in the same way (say, counterclockwise). If $z_n^2 = \sum_i t_{ni}^2$, then

$$\mathfrak{z}^2 = (z_1^2, \cdots, z_k^2, \cdots)$$

is a Γ-cycle of Φ.

REMARK 1. In this chapter, as previously, we regard all chains as linear forms. We recall that two chains (perhaps of two different complexes K

and K') are to be considered as identical if they coincide as linear forms, that is, if the same simplexes have the same nonzero coefficients in both forms. It is of course assumed that all simplexes with nonzero coefficients in the chains being compared belong to both complexes K and K'. Another way of expressing this convention is that every chain x^r of a complex K is also a chain of any other complex K' which contains all the simplexes occurring in x^r with nonvanishing coefficients.

Accordingly, we shall say that a Γ-cycle

$$\mathfrak{z}^r = (z_1^r, z_2^r, \cdots, z_k^r, \cdots)$$

of Φ is equal to the Γ-cycle

$$\mathfrak{z}'^r = (z_1'^r, z_2'^r, \cdots, z_k'^r, \cdots)$$

of Φ' if

$$z_k^r = z_k'^r$$

for every k. In the same way, we shall term

$$\mathfrak{z}^r = (z_1^r, z_2^r, \cdots, z_k^r, \cdots)$$

both a Γ-cycle of Φ and Γ'-cycle of Φ' if z_k^r is a chain of both complexes $K(\Gamma, \delta_k)$ and $K(\Gamma', \delta_k)$ and if the relations

$$z_p^r \sim z_q^r \quad \text{in } K(\Gamma, \epsilon),$$

$$z_p^r \sim z_q^r \quad \text{in } K(\Gamma', \epsilon)$$

hold for arbitrary $\epsilon > 0$ and all sufficiently large p and q.

EXAMPLE. Let Φ and Φ' be two congruent squares lying in two perpendicular planes of three-dimensional space and having a diagonal Φ_0 in common. Let Γ (Γ') be the interior, and Ψ (Ψ') the boundary of Φ (Φ'); and denote by Ψ_0 the set consisting of the two endpoints of the common diagonal Φ_0. Finally, let $\Gamma_0 = \Phi_0 \setminus \Psi_0$.

Suppose that K is the complex whose elements are the common diagonal of the two squares and its two endpoints and that K_n is the nth order barycentric subdivision of K (that is, the subdivision of the diagonal into 2^n equal segments). Assuming a definite direction on Φ_0 and denoting by $t_{n1}^1, \cdots, t_{n\rho(n)}^1$, where $\rho^n = 2^n$, the segments of K_n oriented in the given direction, we set

$$z_k^1 = \sum_{i=1}^{\rho(k)} t_{ki}^1, \qquad \mathfrak{z}^1 = (z_1^1, z_2^1, \cdots, z_n^1, \cdots).$$

Then $\mathfrak{z}^1$ is a one-dimensional Γ-cycle of Φ, Γ'-cycle of Φ', and Γ_0-cycle of Φ_0.

We make particular note of one special case pertaining to Remark 1.

2.10. If Φ_0 is a closed subset of Φ and $\Psi_0 \subseteq \Phi_0$, $\Gamma_0 = \Phi_0 \setminus \Psi_0$, $\Gamma = \Phi \setminus \Psi_0$, then every Γ_0-cycle of Φ_0 is also a Γ-cycle of Φ.

REMARK 2. If there is no need to indicate the set Γ over which a given Γ-cycle is to be taken or if Γ is a variable (as, for instance, if the existence of a Γ-cycle with certain properties is in question), it is preferable to replace the term "Γ-cycle" with the term "relative cycle"; hence

$$\mathfrak{z}^r = (z_1^r, z_2^r, \cdots, z_k^r, \cdots)$$

is a relative cycle of Φ *if* Φ *contains an open set* Γ *such that* $\mathfrak{z}^r$ *is a* Γ*-cycle of* Φ.

The notion of relative cycle in all its various forms (both combinatorial and set-theoretic) was first introduced by Lefschetz. This idea has, in the last two decades, influenced the development of combinatorial techniques and made possible the discovery of a whole set of new topological theorems which it would be impossible to formulate without this concept.

DEFINITION 2.110. If $\mathfrak{z}^r$ is a relative cycle of Φ, every closed subset Φ_0 of Φ with the property that $\mathfrak{z}^r$ is a relative cycle of Φ_0 is called a *carrier* of $\mathfrak{z}^r$ in Φ.

Therefore, if $\mathfrak{z}^r$ is a Γ-cycle of Φ and $\Phi_0 \subset \Phi$ is a carrier of $\mathfrak{z}^r$, there eixsts a closed set $\Psi_0 \subset \Phi_0$ such that $\mathfrak{z}^r$ is a $(\Phi_0 \setminus \Psi_0)$-cycle of Φ_0.

DEFINITION 2.12. A Γ-cycle of Φ is said to be Γ-*homologous to zero in* Φ:

$$\mathfrak{z}^r(\Gamma\text{-}\sim)0 \qquad \text{in } \Phi,$$

if for every $\epsilon > 0$ there exists a $k(\epsilon)$ such that

$$z_k^r \sim 0 \qquad \text{in } K(\Gamma, \epsilon)$$

for all $k > k(\epsilon)$.

EXAMPLE. Once more, let Φ be a square with interior Γ. If n is a natural number and $t_{n1}^1, \cdots, t_{n\sigma}^1$, $\sigma = 2^n$, are the segments obtained by dividing an oriented diagonal of the square into 2^n equal parts, set $z_n^1 = \sum_i t_{ni}^1$. Then

$$\mathfrak{z}^1 = (z_1^1, z_2^1, \cdots, z_k^1, \cdots)$$

is a one-dimensional Γ-cycle of Φ, Γ-homologous to zero in Φ.

REMARK 3. If Φ_0 and $\Psi \subseteq \Phi_0$ are closed subsets of Φ and $\mathfrak{z}^r$ is a $(\Phi_0 \setminus \Psi)$-cycle of Φ_0, then $\mathfrak{z}^r$ is a $(\Phi \setminus \Psi)$-cycle of Φ; if in addition $\mathfrak{z}^r(\Phi_0 \setminus \Psi\text{-}\sim)0$ in Φ_0, then $\mathfrak{z}^r(\Phi \setminus \Psi\text{-}\sim)\ 0$ in Φ. The proof is left to the reader.

§2.2. The groups $Z_\Phi^r(\Gamma, \mathfrak{A})$, $H_\Phi^r(\Gamma, \mathfrak{A})$, $\Delta_\Phi^r(\Gamma, \mathfrak{A})$. If

$$\mathfrak{z}_1^r = (z_{11}^r, z_{12}^r, \cdots, z_{1h}^r, \cdots),$$

$$\mathfrak{z}_2^r = (z_{21}^r, z_{22}^r, \cdots, z_{2h}^r, \cdots)$$

are two Γ-cycles of Φ, then

$$\mathfrak{z}_1^r + \mathfrak{z}_2^r = (z_{11}^r + z_{21}^r, z_{12}^r + z_{22}^r, \cdots, z_{1h}^r + z_{2h}^r, \cdots)$$

is also a Γ-cycle of Φ, the *sum* of $\mathfrak{z}_1^r$ and $\mathfrak{z}_2^r$. It is easy to see that this definition of addition converts the set of all r-dimensional Γ-cycles of Φ (over a given coefficient domain $\mathfrak{A}$) into a group, denoted by $Z_\Phi^r(\Gamma, \mathfrak{A})$. The group $Z_\Phi^r(\Gamma, \mathfrak{A})$ contains the subgroup $H_\Phi^r(\Gamma, \mathfrak{A})$ consisting of all the elements of $Z_\Phi^r(\Gamma, \mathfrak{A})$ which are Γ-homologous to zero in Φ. The factor group

$$Z_\Phi^r(\Gamma, \mathfrak{A})/H_\Phi^r(\Gamma, \mathfrak{A})$$

is denoted by $\Delta_\Phi^r(\Gamma, \mathfrak{A})$:

$$\Delta_\Phi^r(\Gamma, \mathfrak{A}) = Z_\Phi^r(\Gamma, \mathfrak{A})/H_\Phi^r(\Gamma, \mathfrak{A}).$$

§2.3. Canonical and infinitesimal displacements. Isomorphism of the groups $\Delta_{\Phi_0}^r(\Gamma, \mathfrak{A})$ and $\Delta_\Phi^r(\Gamma, \mathfrak{A})$, $\Gamma \subseteq \Phi_0 \subseteq \Phi$. It follows from 1.41 that a canonical displacement S_α^Φ of Φ relative to a closed covering α of Φ induces a homomorphism S_α^Γ of $Z_\Phi^r(\Gamma, \mathfrak{A})$ into $Z^r(K_\alpha \setminus K_{\Psi\alpha}, \mathfrak{A})$ which maps $H_\Phi^r(\Gamma, \mathfrak{A})$ into $H^r(K_\alpha \setminus K_{\Psi\alpha}, \mathfrak{A})$. Hence

2.31. A canonical displacement S_α^Φ of Φ relative to a closed covering α of Φ induces a homomorphism S_α^Γ of $\Delta_{\Phi,\Gamma}^r = \Delta_\Phi^r(\Gamma, \mathfrak{A})$ into $\Delta_{\Gamma\alpha}^r = \Delta^r(K_\alpha \setminus K_{\Psi\alpha}, \mathfrak{A})$.

2.32. Let

$$\mathfrak{z}^r = (z_1^r, z_2^r, \cdots, z_k^r, \cdots), \tag{2.3}$$

z_k^r a δ_k-cycle of $K(\Gamma, \delta_k)$, be a Γ-cycle of Φ and let $S_k z_k^r = z_k'^r$ be an (ϵ_k, Ψ)-displacement of the chain z_k^r, with $\lim_{k\to\infty} \epsilon_k = 0$. Since $z_k'^r$ is in this case a cycle of $K(\Gamma, \delta_k + 2\epsilon_k)$, homologous to z_k^r in this complex, it follows that

$$\mathfrak{z}'^r = (z_1'^r, z_2'^r, \cdots, z_k'^r, \cdots)$$

is a Γ-cycle Γ-homologous to $\mathfrak{z}^r$.

The passage from $\mathfrak{z}^r$ to $\mathfrak{z}'^r$, as well as the Γ-cycle $\mathfrak{z}'^r$ itself, is called an *infinitesimal displacement of* $\mathfrak{z}^r$.

We shall use the method of infinitesimal displacements to prove the isomorphism noted in the title of this subsection.

Set $\Phi_0 \setminus \Gamma = \Phi_0 \cap \Psi = \Psi_0$ and let p be an arbitrary point of Φ. If $p \in \Phi_0$, put $Sp = p$; if $p \notin \Phi_0$, then $p \in \Psi \setminus \Psi_0$, and Sp will denote any (definite) point $q \in \Psi_0$ for which $\rho(p, q)$ assumes its minimum.

If (2.3) is any Γ-cycle of Φ and ϵ_k is the maximum of $\rho(p, Sp)$, p a vertex of the complex $|z_k^r|$, we shall prove that $\lim_{k\to\infty} \epsilon_k = 0$. In the contrary case, there would exist a convergent sequence

$$p_1, p_2, \cdots, p_n, \cdots,$$

with $p_n \in \Psi \setminus \Psi_0$ a vertex of $|z_{k(n)}^r|$ and $\rho(p_n, Sp_n)$, and therefore also $\rho(p_n, \Psi_0)$, greater than some positive ϵ independent of n. But p_n is a vertex of a simplex $T_n^r \in |z_{k(n)}^r|$ with at least one vertex in Γ and hence in Φ_0.

Therefore, $p = \lim_{n\to\infty} p_n$, contained in Ψ (because of the fact that $p_n \in \Psi$), is also contained in Φ_0, that is, in Ψ_0. This contradicts the fact that

$$\epsilon < \rho(p_n, \Psi_0)$$

for all n.

Therefore the assignment to each vertex $p_k \in |z_n^r|$ of the point Sp_k yields an infinitesimal displacement of the Γ-cycle $\mathfrak{z}^r$ of Φ into the Γ-cycle $S\mathfrak{z}^r$ of Φ_0. The displacement induces a homomorphism S of $Z_{\Phi,\Gamma}^r$ onto $Z_{\Phi_0,\Gamma}^r$ (*onto* because S maps every Γ-cycle of Φ_0 into itself).

The homomorphism S maps $H_{\Phi,\Gamma}^r$ onto $H_{\Phi_0,\Gamma}^r$ and therefore induces a homomorphism of $\Delta_{\Phi,\Gamma}^r$ onto $\Delta_{\Phi_0,\Gamma}^r$. Since the cycles $\mathfrak{z}^r$ and $S\mathfrak{z}^r$ are obviously Γ-homologous in Φ, $S\mathfrak{z}^r(\Gamma\text{-}\sim)0$ in Φ_0 implies $\mathfrak{z}^r(\Gamma\text{-}\sim)0$ in Φ. Hence S is an isomorphism of $\Delta_{\Phi,\Gamma}^r$ onto $\Delta_{\Phi_0,\Gamma}^r$. This completes the proof.

§2.4. The groups $\Delta_\Phi^r(\Gamma, \mathfrak{A})$ and the dimension of Φ. The method of infinitesimal displacements affords an easy proof of the following important theorem:

2.4. *If a compactum Φ has dimension n, then $\Delta_\Phi^r(\Gamma, \mathfrak{A})$ is the null group for $r > n$ and for arbitrary coefficient domain $\mathfrak{A}$ and open set $\Gamma \subseteq \Phi$.*

This may also be expressed by saying that every r-dimensional Γ-cycle $\mathfrak{z}^r$ of an n-dimensional compactum Φ is Γ-homologous to zero in Φ for $r > n$.

Proof. For every m construct a $1/2^m$-covering $\alpha_m = \{A_1^m, \cdots, A_{s(m)}^m\}$ of Φ having order $n + 1$. Denote by ϵ_m a Lebesgue number of α_m and by K_m the nerve of α_m realized in Φ in such a way that the vertices of K_m corresponding to the elements A_i^m of α_m which intersect Ψ are points of Ψ. Every subsequence of a Γ-cycle $\mathfrak{z}^r$ is Γ-homologous to all of $\mathfrak{z}^r$ in Φ; hence to show that $\mathfrak{z}^r(\Gamma\text{-}\sim)0$ in Φ it is enough to prove that the analogous relation holds for a subsequence of $\mathfrak{z}^r$. Consider a subsequence of $\mathfrak{z}^r$ whose kth element is a cycle of $K(\Gamma, \epsilon_k)$ and write it also as

$$\mathfrak{z}^r = (z_1^r, z_2^r, \cdots, z_k^r, \cdots).$$

Now let $z_k'^r = S_k z_k^r$ be a canonical displacement of z_k^r relative to α_k. Then

$$S\mathfrak{z}^r = (z_1'^r, z_2'^r, \cdots, z_k'^r, \cdots)$$

is an infinitesimal displacement of $\mathfrak{z}^r$; hence $\mathfrak{z}^r(\Gamma\text{-}\sim)S\mathfrak{z}^r$. But $z_k'^r$ is an r-chain of the n-complex K_k; since $r > n$, $z_k'^r = 0$ so that $\mathfrak{z}^r(\Gamma\text{-}\sim)0$ in Φ.

§2.5. Remark. If a closed set Ψ of Φ is a single point, then the groups $Z_\Phi^r(\Gamma)$, $H_\Phi^r(\Gamma)$, $\Delta_\Phi^r(\Gamma)$ coincide with the groups $Z^r(\Phi)$, $H^r(\Phi)$, $\Delta^r(\Phi)$, respectively, for $r \geq 1$.

Proof. For arbitrary $\epsilon > 0$, $L_{\Psi,\epsilon}^r$, $r \geq 1$, and $H_{\Psi,\epsilon}^0$ are null groups, since $K(\Psi, \epsilon)$ consists of one point. Moreover, the set of r-simplexes ($r \geq 1$) of $K(\Phi, \epsilon)$ and $K(\Gamma, \epsilon)$ coincide; whence it easily follows that not only $L_{\Phi,\epsilon}^r$ and $L_{\Gamma,\epsilon}^r$, but also $Z_{\Phi,\epsilon}^r$ and $Z_{\Gamma,\epsilon}^r$, are identical.

In fact, if $x^r \in Z_{\Gamma,\epsilon}{}^r$, then $\Delta x^r \in H_{\Psi,\epsilon}{}^{r-1}$; and since $H_{\Psi,\epsilon}{}^{r-1}$ is the null group, $\Delta x^r = 0$, that is, $x^r \in Z_{\Phi,\epsilon}{}^r$.

We therefore need only prove that $H_{\Phi,\epsilon}{}^r$ and $H_{\Gamma,\epsilon}{}^r$ are also identical.

It is clear that $z^r \in H_{\Phi,\epsilon}{}^r$ implies $z^r \in H_{\Gamma,\epsilon}{}^r$. If $z^r \in H_{\Gamma,\epsilon}{}^r$, there exists an element x^{r+1} of $L_{\Gamma,\epsilon}{}^{r+1} = L_{\Phi,\epsilon}{}^{r+1}$ such that $\Delta_\Gamma x^{r+1} = z^r$. But

$$\Delta x^{r+1} - \Delta_\Gamma x^{r+1} \in L_{\Psi,\epsilon}{}^r$$

and is therefore equal to zero. Hence

$$\Delta x^{r+1} = \Delta_\Gamma x^{r+1} = z^r,$$

so that $z^r \in H_{\Phi,\epsilon}{}^r$.

It now follows that the proper r-cycles of Φ are identical with the r-dimensional Γ-cycles, that is, that $Z^r(\Phi) = Z_\Phi{}^r(\Gamma)$; and that a proper cycle is homologous to zero in Φ if, and only if, it is Γ-homologous to zero. Hence $H^r(\Phi, \mathfrak{A}) = H_\Phi{}^r(\Gamma, \mathfrak{A})$ and $\Delta^r(\Phi, \mathfrak{A}) = \Delta_\Phi{}^r(\Gamma, \mathfrak{A})$.

PROBLEM. Suppose $\dim \Psi = p$, $\Psi \subset \Phi$. Prove that the groups $\Delta_\Phi{}^r(\Gamma)$ and $\Delta^r(\Phi)$ are isomorphic for $r > p + 1$.

§3. The homomorphism of $\Delta_\Phi{}^r(\Gamma, \mathfrak{A})$ into $\Delta_{\Phi'}{}^r(\Gamma', \mathfrak{A})$ induced by a (Ψ, Ψ')-mapping $C_{\Phi'}{}^\Phi$

§3.1. The homomorphism $C_{\Phi'}{}^\Phi$. Let Φ and Φ' be compacta, let Γ, Γ' be open subsets of Φ, Φ', respectively, and set $\Psi = \Phi \setminus \Gamma$, $\Psi' = \Phi' \setminus \Gamma'$.

A continuous mapping $C_{\Phi'}{}^\Phi$ of Φ into Φ' with the property that the image of the set Ψ is contained in Ψ':

$$C_{\Phi'}{}^\Phi \Psi \subseteq \Psi', \tag{3.11}$$

is called a (Ψ, Ψ')-*mapping*.

Condition (3.11) implies that the simplicial mapping $C_{\Phi'}{}^\Phi$ of the complex $K(\Phi, \epsilon)$ into the complex $K(\Phi', \epsilon_C)$ induced by the continuous mapping $C_{\Phi'}{}^\Phi$ takes the complex $K(\Psi, \epsilon)$ into $K(\Psi', \epsilon_C)$ (see XI, 3.1). Hence (VII, 8.4) the mapping $C_{\Phi'}{}^\Phi$ induces a homomorphism $\Gamma' C_{\Phi'}{}^\Phi = K(\Gamma', \epsilon_C) C_{\Phi'}{}^\Phi$ of $L_{\Gamma,\epsilon}{}^r$ into $L_{\Gamma,\epsilon_C}{}^r$ which commutes with Δ in the sense that

$$\Delta_{\Gamma'} \Gamma' C_{\Phi'}{}^\Phi x^r = \Gamma' C_{\Phi'}{}^\Phi \Delta_\Gamma x^r \tag{3.12}$$

for every $x^r \in L_{\Gamma,\epsilon}{}^r$.

Since the mapping $C_{\Phi'}{}^\Phi$ of Φ into Φ' is uniformly continuous, so that ϵ_C approaches zero as ϵ approaches zero, the mapping $\Gamma' C_{\Phi'}{}^\Phi$ assigns to every Γ-cycle

$$\mathfrak{z}^r = (z_1{}^r, z_2{}^r, \cdots, z_k{}^r, \cdots)$$

of Φ the Γ'-cycle

$$\Gamma' C_{\Phi'}{}^\Phi \mathfrak{z}^r = (\Gamma' C_{\Phi'}{}^\Phi z_1{}^r, \Gamma' C_{\Phi'}{}^\Phi z_2{}^r, \cdots, \Gamma' C_{\Phi'}{}^\Phi z_k{}^r, \cdots)$$

of Φ'. Hence the continuous mapping $C_{\Phi'}{}^{\Phi}$ induces a homomorphism $C_{\Gamma'}{}^{\Gamma} = \Gamma' C_{\Phi'}{}^{\Phi}$ of $Z_{\Phi}{}^{r}(\Gamma)$ into $Z_{\Phi'}{}^{r}(\Gamma')$.

Because of the uniform continuity of the mapping and (3.11) it now follows that $C_{\Gamma'}{}^{\Gamma}$ maps $H_{\Phi}{}^{r}(\Gamma)$ into $H_{\Phi'}{}^{r}(\Gamma')$. Consequently the homomorphism $C_{\Gamma'}{}^{\Gamma}$ of $Z_{\Phi}{}^{r}(\Gamma)$ into $Z_{\Phi'}{}^{r}(\Gamma')$ induces a homomorphism (denoted by the same symbol) of $\Delta_{\Phi}{}^{r}(\Gamma)$ into $\Delta_{\Phi'}{}^{r}(\Gamma')$ which we shall call the *homomorphism induced by the* (Ψ, Ψ')-*mapping* $C_{\Phi'}{}^{\Phi}$.

§3.2. (Ψ, Ψ')-homologous and (Ψ, Ψ')-homotopic mappings; (Ψ, Ψ')-deformations. Let us consider again compacta Φ, Φ' and closed subsets $\Psi \subseteq \Phi$, $\Psi' \subseteq \Phi'$, with $\Gamma = \Phi \setminus \Psi$, $\Gamma' = \Phi' \setminus \Psi'$. In complete analogy with the definitions of Chapter XI, we shall say that two (Ψ, Ψ')-mappings C_0 and C_1 of Φ into Φ' are (Ψ, Ψ')-*homologous* if they induce identical homomorphisms of $\Delta_{\Phi}{}^{r}(\Gamma, \mathfrak{A})$ into $\Delta_{\Phi'}{}^{r}(\Gamma', \mathfrak{A})$ for all r and all $\mathfrak{A}$.

DEFINITION 3.21. A deformation C_θ, $0 \leq \theta \leq 1$, of a continuous (Ψ, Ψ')-mapping C_0 into a continuous (Ψ, Ψ')-mapping C_1 is said to be a (Ψ, Ψ')-*deformation* if C_θ is a (Ψ, Ψ')-mapping for all θ, $0 \leq \theta \leq 1$.

Two (Ψ, Ψ')-mappings are (Ψ, Ψ')-*homotopic* if there is a (Ψ, Ψ')-deformation taking one into the other.

THEOREM 3.22. *Two (Ψ, Ψ')-homotopic (Ψ, Ψ')-mappings C_0 and C_1 are (Ψ, Ψ')-homologous.*

This theorem follows from

3.23. If C_0 and C_1 are (Ψ, Ψ')-homotopic and

$$\mathfrak{z}^r = (z_1^r, z_2^r, \cdots, z_k^r, \cdots)$$

is a Γ-cycle of Φ, then the Γ'-cycles $\Gamma' C_0 \mathfrak{z}^r$ and $\Gamma' C_1 \mathfrak{z}^r$ of Γ' are Γ'-homologous.

We shall give a proof of Theorem 3.23 completely analogous to that of Theorem 3.3 of XI. It is required to show that for every $\epsilon > 0$ there exists a $k(\epsilon)$ such that

$$\Gamma' C_0 z_k^r \sim \Gamma' C_1 z_k^r \qquad \text{in } K(\Gamma', \epsilon)$$

for $k > k(\epsilon)$. To this end, as in XI, we consider a (Ψ, Ψ')-deformation C_θ of C_0 into C_1 and define $\delta > 0$ so that

$$\rho(p', p'') < \delta \quad \text{in } \Phi \quad \text{and} \quad |\theta' - \theta''| < \delta$$

imply

$$\rho(C_{\theta'} p', C_{\theta''} p'') < \epsilon/3.$$

Let us now choose $k(\epsilon)$ large enough so that the mesh of the complex $|z_k^r|$ is $< \delta$ for $k > k(\epsilon)$.

Take a definite $k > k(\epsilon)$ and index the vertices of $|z_k^r|$ in a definite order:

$$e_1, \cdots, e_\mu.$$

We shall divide the segment $0 \leq \theta \leq 1$ into segments of equal length $<\delta$ by means of the points

$$0 = \theta_0, \theta_1, \cdots, \theta_s = 1$$

and prove the relation

$$\Gamma' C_{\theta(i+1)} z_k^r - \Gamma' C_{\theta(i)} z_k^r \sim 0 \quad \text{in } K(\Gamma', \epsilon) \tag{3.24}$$

for arbitrary $i = 0, 1, \cdots, s - 1$. Theorem 3.23 will then follow.

To prove (3.24) we construct the prism Π_k over the skeleton complex $| z_k^r |$, and map Π_k into $K(\Phi', \epsilon)$ by means of a simplicial mapping S defined as follows on the vertices p_m and q_m of the lower and upper bases of Π_k :

$$S(p_m) = C_{\theta(i)}(p_m), \qquad S(q_m) = C_{\theta(i+1)}(p_m).$$

The mapping S transforms the prism Πz_k^r over z_k^r (contained in Π_k) into the chain $S\Pi z_k^r \in L_{\Phi',\epsilon}^r$, with

$$\Delta S\Pi z_k^r = S\Delta\Pi z_k^r = C_{\theta(i+1)} z_k^r - C_{\theta(i)} z_k^r - S\Pi\Delta z_k^r, \tag{3.25}$$

where $\Pi\Delta z_k^r$ is the prism (in the complex Π_k) over the chain $\Delta z_k^r \in Z_{\Psi,\delta}^{r-1}$.

S maps all the vertices of $\Pi\Delta z_k^r$ into points of Ψ', so that

$$S\Pi\Delta z_k^r \in Z_{\Psi',\epsilon}^{r-1}. \tag{3.26}$$

From (3.25), (3.26) we get

$$\Gamma'\Delta S\Pi z_k^r = \Gamma' C_{\theta(i+1)} z_k^r - \Gamma' C_{\theta(i)} z_k^r.$$

Since $\Delta_{\Gamma'}\Gamma' = \Gamma'\Delta$ (see VII, Theorem 6.51), it follows that

$$\Delta_{\Gamma'}\Gamma' S\Pi z_k^r = \Gamma' C_{\theta(i+1)} z_k^r - \Gamma' C_{\theta(i)} z_k^r.$$

This proves (3.24) and hence 3.23 and 3.22.

§3.3. Deformation of a relative cycle of Φ. Let

$$\mathfrak{z}^r = (z_1^r, z_2^r, \cdots, z_k^r, \cdots)$$

be a Γ-cycle of Φ and consider an arbitrary (Ψ, Ψ')-deformation C_θ, $0 \leq \theta \leq 1$, of the identity mapping C_0 of Φ. We shall call the family of Γ-cycles $C_\theta \mathfrak{z}^r$ of Φ indexed by the parameter θ a *deformation of the Γ-cycle $\mathfrak{z}^r$* in Φ.

Theorem 3.23 implies that the Γ-cycles $\mathfrak{z}^r = C_0\mathfrak{z}^r$ and $C_1\mathfrak{z}^r$ are Γ-homologous in Φ:

3.31. *A deformation in a compactum Φ maps every Γ-cycle $\mathfrak{z}^r$ of Φ into a Γ-cycle of Φ which is Γ-homologous to $\mathfrak{z}^r$ in Φ.*

COROLLARY. If there exists a (Ψ, Ψ')-deofrmation C_θ of Φ into itself such that $C_1(\Phi) \subseteq \Psi$, then every Γ-cycle $\mathfrak{z}^r$ of Φ is Γ-homologous to zero in Φ.

Indeed, $C_1(\Phi)$, and hence Ψ, is a carrier of $C_1\mathfrak{z}^r$. Therefore $C_1\mathfrak{z}^r = 0$; but $\mathfrak{z}^r(\Gamma\text{-}\sim)C_1\mathfrak{z}^r$, and this is what we were to prove.

§4. The groups $\Delta_\Phi^r(\Gamma)$ of polyhedra Φ and Ψ

§4.1. Introductory remarks. Suppose that Φ and Ψ are polyhedra, $\Psi \subset \Phi$, and that Ψ is the body of a complex $K_{\Psi\alpha}$ which is a subcomplex of a triangulation K_α of Φ:

$$\Phi = \| K_\alpha \|, \qquad \Psi = \| K_{\Psi\alpha} \|, \qquad K_{\Psi\alpha} \subset K_\alpha .$$

As usual, we set $\Gamma = \Phi \setminus \Psi$.

We shall denote the barycentric subdivision of K_α by $K_{\alpha 1}$ and the barycentric covering of Φ corresponding to K_α by α.

If e is a vertex of $K_{\Psi\alpha}$, it is also a vertex of K_α. Let $T^*(e)$ be the barycentric star of e relative to K_α and denote by $T_\Psi{}^*(e)$ the barycentric star of e relative to $K_{\Psi\alpha}$.

The complex $T_\Psi{}^*(e)$ is made up of all the simplexes of $K_{\alpha 1}$ with e as their last vertex (see IV, 2.2, Remark 1) which are contained in simplexes of $K_{\Psi\alpha}$; hence if $\overline{T}_\Psi{}^*(e)$ and $\overline{T}^*(e)$ are the corresponding closed barycentric stars (IV, Def. 5.31),

$$\overline{T}_\Psi{}^*(e) = \Psi \cap \overline{T}^*(e),$$

where $e \in K_{\Psi\alpha}$.

On the other hand, IV, Theorem 5.42 implies that if $\Psi \cap \overline{T}^*(e) \neq 0$, then e is a vertex of $K_{\Psi\alpha}$; so that by the above, $\overline{T}_\Psi{}^*(e) = \Psi \cap \overline{T}^*(e)$. Hence

4.11. *The covering $\Psi\alpha$ consisting of the (nonempty) intersections of the elements of the covering α with Ψ is the barycentric covering corresponding to $K_{\Psi\alpha}$.*

§4.2. The fundamental theorem.

4.2. *The group $\Delta_\Phi^r(\Gamma) = \Delta_\Gamma^r$ is isomorphic to the group $\Delta^r(K_\alpha \setminus K_{\Psi\alpha}) = \Delta_{\Gamma\alpha}^r$.*

Proof. Let $\delta > 0$ be a Lebesgue number of both coverings α and $\Psi\alpha$.

It follows from Theorem 2.31 that a canonical displacement S_α^Φ induces a homomorphism $S_\alpha^\Gamma = (K_\alpha \setminus K_{\Psi\alpha})S_\alpha^\Phi$ of Δ_Γ^r into $\Delta_{\Gamma\alpha}^r$.

a) *The homomorphism S_α^Γ is a mapping onto $\Delta_{\Gamma\alpha}^r$.* In fact, if $z_\alpha^r \in \mathfrak{z}_\alpha^r \in \Delta_{\Gamma\alpha}^r$ and $z_{\alpha 1}^r, z_{\alpha 2}^r, \cdots, z_{\alpha h}^r, \cdots$ are the consecutive barycentric subdivisions of z_α^r, then

$$\mathfrak{z}^r = (z_{\alpha 1}^r, z_{\alpha 2}^r, \cdots, z_{\alpha h}^r, \cdots)$$

is a Γ-cycle and

$$S_\alpha^\Phi \mathfrak{z}^r = z_\alpha^r;$$

hence

$$S_\alpha^\Gamma \mathfrak{z}^r = \mathfrak{z}_\alpha^r,$$

where $\mathfrak{z}^r$ is the element (homology class) of Δ_Γ^r containing $\mathfrak{z}^r$.

b) *The homomorphism* S_α^Γ *is an isomorphism.* To show this it is enough to prove the following:

If

$$\mathfrak{z}^r = (z_1^r, z_2^r, \cdots, z_h^r, \cdots) \tag{4.20}$$

is a Γ-cycle of a polyhedron Φ and $(K_\alpha \setminus K_{\Psi\alpha})S_\alpha^\Phi \mathfrak{z}^r \sim 0$ in $K_\alpha \setminus K_{\Psi\alpha}$, then $\mathfrak{z}^r \in H_\Gamma^r$, that is, for every $\epsilon > 0$ there is a $k(\epsilon)$ such that $z_k^r \sim 0$ in $K(\Gamma, \epsilon)$ for $k > k(\epsilon)$.

The proof proceeds in complete analogy with the reasoning of XI, 4.4. Let $K_{\alpha h}$, $K_{\Psi\alpha h}$ denote the barycentric subdivisions of the complexes K_α, $K_{\Psi\alpha}$ of order h. For a prescribed $\epsilon > 0$ let $\delta_\alpha < \epsilon$ be a Lebesgue number of both K_α and $K_{\Psi\alpha}$ and choose h so that $K_{\alpha h}$ is a δ_α-complex. Let $\delta_h < \delta_\alpha$ be a Lebesgue number of both $K_{\alpha h}$ and $K_{\Psi\alpha h}$ and choose $k(\epsilon)$ so that all the cycles z_k^r are δ_h-cycles for $k \geq k(\epsilon)$.

Suppose $k > k(\epsilon)$ and consider a canonical displacement $S_{\alpha h}^\Phi$ relative to $K_{\alpha h}$. Then

$$z_k^r \sim (K_\alpha \setminus K_{\Psi\alpha})S_{\alpha h}^\Phi z_k^r \qquad \text{in } K(\Gamma, \delta_\alpha),$$

$$S_{\alpha h}^\Phi z_k^r \in Z_{\Gamma\alpha}^r, \quad \text{with} \quad Z_{\Gamma\alpha}^r = Z^r(K_\alpha \setminus K_{\Psi\alpha}).$$

According to X, Theorem 2.35 there exists a cycle $z_\alpha^r \in Z_{\Gamma\alpha}^r$ for which

$$S_{\alpha h}^\Phi z_k^r \sim s_{\alpha h}^\alpha z_\alpha^r \qquad \text{in } K_{\alpha h} \setminus K_{\Psi\alpha h}.$$

Hence

$$z_k^r \sim s_{\alpha h}^\alpha z_\alpha^r \qquad \text{in } K(\Gamma, \delta_\alpha). \tag{4.21}$$

Applying a canonical displacement S_α^Φ relative to K_α to (4.21), we get

$$(K_\alpha \setminus K_{\Psi\alpha})S_\alpha^\Phi z_k^r \in Z^r(K_\alpha \setminus K_{\Psi\alpha})$$

on the left, and

$$(K_\alpha \setminus K_{\Psi\alpha})S_\alpha^\Phi s_{\alpha h}^\alpha z_\alpha^r = z_\alpha^r$$

on the right. Consequently

$$(K_\alpha \setminus K_{\Psi\alpha})S_\alpha^\Phi z_k^r \sim z_\alpha^r \qquad \text{in } K_\alpha \setminus K_{\Psi\alpha}.$$

But, by assumption,

$$(K_\alpha \setminus K_{\Psi\alpha})S_\alpha^\Phi z_k^r \sim 0 \qquad \text{in } K_\alpha \setminus K_{\Psi\alpha},$$

so that

$$z_\alpha^r \sim 0 \qquad \text{in } K_\alpha \setminus K_{\Psi\alpha}$$

and

$$s_{\alpha h}^\alpha z_\alpha^r \sim 0 \qquad \text{in } K_{\alpha h} \setminus K_{\Psi\alpha h} \subseteq K(\Gamma, \delta_\alpha).$$

Hence, by (4.21),

$$z_k^r \sim 0 \qquad \text{in } K(\Gamma, \delta_\alpha) \subseteq K(\Gamma, \epsilon).$$

This completes the proof.

§4.3. The homomorphism $C_{\alpha'}{}^{\alpha}$ of $\Delta_{\Gamma\alpha}{}^{r} = \Delta^{r}(K_{\alpha} \setminus K_{\Psi\alpha}, \mathfrak{A})$ into $\Delta_{\Gamma'\alpha'}{}^{r} = \Delta^{r}(K_{\alpha'} \setminus K_{\Psi'\alpha'}, \mathfrak{A})$ induced by a (Ψ, Ψ')-mapping $C_{\Phi'}{}^{\Phi}$. *Given* polyhedra $\Phi = \| K_{\alpha} \|$, $\Psi = \| K_{\Psi\alpha} \|$, $K_{\Psi\alpha} \subseteq K_{\alpha}$, and $\Phi' = \| K_{\alpha'} \|$, $\Psi' = \| K_{\Psi'\alpha'} \|$, $K_{\Psi'\alpha'} \subseteq K_{\alpha'}$, a (Ψ, Ψ')-mapping $C_{\Phi'}{}^{\Phi}$ induces a homomorphism $C_{\alpha'}{}^{\alpha}$ of $\Delta_{\Gamma\alpha}{}^{r}$ into $\Delta_{\Gamma'\alpha'}{}^{r}$ given by the following rule (see XI, 4.7):

4.31. Let $\delta' > 0$ be a Lebesgue number of both α' and $\Psi'\alpha'$ and choose δ sufficiently small so that $\delta_{C} < \delta'$ (see XI, 3.1; for convenience of notation we have used C as a subscript for δ in place of $C_{\Phi'}{}^{\Phi}$).

Now choose a subdivision $K_{\alpha h}$ of K_{α} of mesh $<\delta$ and let $\mathfrak{z}_{\alpha}{}^{r} \in \Delta_{\Gamma\alpha}{}^{r}$ be arbitrary. To obtain $C_{\alpha'}{}^{\alpha}\mathfrak{z}_{\alpha}{}^{r} \in \Delta_{\Gamma'\alpha'}{}^{r}$ choose any $z_{\alpha}{}^{r} \in \mathfrak{z}_{\alpha}{}^{r}$, apply the subdivision operator $s_{\alpha h}{}^{\alpha}$ to it, then take the image of the cycle $s_{\alpha h}{}^{\alpha} z_{\alpha}{}^{r}$ under the mapping $\Gamma' C_{\Phi'}{}^{\Phi}$, that is, the cycle $\Gamma' C_{\Phi'}{}^{\Phi} s_{\alpha h}{}^{\alpha} z_{\alpha}{}^{r}$ of the complex $K(\Gamma', \delta')$, and subject it to a canonical displacement $(K_{\alpha'} \setminus K_{\Psi'\alpha'}) S_{\alpha'}{}^{\Phi'}$. The result is a cycle $(K_{\alpha'} \setminus K_{\Psi'\alpha'}) S_{\alpha'}{}^{\Phi'} \Gamma' C_{\Phi'}{}^{\Phi} s_{\alpha h}{}^{\alpha} z_{\alpha}{}^{r}$ of the complex $K_{\alpha'} \setminus K_{\Psi'\alpha'}$ whose homology class is the desired element

$$C_{\alpha'}{}^{\alpha}\mathfrak{z}_{\alpha}{}^{r} \in \Delta_{\Gamma'\alpha'}{}^{r}.$$

We now consider the simplicial mapping $S_{\alpha'}{}^{\alpha h}$ of $K_{\alpha h}$ into $K_{\alpha'}$ defined by the rule in XI, 5.1. (In XI, 5.1, σ and Ψ are the α' and Φ' of the present section.) This mapping is a simplicial approximation to the continuous mapping $C_{\Phi'}{}^{\Phi}$. It is easy to see that $S_{\alpha'}{}^{\alpha h}$ maps $K_{\Psi\alpha h}$ into $K_{\Psi'\alpha'}$ and thus induces a homomorphism $(K_{\alpha'} \setminus K_{\Psi'\alpha'}) S_{\alpha'}{}^{\alpha h}$ of $\Delta_{\Gamma\alpha}{}^{r}$ into $\Delta_{\Gamma'\alpha'}{}^{r}$. Exactly as in XI, 5 we get

4.32. *$C_{\alpha'}{}^{\alpha}$ and $(K_{\alpha'} \setminus K_{\Psi'\alpha'}) S_{\alpha'}{}^{\alpha h} s_{\alpha h}{}^{\alpha}$ are identical homomorphisms of $\Delta_{\Gamma\alpha}{}^{r}$ into $\Delta_{\Gamma'\alpha'}{}^{r}$.*

§4.4. Definition of the homology dimension of a polyhedron. Another proof of the invariance of the dimension number. If Φ is the body of an n-dimensional triangulation K^{n} and $T^{n} \in K^{n}$, then $\| T^{n} \|$ is an open subset of Φ and according to Theorem 4.2 $\Delta_{\Phi}{}^{n}(\| T^{n} \|, \mathfrak{A})$ is isomorphic to the nth Betti group of the complex consisting of the single element T^{n}, that is, it is isomorphic to $\mathfrak{A}$ and therefore it is not the null group.

Hence for every n-dimensional polyhedron Φ it is possible to find an open set $\Gamma = \| T^{n} \|$ for which the group $\Delta_{\Phi}{}^{n}(\Gamma, \mathfrak{A})$ is different from zero for arbitrary $\mathfrak{A}$. On the other hand, the group $\Delta_{\Phi}{}^{r}(\Gamma, \mathfrak{A})$ is the null group for $r > n$ and arbitrary Γ and $\mathfrak{A}$ (Theorem 2.4). Thus

4.4. *The dimension number of a polyhedral complex K (the dimension of the polyhedron $\| K \|$) is the maximum number n such that $\| K \|$ contains an open set Γ for which $\Delta_{\|K\|}{}^{n}(\Gamma, \mathfrak{A})$ is different from zero.*

Here $\mathfrak{A}$ is any coefficient domain.

Theorem 4.4 gives a new invariant definition of the dimension number, known as the homology dimension, of a triangulation of a polyhedron and thereby furnishes another proof of its invariance.

§4.5. (See Aleksandrov [*d*, *i*], Bibliography, Vol. 1.) **The definition of the homology dimension of a compactum.** A natural consequence of Theorem 4.4 is the following definition:

DEFINITION 4.5. *The homology dimension* $d(\Phi, \mathfrak{A})$ *of a compactum* Φ *over the coefficient domain* $\mathfrak{A}$ *is the maximum number* n *for which there exists an open set* Γ *with* $\Delta_\Phi{}^n(\Gamma, \mathfrak{A}) \neq 0$. *If there is no maximum* n *with this property, then*

$$d(\Phi, \mathfrak{A}) = \infty$$

by definition. In the latter case, then, for every n *there is a* $\Gamma_n \subset \Phi$ *such that* $\Delta_\Phi{}^n(\Gamma, \mathfrak{A}) \neq 0$.

Theorem 2.4 implies

4.51. *For an arbitrary compactum* Φ *and coefficient domain* $\mathfrak{A}$,

$$d(\Phi, \mathfrak{A}) \leq \dim \Phi.$$

From Theorem 4.4 we get

4.52. *For an arbitrary polyhedron* Φ *and coefficient domain* $\mathfrak{A}$,

$$d(\Phi, \mathfrak{A}) = \dim \Phi.$$

The proof of the following fundamental result is beyond the scope of this book:

4.53. *For an arbitrary compactum* Φ,

$$\dim \Phi = d(\Phi, \mathfrak{R}_1)$$

(*where* $\mathfrak{R}_1$ *is the additive group of rationals* (mod 1)).

Hence dim Φ is also one of the homology dimensions and, because of Theorem 4.51, it is the maximum homology dimension.

L. S. Pontryagin has given an example of a compactum Φ in R^4 whose homology dimension over a certain coefficient domain is different from dim Φ. It is natural to refer to compacta Φ with the property that

$$d(\Phi, \mathfrak{A}) = \dim \Phi$$

for arbitrary coefficient domain $\mathfrak{A}$ as *dimensionally full-valued.* All polyhedra are dimensionally full-valued (by Theorem 4.52). It can be proved that all compacta in R^3 are also dimensionally full-valued.

§5. Pseudomanifolds with boundary

§5.1. Orientation of a pseudomanifold with boundary. Let $\Phi = \| K_\alpha \|$ be an n-dimensional pseudomanifold with boundary $\Psi = \| K_{\Psi\alpha} \|$; as usual, we set $\Gamma = \Phi \setminus \Psi$. For every triangulation K_α of Φ, the group $\Delta_{\Gamma\alpha}{}^n$ (see 4.2) is isomorphic to $\Delta_\Gamma{}^n$. From this we deduce, as in XI, 4.8, the nvariance of the orientability of a pseudomanifold with boundary:

5.11. *If Φ is an n-dimensional pseudomanifold with boundary Ψ, $\Gamma = \Phi \setminus \Psi$, only two cases can occur: either $\Delta_\Phi^n(\Gamma, J)$ is infinite cyclic or $\Delta_\Phi^n(\Gamma, J)$ is the null group. In the first case, all the triangulations of Φ are orientable combinatorial pseudomanifolds with boundary and Φ is said to be orientable. In the second case, all the triangulations of Φ are nonorientable combinatorial pseudomanifolds with boundary and Φ is said to be nonorientable.*

Each of the two generators $\pm\mathfrak{z}_\Phi^n$ of the infinite cyclic group $\Delta_\Phi^n(\Gamma, J)$ is called an *orientation* of the orientable pseudomanifold Φ. Repeating the arguments of XI, 4.8, it is easy to see that the orientations of Φ correspond (1–1) in a perfectly definite way to the orientations $\pm\mathfrak{z}_\alpha^n$ of an arbitrary triangulation K_α of Φ.

§5.2. Introductory remarks; definition of the degree of a continuous mapping of a pseudomanifold with boundary. Let Φ and Φ' be orientable pseudomanifolds, either with boundary or closed.

Suppose that C is a continuous mapping of Φ into Φ' satisfying the conditions stated below.

If Φ or Φ' is a pseudomanifold with boundary we denote the boundary by Ψ or Ψ'. If Φ is closed, we set $\Psi = 0$ and if Φ and Φ' are both closed, we put $\Psi = 0$, $\Psi' = 0$ (this case was discussed in XI, 6). For the case of Φ a pseudomanifold with boundary and Φ' closed, Ψ' will be defined below.

With regard to the mapping C we shall make the following assumptions:

If Φ' is a pseudomanifold with boundary Ψ', then

$$C(\Psi) \subseteq \Psi';$$

but if Φ' is closed and Φ is not, then we shall assume that $C(\Psi)$ is a single point and this point will be denoted by Ψ'. (It is enough to require that $\dim C(\Psi) \leq n - 2$; the remainder of the argument is in this case left to the reader as an exercise; in this connection see the Problem proposed at the end of §2; this problem must be solved first.)

In all cases, we set

$$\Gamma = \Phi \setminus \Psi, \qquad \Gamma' = \Phi' \setminus \Psi'.$$

Now choose definite orientations $\mathfrak{z}^n$ and $\mathfrak{z}'^n$ of Φ and Φ', and proceed exactly as in XI, 6, substituting $\Delta_\Phi^n(\Gamma, J)$, $\Delta_{\Phi'}^n(\Gamma', J)$ for $\Delta_0^n(\Phi)$, $\Delta_0^n(\Phi')$, Γ-cycles for proper cycles, and Γ-homologoies for homologies. This brings us to the notion of the degree of the mapping C: C induces a homomorphism (denoted by the same symbol) of $\Delta_\Phi^n(\Gamma)$ into $\Delta_{\Phi'}^n(\Gamma')$ such that

$$C(\mathfrak{z}^n) = \gamma\mathfrak{z}'^n,$$

where γ is an integer, the *degree* of the mapping.

Using the same notation as in 4.3, we see that for sufficiently large h, a simplicial mapping $S_{\alpha'}^{\alpha h}$ of $K_{\alpha h}$ into $K_{\alpha'}$ maps $K_{\Psi\alpha h}$ into $K_{\Psi'\alpha'}$ and is a simplicial approximation to C. Further, it can be proved in exactly the same way as in XI, 6.3 that the degree of C is equal to the degree of the simplicial mapping $S_{\alpha'}{}^{\alpha h}$ of the combinatorial pseudomanifold $K_{\alpha h}$ into the combinatorial pseudomanifold $K_{\alpha'}$ (see VIII, 5 (especially 5.22)).

§5.3. Some properties of the degree of a mapping. Retaining the notation and assumptions of the beginning of this section, let us suppose that K_α is divided into a finite number of n-dimensional combinatorial pseudomanifolds $K_{\alpha\lambda}$, $\lambda = 1, \cdots, s$, with boundaries $K_{\Psi\alpha\lambda}$ and that

$$K_{\alpha\lambda'} \cap K_{\alpha\lambda''} = K_{\Psi\alpha\lambda'} \cap K_{\Psi\alpha\lambda''}$$

for arbitrary λ', λ''.

Let $\sum_i t_{\alpha i}{}^n$ be a prescribed orientation of K_α and for each $K_{\alpha\lambda}$ choose the orientation $\sum_i t_{\alpha\lambda i}{}^n$, with the summation extended over all

$$| t_{\alpha\lambda i}{}^n | \in K_{\alpha\lambda} .$$

Suppose that C maps each of the pseudomanifolds $\Phi_\lambda = \| K_{\alpha\lambda} \|$ into Φ' in such a way that the boundary $\Psi_\lambda = \| K_{\Psi\alpha\lambda} \|$ is mapped into Ψ'. Assuming the above orientation for each of the pseudomanifolds $\Phi_\lambda = \| K_{\alpha\lambda} \|$, we can speak of the degree γ_λ of the mapping C of Φ_λ into Φ'. We then arrive at the following *addition theorem*:

$$\gamma = \sum \gamma_\lambda , \tag{5.31}$$

where γ is the degree of the mapping C of Φ into Φ'.

To prove (5.31) we construct, in accordance with the rule of XI, 5.1, a simplicial approximation $S_{\alpha'}{}^{\alpha h}$ to the mapping C of Φ into Φ' which takes $K_{\Psi\alpha}$ and all the $K_{\Psi\alpha\lambda}$ into $K_{\Psi'\alpha'}$. The algebraic number of simplexes of K_α covering an oriented simplex $t_{\alpha'}{}^n$ of $K_{\alpha'}$ (see VIII, Theorem 5.21) is equal to the sum of the algebraic numbers of the simplexes of the complexes $K_{\alpha\lambda}$, $\lambda = 1, \cdots, s$, covering the simplex $t_{\alpha'}{}^n$. Formula (5.31) follows.

The theorems of XI, 6.4 apply word for word to the case of pseudomanifolds with boundary, with the exception that in 6.43 it is of course necessary to assume that the mappings $C_2{}^1$ and $C_3{}^2$ are (Ψ_1, Ψ_2)- and (Ψ_2, Ψ_3)-mappings, respectively.

5.31. *If Φ is a closed pseudomanifold and Φ' is a pseudomanifold with boundary, the degree of C is zero.*

Proof. In this case $\Gamma = \Phi$, every n-dimensional Γ-cycle of Φ is a proper cycle of Φ, and it is mapped onto a proper n-cycle of Φ' which is homologous to zero (Φ' does not contain any nonbounding proper n-cycles). Therefore, if $\mathfrak{z}_\Phi$ and $\mathfrak{z}_{\Phi'}$ are orientations of Φ and Φ' and $z^n \in \mathfrak{z}_\Phi$, the element $\gamma\mathfrak{z}_{\Phi'}$ of the group $\Delta_{\Phi'}{}^n(\Gamma')$ containing $C(z^n)$ is the identity of $\Delta_{\Phi'}{}^n(\Gamma')$, that is, $\gamma = 0$.

§5.4. Examples.

EXAMPLE 1. The polyhedra Φ and Φ' are two 2-spheres, represented by the closed complex planes $w = r(\cos\varphi + i\sin\varphi)$, $w' = r'(\cos\varphi' + i\sin\varphi')$. The mapping $C_{\Phi'}{}^{\Phi}$ assigns to each point $w = r(\cos\varphi + i\sin\varphi)$ of the sphere Φ the point $w' = C(w) = r(\cos n\varphi + i\sin n\varphi)$ of the sphere Φ'.

If the sphere Φ is triangulated in the form of a double tetrahedron, while

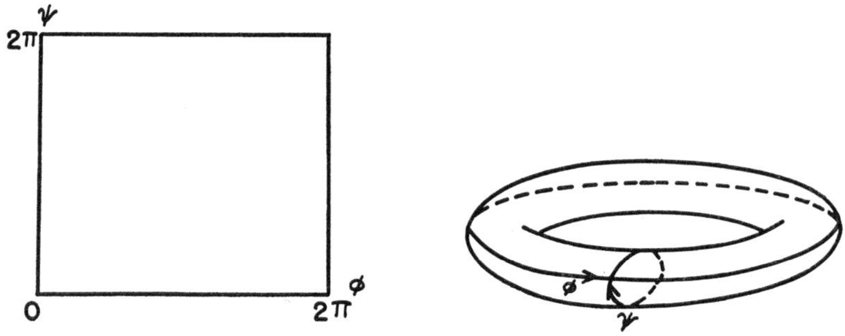

FIG. 134

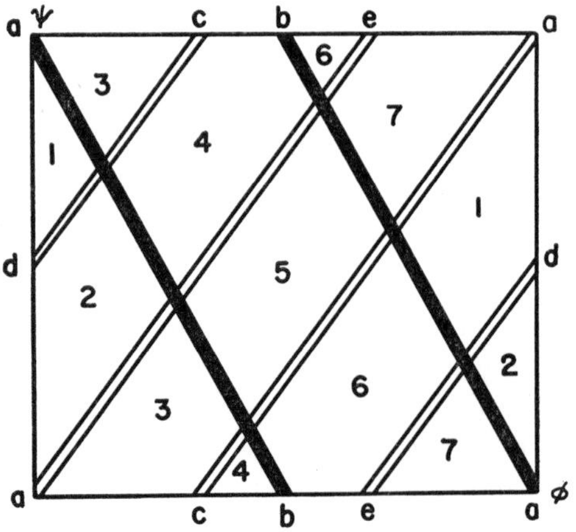

FIG. 135

Φ' is triangulated into a double pyramid with base in the form of a $3n$-gon, it is easy to show that the degree of $C_{\Phi'}{}^{\Phi}$ is n.

EXAMPLE 2. Let Φ and Φ' be tori. Introduce geographic coordinates on Φ, the angles φ and ψ measured as indicated in Fig. 134. On Φ' the analogous coordinates will be denoted by φ', ψ'.

a) Define $C_{\Phi'}{}^{\Phi}$ as follows:

$$C_{\Phi'}{}^{\Phi}(\varphi, \psi) = (n\varphi, \psi),$$

that is,

$$\varphi' = n\varphi, \qquad \psi' = \psi.$$

If Φ is divided into n rings by the meridians $\varphi = 2k\pi/n$ $(k = 0, 1, \cdots, n-1)$, then each ring, under the mapping $C_{\Phi'}{}^{\Phi}$, covers the torus Φ' with degree $+1$. It follows easily (by the Addition Theorem of 5.3 or by triangulating Φ and Φ') that the degree of $C_{\Phi'}{}^{\Phi}$ is n. The mappings

$$C_{\Phi'}{}^{\Phi}(\varphi, \psi) = (\varphi, n\psi),$$

$$C_{\Phi'}{}^{\Phi}(\varphi, \psi) = (n\psi, \varphi)$$

have the same degree, n.

b) Let us consider in general mappings of the form

$$C_{\Phi'}{}^{\Phi}(\varphi, \psi) = (a\varphi + b\psi, c\varphi + d\psi),$$

that is,

$$\varphi' = a\varphi + b\psi, \qquad \psi' = c\varphi + d\psi,$$

with a, b, c, d integers such that the pairs a, b and c, d are relatively prime. In the contrary case $C_{\Phi'}{}^{\Phi}$ is factorable into two mappings, one of which has the form $\varphi' = m\varphi$, $\psi' = p\psi$ and the other the form $\varphi' = a'\varphi + b'\psi$, $\psi' = c'\varphi + d'\psi$, with the pairs a', b' and c', d' relatively prime.

In order for $C_{\Phi'}{}^{\Phi}$ to be a mapping onto Φ' it is obviously necessary that the determinant

$$D = \begin{vmatrix} a & b \\ c & d \end{vmatrix}$$

be different from zero. For convenience we shall term the curves $a\varphi + b\psi = 0$ and $c\varphi + d\psi = 0$ on Φ "spirals" (φ and ψ are defined up to integral multiples of 2π). These two "spirals" divide the torus into curvilinear parallelograms, each of which covers Φ' under the mapping $C_{\Phi'}{}^{\Phi}$. The number of these parallelograms is equal to D (it is left to the reader to prove this, for instance, by using a plane representation of the torus). Hence the degree of $C_{\Phi'}{}^{\Phi}$ is D. Fig. 135 shows the division of the torus by the spirals

$$3\varphi - 2\psi = 0 \quad \text{and} \quad 2\varphi + \psi = 0$$

into $\begin{vmatrix} 3 & -2 \\ 2 & 1 \end{vmatrix} = 7$ parallelograms.

EXAMPLE 3. Φ is a torus, Φ' a 2-sphere. Both surfaces are imbedded in three-dimensional space and each point p of the torus is assigned its projection $C_{\Phi'}{}^{\Phi}(p)$ from the center of the sphere Φ' onto the sphere (that is, $C_{\Phi'}{}^{\Phi}(p)$ is the point of intersection with the sphere of the line joining p to the center of the sphere). $C_{\Phi'}{}^{\Phi}$ is a continuous mapping of Φ onto Φ' if the center of the sphere is not on the torus. The degree of $C_{\Phi'}{}^{\Phi}$ is 1 if the center of the sphere is inside the torus and 0 if it is outside the torus.

REMARK. We have already shown in II, 2.5 (see also XI, 6.4) that there exist mappings of a circumference onto a circumference with arbitrarily prescribed degree. This is also true for an n-sphere for every n; in this connection see XVI, 6.1; it is recommended that XVI, 6.1 be read at this point.

§6. The groups $\Delta_p^r(\Phi)$ (The local Δ^r-groups of a compactum Φ)

§6.1. Definition of the groups $\Delta_p^r(\Phi)$. We shall begin with several auxiliary definitions.

If Γ and Γ_1 are two open sets of a compactum Φ and $x^r = \sum a_i t_i^r$ is a chain of the complex $K(\Gamma_1, \epsilon)$, set

$$\Gamma x^r = K(\Gamma \cap \Gamma_1, \epsilon)x^r.$$

In other words, the chain Γx^r is defined only on $K(\Gamma \cap \Gamma_1, \epsilon)$ and coincides there with x^r. It may also be thought of as obtained from x^r by retaining in the linear form $\sum a_i t_i^r$ only those terms with simplexes t_i^r at least one of whose vertices is contained in $\Gamma \cap \Gamma_1$.

If

$$\mathfrak{z}^r = (z_1^r, z_2^r, \cdots, z_h^r, \cdots)$$

is a Γ_1-cycle of Φ, then

$$\Gamma\mathfrak{z}^r = (\Gamma z_1^r, \Gamma z_2^r, \cdots, \Gamma z_h^r, \cdots)$$

is a $(\Gamma \cap \Gamma_1)$-cycle of Φ.

We are now ready for the fundamental definitions of this section.

DEFINITION 6.13. Let p be an arbitrary but fixed point of Φ and Γ a neighborhood of p (that is, an open set containing p). Every Γ-cycle of Φ is called a *cycle in the point* $p \in \Phi$.

DEFINITION 6.14. Suppose that $\mathfrak{z}_1^r = (z_{11}^r, z_{12}^r, \cdots, z_{1h}^r, \cdots)$, $\mathfrak{z}_2^r = (z_{21}^r, z_{22}^r, \cdots, z_{2h}^r, \cdots)$ are Γ_1- and Γ_2-cycles, respectively, in $p \in \Phi$ and set $\Gamma = \Gamma_1 \cap \Gamma_2$. Then $p \in \Gamma$, so that the Γ-cycle

$$(\Gamma z_{11}^r + \Gamma z_{21}^r, \cdots, \Gamma z_{1h}^r + \Gamma z_{2h}^r, \cdots) \tag{6.14}$$

is a cycle in the point p. The cycle (6.14) is called the *sum of the cycles* $\mathfrak{z}_1^r$ *and* $\mathfrak{z}_2^r$. It is not difficult to see that this definition of addition turns the set of all r-cycles in a point $p \in \Phi$ (over a given coefficient domain $\mathfrak{A}$) into a group, which will be written as $Z_p^r(\Phi, \mathfrak{A})$.

DEFINITION 6.15. A Γ-cycle in $p \in \Phi$:

$$\mathfrak{z}^r = (z_1^r, z_2^r, \cdots, z_h^r, \cdots)$$

is said to be *homologous to zero in* p, $\mathfrak{z}^r(p\text{-}\sim)0$, if there is an open set $\Gamma_1 \subseteq \Gamma$ containing p such that the cycle $\Gamma_1\mathfrak{z}^r$ is Γ_1-homologous to zero in Φ. Two cycles in p are homologous, by definition, if their difference is homologous to zero in p.

REMARK 1. From the definition of homology in a point, we immediately obtain the following proposition:

If $\mathfrak{z}^r$ is a cycle in $p \in \Phi$ and Γ_0 is a neighborhood of p in Φ (an open set containing p), then

$$\mathfrak{z}^r(p\text{ - }\sim)\Gamma_0\mathfrak{z}^r \qquad \text{in } \Phi.$$

REMARK 2. If $\mathfrak{z}^r = (z_1^r, z_2^r, \cdots, z_h^r, \cdots)$ is a Γ-cycle Γ-homologous to zero in Φ and if Φ_0 is closed, while $\Gamma_0 \subseteq \Gamma \subseteq \Phi_0$ is open in Φ, then

$$\Gamma_0\mathfrak{z}^r(\Gamma_0\text{ - }\sim)0 \qquad \text{in } \Phi_0\,.$$

Proof. Choose chains x_k^{r+1} of $K(\Gamma, \epsilon_k)$, $\lim \epsilon_k = 0$, such that

$$\Delta_\Gamma x_k^{r+1} = z_k^r.$$

Since $K(\Gamma_0, \epsilon)$ is open in $K(\Gamma, \epsilon)$ for every ϵ, it follows that

$$\Gamma_0 z_k^r = \Gamma_0\Delta_\Gamma x_k^{r+1} = \Delta_0\Gamma_0 x_k^{r+1},$$

where Δ_0 is the boundary operator in Γ_0. This completes the proof.

It is easily seen that the cycles in $p \in \Phi$ which are homologous to zero form a subgroup $H_p^r(\Phi, \mathfrak{A})$ of $Z_p^r(\Phi, \mathfrak{A})$.

DEFINITION 6.16. The group $\Delta_p^r(\Phi, \mathfrak{A}) = Z_p^r(\Phi, \mathfrak{A})/H_p^r(\Phi, \mathfrak{A})$ is called the *rth Betti* (or Δ^r-) *group of* Φ *in* p *over* $\mathfrak{A}$. The rank of $\Delta_p^r(\Phi, \mathfrak{R})$ relative to $\mathfrak{R}$ is called the *rth Betti number of* Φ *in* p; that is, the rth Betti number of Φ in p is the maximum number k with the property that there are k elements of $Z_p^r(\Phi, \mathfrak{R})$ such that no nontrivial linear combination of these elements with coefficients in $\mathfrak{R}$ is contained in $H_p^r(\Phi, \mathfrak{R})$.

REMARK 3. We shall write Z_p^r, H_p^r, Δ_p^r in place of $Z_p^r(\Phi, \mathfrak{A})$, $H_p^r(\Phi, \mathfrak{A})$, $\Delta_p^r(\Phi, \mathfrak{A})$.

REMARK 4. If p is considered a variable, the groups Δ_p^r are often referred to as the *local Betti groups of* Φ.

§6.2. The local character of the groups Δ_p^r. In this subsection we shall prove that the groups Δ_p^r really express a certain local property of Φ, that is, a property depending only on the structure of Φ in an arbitrarily small neighborhood of p. This property is formulated in 6.21 and 6.22.

6.21. *If* Φ_0 *is a closed subset of* Φ *containing a neighborhood of* p (*relative to* Φ), *then* $\Delta_p^r(\Phi_0, \mathfrak{A})$ *is isomorphic to* $\Delta_p^r(\Phi, \mathfrak{A})$.

Proof. Let us choose neighborhoods Γ_0 and Γ_1 of p such that

$$\bar{\Gamma}_0 \subset \Gamma_1 \subseteq \Phi_0$$

and assume that

$$\Phi \neq \Phi_0, \qquad \epsilon = \rho(\bar{\Gamma}_0, \Phi \setminus \Phi_0).$$

Then

$$\epsilon = \rho(\bar{\Gamma}_0, \Phi \setminus \Phi_0) \geq \rho(\bar{\Gamma}_0, \Phi \setminus \Gamma_1) > 0.$$

Suppose

$$\mathfrak{z}^r = (z_1^r, z_2^r, \cdots, z_k^r, \cdots) \in Z_p^r(\Phi, \mathfrak{A});$$

$$\Gamma_0\mathfrak{z}^r = (\Gamma_0 z_1^r, \Gamma_0 z_2^r, \cdots, \Gamma_0 z_k^r, \cdots).$$

Only a finite number of terms $| \Gamma_0 z_k^r |$ can contain simplexes with at least one vertex in $\Phi \setminus \Phi_0$ (because such simplexes have diameter $> \epsilon$). If these terms are

$$\Gamma_0 z_1^r, \cdots, \Gamma_0 z_{h-1}^r,$$

set

$$(\Gamma_0)\mathfrak{z}^r = (\Gamma_0 z_h^r, \Gamma_0 z_{h+1}^r, \cdots).$$

Then

$$(\Gamma_0)\mathfrak{z}^r \in Z_{\Phi_0}^r(\Gamma_0, \mathfrak{A}) \subseteq Z_p^r(\Phi_0, \mathfrak{A}).$$

If the cycle $(\Gamma_0)\mathfrak{z}^r$ is made to correspond to each cycle $\mathfrak{z}^r \in Z_p^r(\Phi)$, the result is a homomorphism (Γ_0) of $Z_p^r(\Phi)$ into $Z_p^r(\Phi_0)$. Since $\mathfrak{z}^r(\Gamma\text{-}\sim)0$ in Φ implies $(\Gamma_0)\mathfrak{z}^r(\Gamma_0\text{-}\sim)0$ in Φ_0 (by 6.1, Remark 2), (Γ_0) induces a homomorphism (denoted by the same symbol) of $\Delta_p^r(\Phi)$ into $\Delta_p^r(\Phi_0)$. Because

$$\mathfrak{z}^r(p\text{-}\sim)\ (\Gamma_0)\mathfrak{z}^r \qquad \text{in } \Phi,$$

it follows that $(\Gamma_0)\mathfrak{z}^r(p\text{-}\sim)0$ in Φ_0 implies $\mathfrak{z}^r(p\text{-}\sim)0$ in Φ; that is, (Γ_0) is an isomorphism of $\Delta_p^r(\Phi)$ into $\Delta_p^r(\Phi_0)$.

Suppose that $\mathfrak{z}_0 \in \Delta_p^r(\Phi_0)$ and that

$$\mathfrak{z}^r = (z_1^r, z_2^r, \cdots, z_k^r, \cdots)$$

is a Γ_0-cycle of Φ_0 contained in $\mathfrak{z}_0$. If $\mathfrak{z}^r$ is thought of as an element of $Z_p^r(\Phi)$, we get

$$(\Gamma_0)\mathfrak{z}^r = \mathfrak{z}^r.$$

Hence (Γ_0) maps $\Delta_p^r(\Phi)$ *onto* $\Delta_p^r(\Phi_0)$. This completes the proof.

Theorem 6.21 implies

THEOREM 6.22. *Let* $\Phi_1 \subseteq R$, $\Phi_2 \subseteq R$ *be two compacta in a metric space* R

and suppose that the intersection of Φ_1 and Φ_2 contains a nonempty set Γ open in both Φ_1 and Φ_2.

If $p \in \Gamma$, the groups $\Delta_p^r(\Phi_1, \mathfrak{A})$ and $\Delta_p^r(\Phi_2, \mathfrak{A})$ are isomorphic.

Indeed, both groups are isomorphic, by 6.21, to the group $\Delta_p^r(\bar{\Gamma}, \mathfrak{A})$.

DEFINITION 6.2. (The local Betti groups of a locally compact metric space.) Let R be a locally compact metric space and let p be a point of R. By the group $\Delta_p^r(R, \mathfrak{A})$ (defined up to an isomorphism) we shall mean the group $\Delta_p^r(\bar{\Gamma}, \mathfrak{A})$, where Γ is any neighborhood of p in R with compact closure $\bar{\Gamma}$.

According to 6.21 and 6.22, this definition is independent of the choice of the neighborhood Γ.

§7. The local Δ-groups of polyhedra

§7.1. Notation and introductory remarks. Let $\Phi = \| K_\alpha \|$ be a polyhedron and p an arbitrary point of Φ. As usual, $K_{\alpha h}$ is the barycentric subdivision of K_α of order h; a symbol with index α (or αh) stands for a subcomplex of K_α (or $K_{\alpha h}$); a transition from a subcomplex of K_α (or $K_{\alpha h}$) to its subdivision in $K_{\alpha k}$ is indicated by adding the symbol k on the right of those already present.

O_α ($o_{\alpha h}$) denotes the star of the carrier of p in K_α (in $K_{\alpha h}$).

As usual, a complex surrounded by vertical bars on either side indicates the combinatorial closure of the complex; thus $| O_\alpha |$ is the combinatorial closure of O_α.

In the sequel we shall put (see Fig. 136; in this figure it has been assumed that the carrier of p in K_α is p itself, that is, that p is a vertex of K_α and therefore also of $K_{\alpha h}$; it is left to the reader to draw figures for other cases):

$$Q_\alpha = | O_\alpha | \setminus O_\alpha, \qquad q_{\alpha h} = | o_{\alpha h} | \setminus o_{\alpha h},$$
$$P_{\alpha h} = | O_{\alpha h} | \setminus o_{\alpha h},$$

with the convention that $| O_{\alpha h} | = | O_\alpha |_h$ is the subdivision of $| O_\alpha |$ in $K_{\alpha h}$; in the same way, $Q_{\alpha h}$ is the subdivision of Q_α.

We shall denote the bodies of the complexes

$$O_\alpha,\ o_{\alpha h},\ | O_\alpha |,\ | o_{\alpha h} |,\ Q_\alpha,\ q_{\alpha h},\ P_{\alpha h}$$

by

$$\Gamma_\alpha,\ \gamma_{\alpha h},\ \Phi_\alpha,\ \varphi_{\alpha h},\ \Psi_\alpha,\ \psi_{\alpha h},\ \Pi_{\alpha h},$$

respectively; hence Φ_α, $\varphi_{\alpha h}$, Ψ_α, $\psi_{\alpha h}$, $\Pi_{\alpha h}$, are polyhedra and

$$\Phi_\alpha = \Gamma_\alpha \cup \Psi_\alpha, \qquad \varphi_{\alpha h} = \gamma_{\alpha h} \cup \psi_{\alpha h}, \qquad \Pi_{\alpha h} = (\Phi_\alpha \setminus \varphi_{\alpha h}) \cup \psi_{\alpha h}.$$

Finally we shall use C_θ, $0 \leq \theta \leq 1$, to designate a deformation of the polyhedron $\Pi_{\alpha h}$ defined as follows. Draw the ray py through each point

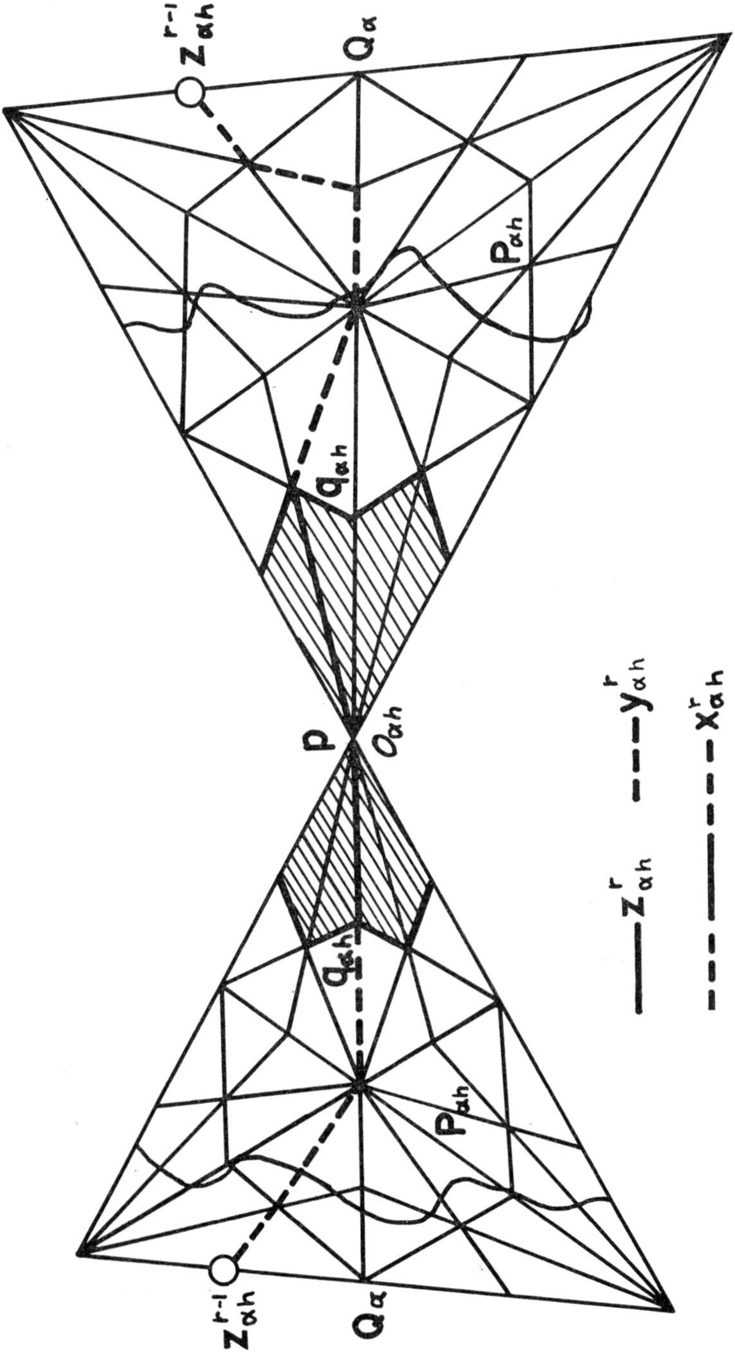

$Z^{r-1}_{\alpha h}$
Q_α
$P_{\alpha h}$
$q_{\alpha h}$
p
$O_{\alpha h}$
$q_{\alpha h}$
$P_{\alpha h}$
$Z^{r-1}_{\alpha h}$
Q_α
$Z^r_{\alpha h}$
$y^r_{\alpha h}$
$x^r_{\alpha h}$

FIG. 136

$y \in \Pi_{\alpha h}$ and let $C_1 y$ be the last point of py (in the direction from p to y) contained in the closure of the carrier of y in K_α. It is clear that $C_1 y \in \Psi_\alpha$. The point $C_\theta y$, $0 \leq \theta < 1$, is, by definition, that point of the segment $[y, C_1 y]$ which divides it in the ratio $\theta:(1 - \theta)$.

The deformation C_θ maps the entire polyhedron $\Pi_{\alpha h}$, and hence $\psi_{\alpha h}$ also, into Ψ_α in such a way that the entire deformation takes place in $\Pi_{\alpha h}$ and all points of Ψ_α remain fixed. Hence every proper cycle of $\psi_{\alpha h}$ is homologous in $\Pi_{\alpha h}$ to a proper cycle of Ψ_α.

Therefore (by XI, 4.61),

7.11. *Every cycle of the complex $q_{\alpha h}$ is homologous in $P_{\alpha h}$ to a cycle of the complex $Q_{\alpha h}$.*

§7.2. The fundamental theorem.

7.2. *The group $\Delta_p^r(\Phi, \mathfrak{A})$ is isomorphic to the group $\Delta^r(O_\alpha, \mathfrak{A})$.*

Proof. Since $\Delta_p^r(\Phi, \mathfrak{A})$ is isomorphic to $\Delta_p^r(\Phi_\alpha, \mathfrak{A})$ (according to 6.21), while $\Delta^r(O_\alpha, \mathfrak{A})$ is isomorphic to $\Delta_{\Phi_\alpha}{}^r(\Gamma_\alpha, \mathfrak{A})$ (by 4.2), it is enough to show that $\Delta_p^r(\Phi_\alpha, \mathfrak{A})$ and $\Delta_{\Phi_\alpha}{}^r(\Gamma_\alpha, \mathfrak{A})$ are isomorphic.

To simplify the notation in the following proof, we shall write Z^r, H^r, Δ^r, Z_p^r, H_p^r, Δ_p^r in place of

$$Z_{\Phi_\alpha}{}^r(\Gamma_\alpha, \mathfrak{A}),\ H_{\Phi_\alpha}{}^r(\Gamma_\alpha, \mathfrak{A}),\ \Delta_{\Phi_\alpha}{}^r(\Gamma_\alpha, \mathfrak{A}),\ Z_p^r(\Phi_\alpha, \mathfrak{A}),\ H_p^r(\Phi_\alpha, \mathfrak{A}),\ \Delta_p^r(\Phi_\alpha, \mathfrak{A}),$$

respectively.

An isomorphic mapping I of Δ^r onto Δ_p^r is constructed as follows:

Every element $\mathfrak{z}^r$ of Z^r is also an element of Z_p^r, so that

$$Z^r \subseteq Z_p^r. \tag{7.21}$$

In addition, $\mathfrak{z}^r(\Gamma_\alpha\text{-}\sim)0$ in Φ_α implies that $\mathfrak{z}^r(p\text{-}\sim)0$ in Φ_α; hence

$$H^r \subseteq H_p^r. \tag{7.22}$$

It follows from (7.21) and (7.22) that the identity mapping of Z^r into Z_p^r induces a homomorphism I of Δ^r into Δ_p^r. We shall prove that I is an isomorphism of Δ^r onto Δ_p^r.

STEP 1. *The homomorphism I is a mapping onto.*

Suppose $\mathfrak{z}_p^r \in \Delta_p^r$. Choose a cycle $\mathfrak{z}_1^r$ in p of the homology class $\mathfrak{z}_p^r$; $\mathfrak{z}_1^r$ is a Γ-cycle for some Γ containing p. Let h be sufficiently large, so that $\gamma_{\alpha h} \subseteq \Gamma$ (see Fig. 136). Then the $\gamma_{\alpha h}$-cycle $\gamma_{\alpha h}\mathfrak{z}_1^r$ is homologous to $\mathfrak{z}_1^r$ in p and is therefore contained in $\mathfrak{z}_p^r$.

Since $\gamma_{\alpha h}\mathfrak{z}_1^r$ is a $\gamma_{\alpha h}$-cycle of the polyhedron $\varphi_{\alpha h}$, the body of the complex $|o_{\alpha h}|$, it is certainly $\gamma_{\alpha h}$-homologous, and therefore homologous in p, to a cycle $\mathfrak{z}^r$ of the form

$$\mathfrak{z}^r = (z_{\alpha h}^r,\ s_{h+1}^h z_{\alpha h}^r,\ \cdots,\ s_k^h z_{\alpha h}^r,\ \cdots), \tag{7.23}$$

where $z_{\alpha h}^r \in Z^r(o_{\alpha h})$ and $s_k^h z_{\alpha h}^r$ is the subdivision of $z_{\alpha h}^r$ in $K_{\alpha k}$.

Because of 7.11, the cycle $\Delta z_{\alpha h}^{\ r}$ is homologous in $P_{\alpha h}$ to a cycle $z_{\alpha h}^{\ r-1}$ of $Q_{\alpha h}$; whence there exists a chain (Fig. 136)

(7.24) $$y_{\alpha h}^{\ r} \in L^r(P_{\alpha h})$$

satisfying the condition

(7.25) $$\Delta y_{\alpha h}^{\ r} = z_{\alpha h}^{\ r-1} - \Delta z_{\alpha h}^{\ r}.$$

Assume that $k > h$; then

(7.26) $$\Delta s_k^{\ h} y_{\alpha h}^{\ r} = s_k^{\ h} z_{\alpha h}^{\ r-1} - \Delta s_k^{\ h} z_{\alpha h}^{\ r}$$

in $P_{\alpha hk}$. We now set

(7.27) $$x_{\alpha h}^{\ r} = z_{\alpha h}^{\ r} + y_{\alpha h}^{\ r}.$$

By (7.27) and (7.26),

(7.28) $$\Delta s_k^{\ h} x_{\alpha h}^{\ r} = \Delta s_k^{\ h} z_{\alpha h}^{\ r} + \Delta s_k^{\ h} y_{\alpha h}^{\ r} = s_k^{\ h} z_{\alpha h}^{\ r-1} \in Z^{r-1}(Q_{\alpha k})$$

and (7.28) implies that

$$\mathfrak{z}'^r = (x_{\alpha h}^{\ r}, s_{h+1}^{\ h} x_{\alpha h}^{\ r}, \cdots, s_k^{\ h} x_{\alpha h}^{\ r}, \cdots) \in Z^r.$$

Finally, since $s_k^{\ h} y_{\alpha h}^{\ r}$ is a chain of $P_{\alpha hk}$ whose body does not intersect $\gamma_{\alpha h}$,

$$\gamma_{\alpha h} s_k^{\ h} y_{\alpha h}^{\ r} = 0.$$

Consequently

$$\gamma_{\alpha h} \mathfrak{z}'^r = (z_{\alpha h}^{\ r}, s_{h+1}^{\ h} z_{\alpha h}^{\ r}, \cdots, s_k^{\ h} z_{\alpha k}^{\ r}, \cdots) = \mathfrak{z}^r,$$

so that $\mathfrak{z}^r \sim \mathfrak{z}'^r$ in p, that is $\mathfrak{z}'^r \in Z_p^{\ r}$. Hence every homology class $\mathfrak{z}_p^{\ r} \in \Delta_p^{\ r}$ contains at least one element of Z^r and this proves that I is a mapping onto.

STEP 2. *The homomorphism I is an isomorphism.*

Suppose $\mathfrak{z}^r \in \Delta^r$ and let $I(\mathfrak{z}^r)$ be the identity of $\Delta_p^{\ r}$. Then every Γ_α-cycle contained in $\mathfrak{z}^r$ is homologous to zero in p. It is required to prove that $\mathfrak{z}^r$ is the identity of Δ^r. To this end, it is enough to exhibit a cycle of $\mathfrak{z}^r$ which is Γ_α-homologous to zero in Φ_α .

The homology class $\mathfrak{z}^r$, as does every element of Δ^r, contains a cycle $\mathfrak{z}_1^{\ r}$ of the form

(7.29) $$\mathfrak{z}_1^{\ r} = (z_\alpha^{\ r}, z_{\alpha 1}^{\ r}, \cdots, z_{\alpha k}^{\ r}, \cdots),$$

with $z_\alpha^{\ r} \in Z^r(O_\alpha)$ and $z_{\alpha h}^{\ r}$ the subdivision of $z_\alpha^{\ r}$ in $K_{\alpha h}$.

We shall prove that $\mathfrak{z}_1^{\ r}(\Gamma_\alpha\text{-}\sim)0$ in Φ_α and the theorem will follow.

Since $\mathfrak{z}_1^{\ r}$ is homologous to zero in p by assumption, there exists a neighborhood Γ of p such that

$$\Gamma \mathfrak{z}_1^{\ r}(\Gamma\text{-}\sim)0 \qquad \text{in } \Phi_\alpha .$$

Choose an h for which $\gamma_{\alpha h} \subseteq \Gamma$ (Fig. 137). Then $\Gamma \mathfrak{z}_1^r(\Gamma\text{-}\sim)0$ in Φ_α implies (because of 6.1, Remark 2) that

$$\gamma_{\alpha h}\mathfrak{z}_1^r(\gamma_{\alpha h}\text{-}\sim)0 \qquad \text{in } \varphi_{\alpha h} .$$

Finally, deleting the first h terms $z_\alpha^r, z_{\alpha 1}^r, \cdots, z_{\alpha,h-1}^r$ of the cycle (7.29), we obtain a Γ_α-cycle $\mathfrak{z}^r$, Γ_α-homologous to the cycle $\mathfrak{z}_1^r$, which may be written in the form

$$\mathfrak{z}^r = (z_{\alpha h}^r, s_{h+1}^h z_{\alpha h}^r, \cdots, s_k^h z_{\alpha h}^r, \cdots) \tag{7.291}$$

with

$$\gamma_{\alpha h}\mathfrak{z}^r(\gamma_{\alpha h}\text{-}\sim)0 \qquad \text{in } \varphi_{\alpha h} . \tag{7.292}$$

It is sufficient to prove that $\mathfrak{z}^r(\Gamma_\alpha\text{-}\sim)0$ *in* Φ_α .

To this end, we note first that

$$\gamma_{\alpha h} s_k^h z_{\alpha h}^r = \| o_{\alpha h} \| s_k^h z_{\alpha h}^r = s_k^h o_{\alpha h} z_{\alpha h}^r .$$

Hence

$$\gamma_{\alpha h}\mathfrak{z}^r = (o_{\alpha h} z_{\alpha h}^r, s_{h+1}^h o_{\alpha h} z_{\alpha h}^r, \cdots, s_k^h o_{\alpha h} z_{\alpha h}^r, \cdots) \tag{7.293}$$

and it follows from

$$\gamma_{\alpha h}\mathfrak{z}^r(\gamma_{\alpha h}\text{-}\sim)0 \qquad \text{in } \varphi_{\alpha h}$$

that

$$o_{\alpha h} z_{\alpha h}^r \sim 0 \qquad \text{in } o_{\alpha h} .$$

Therefore there exists a chain

$$x_{\alpha h}^{r+1} \in L^{r+1}(o_{\alpha h}) \subseteq L^{r+1}(O_{\alpha h})$$

such that

$$\Delta_{\alpha h} x_{\alpha h}^{r+1} = o_{\alpha h} z_{\alpha h}^r ,$$

where $\Delta_{\alpha h}$ is the boundary operator in $o_{\alpha h}$.

Setting

$$y_{\alpha h}^r = \Delta_{\alpha h} x_{\alpha h}^{r+1} - \Delta x_{\alpha h}^{r+1} ,$$

we obtain

$$y_{\alpha h}^r \in L^r(q_{\alpha h}), \tag{7.294}$$

$$\Delta x_{\alpha h}^{r+1} = o_{\alpha h} z_{\alpha h}^r - y_{\alpha h}^r . \tag{7.295}$$

Then

$$0 = \Delta\Delta x_{\alpha h}^{r+1} = \Delta o_{\alpha h} z_{\alpha h}^r - \Delta y_{\alpha h}^r$$

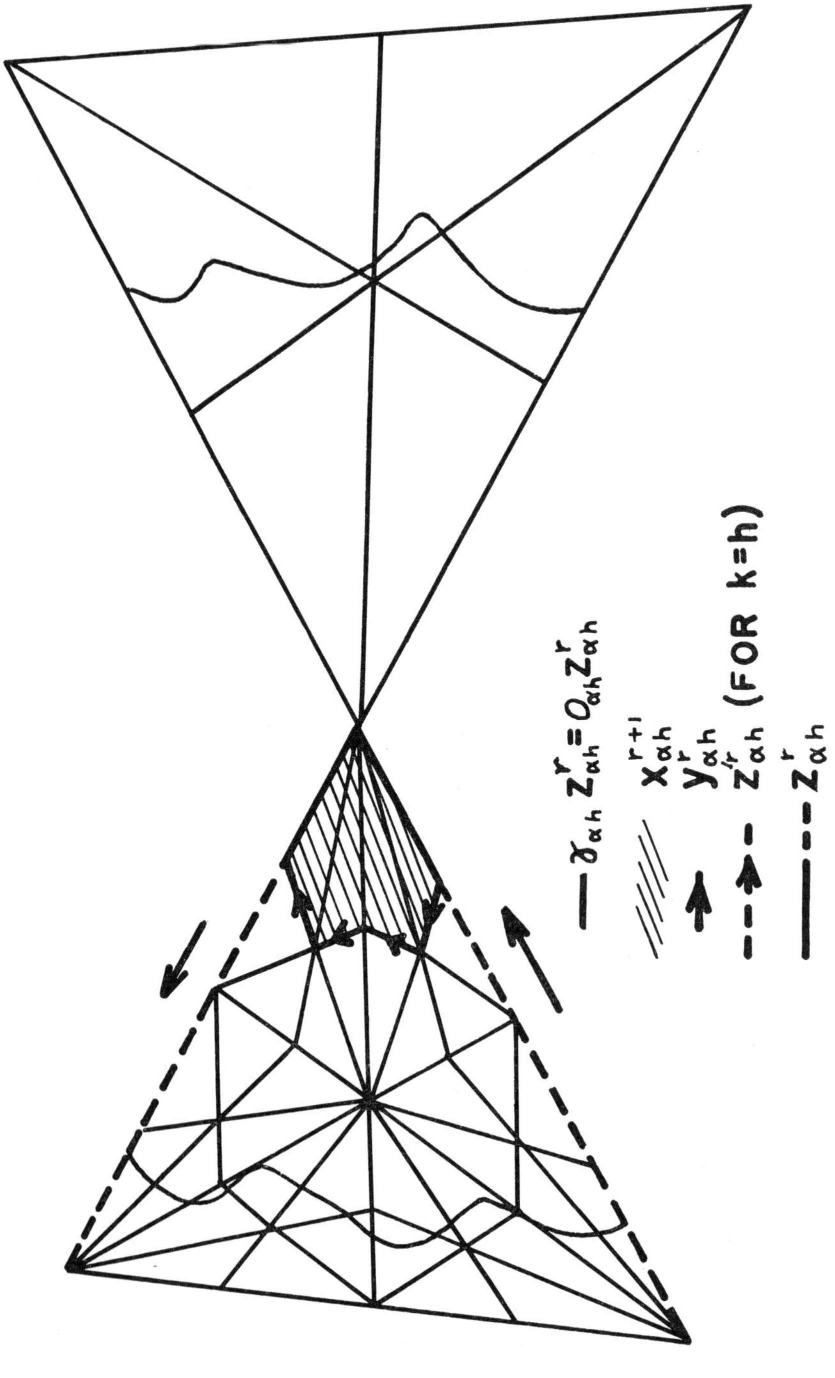
$-\gamma_{\alpha h} Z^r_{\alpha h} = O_{\alpha h} Z^r_{\alpha h}$
$X^{r+1}_{\alpha h}$
$Y^r_{\alpha h}$
$Z'^r_{\alpha h}$ (FOR k=h)
$Z^r_{\alpha h}$

FIG. 137

implies that

(7.296) $$\Delta y_{\alpha h}^{\ r} = \Delta o_{\alpha h} z_{\alpha h}^{\ r}.$$

We now set

(7.297$_h$) $$z_{\alpha k}^{\ \prime r} = s_k^{\ h}[(z_{\alpha h}^{\ r} - o_{\alpha h} z_{\alpha h}^{\ r}) + y_{\alpha h}^{\ r}]$$

for $k = h, h+1, \cdots$ and prove the following three assertions (7.297), (7.298), (7.299), whose intuitive meaning should be clear from Fig. 137:

(7.297) $$\mathfrak{z}'^r = (z_{\alpha h}^{\ \prime r}, z_{\alpha,h+1}^{\ \prime r}, \cdots, z_{\alpha k}^{\ \prime r}, \cdots) \in Z^r,$$

that is, $\mathfrak{z}'^r$ is a Γ_α-cycle of Φ_α ;

(7.298) $$\mathfrak{z}^r(\Gamma_\alpha\text{-}\sim)\mathfrak{z}'^r \qquad \text{in } \Phi_\alpha ;$$

(7.299) $$\mathfrak{z}'^r \in Z_{\alpha h}^{\ r}(\Pi_{\alpha h} \setminus \Psi_\alpha)$$

(where $Z_{\alpha h}^{\ r}(\Pi_{\alpha h} \setminus \Psi_\alpha)$ is the group of $(\Pi_{\alpha h} \setminus \Psi_\alpha)$-cycles of the compactum $\Pi_{\alpha h}$).

Before proving these three assertions, we shall first show that they imply

(7.2991) $$\mathfrak{z}^r(\Gamma_\alpha\text{-}\sim)0 \qquad \text{in } \Phi_\alpha ,$$

from which Theorem 7.2 follows. Since $\mathfrak{z}'^r$ is a $(\Pi_{\alpha h} \setminus \Psi_\alpha)$-cycle of $\Pi_{\alpha h}$ and C_θ is a $(\Psi_\alpha , \Psi_\alpha)$-deformation of $\Pi_{\alpha h}$, with $C_1(\Pi_{\alpha h}) = \Psi_\alpha$, it follows by the Corollary to Theorem 3.31 that

$$\mathfrak{z}'^r(\Pi_{\alpha h} \setminus \Psi_\alpha\text{-}\sim)0 \qquad \text{in } \Pi_{\alpha h} .$$

Therefore

$$\mathfrak{z}'^r(\Phi_\alpha \setminus \Psi_\alpha\text{-}\sim)0 \qquad \text{in } \Phi_\alpha ,$$

that is,

(7.2992) $$\mathfrak{z}'^r(\Gamma_\alpha\text{-}\sim)0 \qquad \text{in } \Phi_\alpha .$$

The required relation (2.7991) is given by (7.298) and (7.2992).

Hence it remains to prove (7.297), (7.298), and (7.299). To prove (7.297), we have (by 7.297$_h$))

$$\Delta z_{\alpha k}^{\ \prime r} = s_k^{\ h}(\Delta z_{\alpha h}^{\ r} - \Delta o_{\alpha h} z_{\alpha h}^{\ r} + \Delta y_{\alpha h}^{\ r}),$$

that is, because of (7.296),

$$\Delta z_{\alpha k}^{\ \prime r} = s_k^{\ h} \Delta z_{\alpha h}^{\ r} \in L^{r-1}(\Psi_{\alpha k}).$$

Since

$$z_{\alpha l}^{\ \prime r} = s_l^{\ k} z_{\alpha k}^{\ \prime r}$$

for $l > k$, it follows that

$$z_{\alpha l}'^{r} - z_{\alpha k}'^{r} \in H^r[K(\Gamma_\alpha, \epsilon)]$$

for sufficiently large k and l. Hence $\mathfrak{z}'^{r}$ is a Γ_α-cycle of Φ_α.

For the proof of (7.298) we have

$$s_k^h z_{\alpha h}^r - z_{\alpha k}'^{r} = s_k^h(o_{\alpha h} z_{\alpha h}^r - y_{\alpha h}^r);$$

that is, by (7.295),

$$s_k^h z_{\alpha h}^r - z_{\alpha k}'^{r} = \Delta s_k^h x_{\alpha h}^{r+1}$$

or

$$s_k^h z_{\alpha h}^r \sim z_{\alpha k}'^{r} \quad \text{in } O_{\alpha k}.$$

Finally, to prove (7.299), we start with

$$(7.294) \qquad \begin{aligned} z_{\alpha h}^r - o_{\alpha h} z_{\alpha h}^r &\in L^r(P_{\alpha h} \setminus Q_{\alpha h}),\\ y_{\alpha h}^r &\in L^r(q_{\alpha h}) \subseteq L^r(P_{\alpha h} \setminus Q_{\alpha h}). \end{aligned}$$

Then (for $k = h + 1, h + 2, \cdots$)

$$z_{\alpha h}'^{r} \in L^r(P_{\alpha h} \setminus Q_{\alpha h}), \qquad z_{\alpha k}'^{r} \in L^r(P_{\alpha hk} \setminus Q_{\alpha hk}),$$

and

$$\mathfrak{z}'^{r} \in Z_{\alpha h}^r(\Pi_{\alpha h} \setminus \Psi_\alpha),$$

which completes the proof.

COROLLARY 1. If R^n is Euclidean n-space and $p \in R^n$, then $\Delta_p^n(R^n, \mathfrak{A})$ is isomorphic to $\mathfrak{A}$, while $\Delta_p^r(R^n, \mathfrak{A})$ is the null group for $0 < r < n$.

To show this, let T^n be an n-simplex containing p. According to Def. 6.2, $\Delta_p^r(R^n, \mathfrak{A})$ is isomorphic to $\Delta_p^r(\overline{T}^n, \mathfrak{A})$ for arbitrary $r > 0$, which is, by Theorem 7.2. the group Δ^n of the complex consisting of a single n-simplex T^n over $\mathfrak{A}$.

COROLLARY 2. The dimension of a polyhedron Φ is equal to the maximum r for which the group $\Delta_p^r(\Phi, \mathfrak{A})$ is different from zero for some $p \in \Phi$. Here, $\mathfrak{A}$ is arbitrary.

COROLLARY 3. If Φ is a polyhedron and p is not an isolated point of Φ, $\Delta_p^0(\Phi)$ is the null group.

This proposition is an immediate consequence of Theorem 7.2 and VIII, Theorem 1.54_0.

We emphasize especially

COROLLARY 4. If Φ is a polyhedron, $p \in \Phi$, and $\Delta_p^r(\Phi, \mathfrak{A})$ is not the null group, then the carrier of p in an arbitrary triangulation K of Φ is a face of an r-simplex $T^r \in K$.

Proof. If the carrier of p in K is an m-simplex, $m > r$, then the complex $O = O_K T^m$ does not contain any r-simplexes. Therefore $\Delta^r(O)$, and by implication Δ_p^r, is the null group. If the carrier T^m of p has dimension $m < r$, but is not a face of any r-simplex of K, then the body of the complex $| O_K T^m |$ is a polyhedron Φ_0 of dimension $<r$. Hence $\Delta_p^r(\Phi_0)$ is the null group. But, because of 6.21, $\Delta_p^r(\Phi)$ is isomorphic to $\Delta_p^r(\Phi_0)$. This proves the proposition.

COROLLARY 5. If the carrier of a point p in a triangulation K^n of an n-dimensional polyhedron Φ is an $(n - 1)$-face of precisely k n-simplexes $T_1^n, \cdots, T_k^n$ of K^n, then $\Delta_p^n(\Phi)$ is the free Abelian group of rank $k - 1$.

This proposition follows immediately from Theorem 7.2 and VIII, Theorem 1.54 $(n, n - 1)$.

§7.3. Application to the invariance of pseudomanifolds (see X, 7). Let Φ be an n-dimensional polyhedron. We shall call a point $p \in \Phi$ a *homological singularity* of Φ if $\Delta_p^n(\Phi, J)$ is not infinite cyclic. Corollaries 1 and 5 imply that a point of a closed n-dimensional pseudomanifold $\| K^n \|$ whose carrier in K^n has dimension $\geq n - 1$ is not a homological singularity.

Hence the set of homological singularities of $\| K^n \|$ has dimension $\leq n - 2$. Conversely, if the set of homological singularities of a polyhedron $\| K^n \|$ has dimension $\leq n - 2$, then every $(n - 1)$-simplex of K^n is the common face of precisely two n-simplexes of K^n. If $\| K^n \|$ is, in addition, a strongly connected polyhedron, it is a closed pseudomanifold.

We have therefore proved the following theorem (see X, 7):

7.3. *A closed n-dimensional pseudomanifold may be characterized as an n-dimensional strongly connected polyhedron Φ whose set of homological singularities has dimension $\leq n - 2$.*

If, in addition, $\Delta_0^n(\Phi) \neq 0$, then Φ is orientable; if $\Delta_0^n(\Phi) = 0$, then Φ is nonorientable.

Thus, we finally arrive at the

INVARIANCE THEOREM FOR PSEUDOMANIFOLDS 7.31. *If a triangulation of a polyhedron Φ is an n-dimensional closed (orientable) combinatorial pseudomanifold, every triangulation of every polyhedron homeomorphic to Φ is also an n-dimensional closed (orientable) pseudomanifold.*

EXERCISE. Formulate and prove the theorem analogous to Theorem 7.3 for pseudomanifolds with boundary.

Appendix 2

ABELIAN GROUPS

We shall assume that the reader is acquainted with only the very elementary notions of group theory [see, for instance, van der Waerden, *Moderne Algebra*, Chapter 2; or Pontryagin, *Topolological Groups*, Chapter 1 (the latter is especially recommended, but contains more material than we shall presuppose)]. We shall therefore assume that the following concepts are familiar to the reader: group, subgroup, homomorphism and isomorphism, factor group, cyclic group. We shall also suppose that the elementary propositions concerning these concepts are known.

In the sequel, all groups are Abelian. Accordingly, the group operation will be referred to as addition.

If the groups A and B are isomorphic, we shall write $A \approx B$.

§1. General remarks

§1.1. Homomorphism. If f is a homomorphism of a group A into a group B, the subgroup of A consisting of all the elements of A mapped by f into the identity of B is called the *kernel* of f. We recall that:

In order that a homomorphism f of A onto B be an isomorphism, it is necessary and sufficient that the kernel of f be the null group, that is, the subgroup consisting of the identity alone.

Let f be a homomorphism of a group A into a group B, and let A_0, B_0 be subgroups of A, B, respectively. If $f(A_0) \subseteq B_0$, f maps every coset of A relative to A_0 into a coset of B relative to B_0. If we assign to each coset $\mathfrak{a} \in A/A_0$ the coset $\mathfrak{b} \in B/B_0$ which contains the image of $\mathfrak{a}$ under the mapping f, we obtain a homomorphism $\mathfrak{f}$ of the factor group A/A_0 into the factor group B/B_0, and we say that $\mathfrak{f}$ *is induced by* f.

General Theorem on Homomorphisms 1.1. *If f is a homomorphism of A onto B, and $f^{-1}(B_0) = A_0$, where A_0, B_0 are subgroups of A, B, respectively, then the induced homomorphism $\mathfrak{f}$ of A/A_0 into B/B_0 is an isomorphism of A/A_0 onto B/B_0.*

Remark. If B_0 is the null group, this theorem becomes

1.10. *If A_0 is the kernel of the homomorphism f of A into B, then f induces an isomorphism of A/A_0 onto B.*

Proof of Theorem 1.1. Since f is *onto* B, $\mathfrak{f}$ is *onto* B/B_0. If the coset $\mathfrak{a} \in A/A_0$ is an element of the kernel of $\mathfrak{f}$, then f maps all the elements of the coset $\mathfrak{a}$ into elements of B_0, whence $\mathfrak{a} = A_0$. Hence the kernel of $\mathfrak{f}$ consists of a single element, the identity $\mathfrak{a} = A_0$ of A/A_0, and this is what we wished to prove.

DEFINITION 1.11. A homomorphism of a group into itself is called an *endomorphism* of the group.

§1.2. Some notable subgroups of a group A. A *proper subgroup* of a group A is any subgroup of A which is neither the group A itself nor the null group (the subgroup of A consisting of the identity alone).

If $a \in A$, the cyclic subgroup of A consisting of all the elements of A of the form ma, m an integer, is called the subgroup of A *generated* by a. Here, by definition,

$$ma = a + a + \cdots + a \quad (m \text{ times})$$

if m is positive, $ma = 0$ if $m = 0$, and

$$ma = (-a) + \cdots + (-a) \quad (|\, m \,| \text{ times})$$

if m is negative.

If the cyclic subgroup generated by a is infinite, then a is said to be an element of *infinite order* of A. In the contrary case, a is said to be an element of *finite order* and the order of a is equal, by definition, to the order (that is, the number of elements) of the cyclic group generated by a.

Let m be a fixed integer. The set of all elements ma, a an arbitrary element of A, is a subgroup of A, denoted by mA.

DEFINITION 1.21. Let A_0 be a subgroup of A. We shall denote by $\hat{A}_0$ the set of all elements $a \in A$ for each of which there is at least one natural number m such that $ma \in A_0$.

The set $\hat{A}_0$ is a subgroup of A called the *division closure of the subgroup A_0 in the group A*.

The subgroup A_0 is *division closed in A*, by definition, if it coincides with its division closure in A.

It is easy to see that the set of all elements of finite order of a given group A is a division closed subgroup of A.

THEOREM 1.22. *A subgroup A_0 of a group A is division closed in A if, and only if, the factor group A/A_0 contains no elements of finite order except the identity.*

The proof of this theorem is left to the reader.

§1.3. Decomposition of a group A into a direct sum of subgroups.

1.3. A group A is said to be a *sum* of subgroups $A_1, \cdots, A_n$ if every $a \in A$ can be represented in at least one way in the form

$$a = a_1 + \cdots + a_n, \qquad a_i \in A_i, \qquad i = 1, \cdots, n.$$

If the representation of every $a \in A$ is *unique*, the sum is called a *direct sum*.

1.31. *Let A be a sum of subgroups A_1 and A_2. The sum is direct if, and only if, $A_1 \cap A_2$ is the identity.*

Proof. 1. If $a \in A_1 \cap A_2$, $a \neq 0$, a may be represented in two distinct ways:

$$a = a + 0, \qquad a = 0 + a.$$

2. Suppose that $A_1 \cap A_2$ is the null group and

$$a = a_1 + a_2 = a'_1 + a'_2,$$

with $a_i \in A_i$, $a'_i \in A_i$, $i = 1, 2$. Then $a_1 - a'_1 = a'_2 - a_2$ is in $A_1 \cap A_2$ and is consequently the identity. Hence $a_1 = a'_1$, $a_2 = a'_2$. This proves the theorem.

§1.4. The direct sum of given groups. Let $A_1, \cdots, A_n$ be a finite number of groups, and let us consider the set A of all possible finite combinations of the form $(a_1, \cdots, a_n)$, where a_i is any element of A_i, $i = 1, \cdots, n$.

Let us define the operation of addition in the set of all these combinations as:

$$(a_1, \cdots, a_n) + (a'_1, \cdots, a'_n) = (a_1 + a'_1, \cdots, a_n + a'_n).$$

This definition converts the set A into a group, called the direct sum of the given groups $A_1, \cdots, A_n$.

Although this definition differs from that of the decomposition of a group into a direct sum of subgroups (see 1.3), it is closely connected with the latter. Indeed, if A_i^0 is the subgroup of the group A just defined which consists of all combinations of the form

$$(0, \cdots, a_i, \cdots, 0), \qquad a_i \in A_i,$$

it is easy to see that A can be decomposed into a direct sum of its subgroups A_i^0. In most cases we may identify each of the groups A_i with the corresponding A_i^0 by identifying the elements a_i and

$$(0, \cdots, a_i, \cdots, 0)$$

of these groups. If A is the direct sum of subgroups $A_1, \cdots, A_n$, we shall write

$$A = A_1 + \cdots + A_n.$$

§1.5. Direct sums and factor groups.

1.51. *If* $A = A_1 + A_2$, *the factor group* A/A_1 *is isomorphic to* A_2.

Proof. If $\mathfrak{a} \in A/A_1$ and $a = a_1 + a_2 \in \mathfrak{a}$, $a_1 \in A_1$, $a_2 \in A_2$, then

$$a_2 = a - a_1 \in \mathfrak{a}.$$

Since the difference of two elements of A_2 is contained in A_2 and consequently can be contained in A_1 only if it is the identity, a_2 is the only element of A_2 in the coset $\mathfrak{a}$. Therefore, setting $f(\mathfrak{a}) = a_2$ yields a $(1-1)$ mapping which is an isomorphism. This proves the theorem.

1.52. *Let*

$$A = A_1 + \cdots + A_n ,$$

and let A'_i *be a subgroup of* A_i $(i = 1, \cdots, n)$; *the elements of the form* $a' = a'_1 + \cdots + a'_n$ *form a subgroup* A' *of* A (A' *is the direct sum of the groups* A'_i). *Under these conditions the factor group* A/A' *is isomorphic to the direct sum of the factor groups* A_i/A'_i $(i = 1, \cdots, n)$:

$$A/A' \approx A_1/A'_1 + \cdots + A_n/A'_n ,$$

where the direct sum is to be taken in the sense of Def. 1.4.

Proof. The elements of the group $\mathfrak{B} = A_1/A'_1 + \cdots + A_n/A'_n$ have the form $\mathfrak{b} = (\mathfrak{a}_1, \cdots, \mathfrak{a}_n)$, with $\mathfrak{a}_i$ taking all values in the group A_i/A'_i. Let us assign to each element

$$a = a_1 + \cdots + a_n \in A, \qquad a_i \in A_i ,$$

the element

$$\mathfrak{b} = f(a) = (\mathfrak{a}_1, \cdots, \mathfrak{a}_n) \in \mathfrak{B}$$

where $\mathfrak{a}_i$ is defined by the condition that $a_i \in \mathfrak{a}_i$. The resulting mapping f of A onto $\mathfrak{B}$ is a homomorphism. To show that f induces an isomorphism $\mathfrak{f}$ of A/A' onto $\mathfrak{B}$, it is enough, by 1.1, to prove that the kernel of f is A'.

The last assertion is proved as follows: since the identities of the groups A_i/A'_i are the groups A'_i,

$$a = a_1 + \cdots + a_n \in A$$

is mapped by f into the identity of $\mathfrak{B}$ only if $a_i \in A'_i$ for all i, that is, if $a \in A'$.

§2. Free Abelian groups

§2.1. Definition. A group X is called a *free* (Abelian) *group of "rank"* n if it contains elements $x_1, \cdots, x_n$ with the property that every $x \in X$ can be uniquely represented in the form

$$x = c_1x_1 + \cdots + c_nx_n , \tag{2.1}$$

c_i an integer. Every set of elements $x_1, \cdots, x_n$ of X with this property is called a *basis* of the free group X. A free group X with a basis consisting of the elements $x_1, \cdots, x_n$ will be denoted by $X = [x_1, \cdots, x_n]$ or simply by X^n. X^n is obviously the direct sum of the infinite cyclic groups X_i generated by the corresponding elements x_i.

A free group of "rank" n is isomorphic to the group of all linear forms $c_1x_1 + \cdots + c_nx_n$ in n variables with integral coefficients.

2.1. *The free group* $X = [x_1, \cdots, x_n]$ *has no elements of finite order except the identity.*

Indeed, if $x = a_1x_1 + \cdots + a_nx_n$ is of order c, then

$$ca_1x_1 + \cdots + ca_nx_n = 0,$$

so that $ca_1 = 0, \cdots, ca_n = 0$, i.e., $a_1 = \cdots = a_n = 0$.

§2.2. The rank of an Abelian group. An expression

$$c_1x_1 + \cdots + c_nx_n,$$

where $x_1, \cdots, x_n$ are elements of a group A and $c_1, \cdots, c_n$ are integers, is called a *linear combination* of the elements $x_1, \cdots, x_n$ of A. A linear combination is said to be *trivial* if all the c_i are zero; in the contrary case it is said to be *nontrivial*. The elements $x_1, \cdots, x_n$ of A are said to be *linearly dependent* if some nontrivial linear combination of these elements vanishes. In the contrary case, the elements are called *linearly independent*.

The maximum m for which A has m linearly independent elements is called the *rank* of A. If A has m linearly independent elements for every m, the rank of A is infinite. We shall denote the rank of A by ρA.

The following proposition shows that this definition of the rank of a group is equivalent to that of Def. 2.1:

THEOREM 2.2. *The rank (in the sense of* 2.2) *of a free group of "rank"* n *(in the sense of* 2.1) *is* n.

Proof. The elements $x_1, \cdots, x_n$ of a basis of a free group are linearly independent. Indeed, since the element 0 can be represented in the form (2.1) uniquely, it follows that

$$c_1x_1 + \cdots + c_nx_n = 0$$

implies that

$$c_1 = \cdots = c_n = 0.$$

It remains to prove that every $n + 1$ elements $y_1, \cdots, y_{n+1}$ of a free group $X = [x_1, \cdots, x_n]$ are linearly dependent. The proof is by induction over n. The theorem is obvious for $n = 1$: the elements $y_1 = ax_1$ and $y_2 = bx_1$ are linearly dependent. (If $a = 0$, $1 \cdot y_1 + 0 \cdot y_2 = 0$; if $a \neq 0$, $-b \cdot y_1 + a \cdot y_2 = 0$.)

Suppose the theorem valid for $n = k - 1$; we shall prove it for $n = k$. Let

$$y_h = \sum_{i=1}^{k} a_{hi}x_i, \qquad h = 1, \cdots, k + 1, \tag{2.2}$$

be $k + 1$ elements of $[x_1, \cdots, x_k]$. Then the k elements

$$\bar{y}_h = y_h - a_{h1}x_1, \qquad h = 1, \cdots, k,$$

are contained in the free group $[x_2, \cdots, x_k]$ of "rank" $k - 1$ (in the sense of 2.1). Hence they are linearly dependent by the inductive hypothesis. Therefore there exist integers $c'_1, \cdots, c'_k$, not all equal to zero, such that

$$\textstyle\sum_{h=1}^{k} c'_h \bar{y}_h = 0;$$

then

$$\textstyle\sum_{h=1}^{k} c'_h y_h = b_1 x_1, \tag{2.2$_1$}$$

where

$$b_1 = \textstyle\sum_{h=1}^{k} c'_h a_{h1}.$$

In exactly the same way, considering the elements

$$\bar{y}_h = y_h - a_{hi} x_i, \qquad h = 1, \cdots, k; i = 1, \cdots, k,$$

for arbitrary fixed i, we obtain

$$\textstyle\sum_{h=1}^{k} c'_{ih} y_h = b_i x_i, \qquad i = 1, \cdots, k. \tag{2.2$_i$}$$

If $b_i = 0$ for at least one value of i, then according to (2.2$_i$) the elements $y_1, \cdots, y_k$, and consequently the elements $y_1, \cdots, y_{k+1}$, are linearly dependent. If every b_i is different from zero, it is easy to see that according to (2.2) for $h = k + 1$ and (2.2$_i$) the element

$$b_1 b_2 \cdots b_k y_{k+1}$$

is a linear combination of $y_1, \cdots, y_k$, with the coefficient of y_{k+1} different from zero. Consequently, $y_1, \cdots, y_{k+1}$ are linearly dependent. This proves the theorem.

Having proved the equivalence for free groups of the two definitions of rank, we now have:

2.21. *Every basis of the free group* $[x_1, \cdots, x_n]$ *consists of* n *elements.*

2.22. *If two free groups* $[x_1, \cdots, x_n]$ *and* $[y_1, \cdots, y_m]$ *are isomorphic, then* $m = n$.

The converse, i.e., that the two free groups $[x_1, \cdots, x_n]$ and $[y_1, \cdots, y_n]$ are isomorphic, is obvious.

§2.3. Linear equations. An immediate consequence of Theorem 2.2 is the well known

THEOREM 2.3. *A system*

$$\textstyle\sum_{k=1}^{n+1} a_{ik} \xi_k = 0, \qquad i = 1, \cdots, n, \tag{2.3}$$

of n *linear equations in* $n + 1$ *unknowns* ξ_k *with integral coefficients* a_{ik} *always has nontrivial integral solutions.*

Proof. According to Theorem 2.2, $n + 1$ arbitrary elements

$$y_k = \textstyle\sum_{i=1}^{n} a_{ik} x_i, \qquad k = 1, \cdots, n + 1,$$

of the free group $[x_1, \cdots, x_n]$ are linearly dependent. Hence there exist integers $\xi_1, \cdots, \xi_{n+1}$ not all equal to zero such that

$$0 = \sum_{k=1}^{n+1} \xi_k y_k = \sum_{i=1}^{n} \Big(\sum_{k=1}^{n+1} a_{ik}\xi_k\Big) x_i ,$$

i.e., $\sum_{k=1}^{n+1} a_{ik}\xi_k = 0$ for all $i = 1, \cdots, n$; consequently, the numbers ξ_k are solutions of the system (2.3).

Since we may cancel from the ξ_k any factors which they may have in common, the conclusion of the theorem can be strengthened by requiring that the numbers $\xi_1, \cdots, \xi_{n+1}$ be *relatively prime.*

§2.4. Bases of a free group. The elements $x_1, \cdots, x_n$ are not a unique basis for the group $[x_1, \cdots, x_n]$.

The definition of basis implies that a system of elements $y_1, \cdots, y_n$ of $[x_1, \cdots, x_n]$ forms a basis for $[x_1, \cdots, x_n]$ if, and only if, the system satisfies the following conditions:

1) Every $x \in [x_1, \cdots, x_n]$ may be written as

$$x = \sum_{i=1}^{n} a_i y_i .$$

2) The coefficients a_i are uniquely determined by x.

According to the following theorem Condition 2) follows from 1):

THEOREM 2.4. *If the elements $y_1, \cdots, y_n$ of the group $X = [x_1, \cdots, x_n]$ have the property that every element of X is a linear combination of the elements $y_1, \cdots, y_n$, then the latter form a basis for X.*

Proof. Let $Z = [z_1, \cdots, z_n]$ be a free group and assign to every element $z = \sum_{i=1}^{n} a_i z_i \in Z$ the element $f(z) = \sum_{i=1}^{n} a_i y_i \in X$. By the hypothesis of Theorem 2.4 the mapping f is a homomorphism of Z onto X, with $f(z_i) = y_i$ $(i = 1, \cdots, n)$. Since the elements z_i form a basis for Z, the proof of the theorem will follow if we show that f is an isomorphism. To show this, it is enough to show that the kernel of f is the null group, i.e., that $f(z) = 0$ implies that $z = 0$.

Hence, suppose $z \in Z$ and $f(z) = 0$. Choose elements $z'_1, \cdots, z'_n$ of Z such that $f(z'_i) = x_i$. The elements $z, z'_1, \cdots, z'_n$ are linearly dependent according to Theorem 2.2, that is, there exist numbers $b, b_1, \cdots, b_n$, not all equal to zero, such that

$$bz + b_1 z'_1 + \cdots + b_n z'_n = 0.$$

Since f is a homomorphism and $f(z) = 0, f(z'_i) = x_i$, it follows that

$$b_1 x_1 + \cdots + b_n x_n = 0;$$

whence

$$b_1 = \cdots = b_n = 0,$$

so that

$$bz = 0; \qquad b \neq 0.$$

Since the only element of finite order of the free group Z is the identity, it follows that $z = 0$, q.e.d.

We now have

2.40. *The elements* $y_1, \cdots, y_n$ *form a basis for* $X = [x_1, \cdots, x_n]$ *if, and only if,* x_i *can be written as*

$$x_i = \sum_{k=1}^{n} a_{ik} y_k, \qquad i = 1, \cdots, n. \tag{2.40}$$

For, (2.40) implies that every element of X, as a linear combination of the elements x_i, can also be written as a linear combination of the elements y_i; whence it follows that the elements y_i form a basis for X.

Example. Let

$$y_1 = x_1 + c_2 x_2 + \cdots + c_n x_n,$$

where the c_i are arbitrary integers. Then

$$y_1, x_2, \cdots, x_n$$

are a basis for X. Indeed,

$$x_1 = y_1 - c_2 x_2 - \cdots - c_n x_n.$$

§2.5. Subgroups of free groups. All the applications of group theory in this book, as well as the entire theory of Abelian groups with a finite number of generators, are based essentially on the following theorem:

Theorem 2.51. *Let* X^n *be a free Abelian group of rank* n*. Every non-null subgroup* U *of* X^n *is itself a free Abelian group with rank* $k \leq n$*. Moreover, there is a basis* $y_1, \cdots, y_n$ *of* X^n *and a basis* $u_1, \cdots, u_k$*,* $k \leq n$*, of* U *such that*

$$u_i = \theta_i y_i, \qquad i = 1, \cdots, k,$$

where $\theta_1, \cdots, \theta_k$ *are natural numbers and* θ_{i+1} *is divisible by* θ_i*,* $i = 1, \cdots, k - 1$.

Proof (borrowed from A. G. Kuroš, *Group Theory*). For $n = 1$ the theorem follows immediately from the fact that every subgroup of an infinite cyclic group with generator x is an infinite cyclic group with generator θx, θ a natural number.

Assume that Theorem 2.51 is valid for X^{n-1}. If U is a non-null subgroup of X^n, every basis of X^n determines a unique positive number, namely, the least of the positive numbers which occur as coefficients in the linear forms, relative to the given basis, representing the elements of U. This minimal coefficient changes, in general, with the basis of X^n. We now seek a basis of X^n for which this coefficient assumes its smallest possible value. Let

$$\{x_1, \cdots, x_n\} \tag{2.511}$$

be such a basis.

Let θ_1, $\theta_1 \geq 1$, be the smallest positive coefficient corresponding to this basis, and let

$$u_1 = \theta_1 x_1 + a_2 x_2 + \cdots + a_n x_n$$

be one of those elements of U whose representation relative to the basis (2.511) contains θ_1 as one of its coefficients. (The assumption that θ_1 is a coefficient for u_1 is legitimate, since the basis of X^n is not ordered.)

We shall prove, first, that every coefficient $a_2, \cdots, a_n$ is divisible by θ_1. For, let $a_2 = \theta_1 q + r$, $0 < r < \theta_1$. Let us transform the basis (2.511), replacing the element x_1 by $x_1 + qx_2$. The element u_1 may now be written as

$$u_1 = \theta_1 x'_1 + rx_2 + a_3 x_3 + \cdots + a_n x_n$$

relative to the new basis

$$x'_1 = x_1 + qx_2, x_2, \cdots, x_n.$$

We have thus found a basis of X^n relative to which one of the elements of U contains a positive coefficient less than θ_1. This contradicts the choice of θ_1 and proves the assertion.

Let $a_i = \theta_1 q_i$, $i = 2, \cdots, n$. Transform the basis (2.511) of X^n by replacing x_1 with

$$y_1 = x_1 + q_2 x_2 + \cdots + q_n x_n.$$

It is clear that $u_1 = \theta_1 y_1$. Furthermore, let us collect all the elements of U in whose representations relative to the new basis the coefficient of y_1 is zero. These elements form a subgroup U' of U, whose intersection with the cyclic subgroup $[u_1]$ generated by u_1 is the null group. We shall prove that *U is the sum of the subgroups $[u_1]$ and U'.*

Let

$$u = b_1 y_1 + b_2 x_2 + \cdots + b_n x_n$$

be an arbitrary element of U. If $b_1 = \theta_1 q + r$, $0 \leq r < \theta_1$, then the coefficient of y_1 in the element

$$u' = u - qu_1 = ry_1 + b_2 x_2 + \cdots + b_n x_n$$

of U is less than θ_1; hence, because of the definition of θ_1, $r = 0$. Therefore u' is contained in U' and

$$u = qu_1 + u'$$

is an element of the sum of the subgroups $[u_1]$ and U'.

It now follows that if $U' = 0$, then $U = [u_1]$ and the theorem is proved, If $U' \neq 0$, we obtain a decomposition of U as the direct sum

$$U = [u_1] + U'.$$

The subgroup U' is contained in the subgroup $X' = [x_2, \cdots, x_n]$, which is a free group of rank $n - 1$. Hence, by the inductive hypothesis, U' is a free group. Moreover, there exist bases $y_2, \cdots, y_n$ of X' and $u_2, \cdots, u_k$ of U' such that $k - 1 \leq n - 1$ and $u_i = \theta_i y_i$, with $\theta_i > 0$ $(2 \leq i \leq k)$ and θ_{i+1} divisible by θ_i, $i = 2, 3, \cdots, k - 1$.

We have thus proved that U is a free group of rank k, $k \leq n$. To prove that the bases

$$\{y_1, y_2, \cdots, y_n\} \tag{2.512}$$

of X^n and

$$\{u_1, u_2, \cdots, u_k\} \tag{2.513}$$

of the subgroup $U \subseteq X^n$ have the properties required by the theorem, it remains to prove merely that θ_2 is divisible by θ_1. Let $\theta_2 = \theta_1 q_0 + r_0$, $0 \leq r_0 \leq \theta_1$. Let us replace y_1 by

$$y'_1 = y_1 - q_0 y_2$$

in the basis (2.512) of X^n. The element $u_2 - u_1$ of U may now be written as

$$u_2 - u_1 = (-\theta_1)y'_1 + r_0 y_2$$

relative to the new basis; whence again, because of the choice of θ_1, it follows that $r_0 = 0$. This proves Theorem 2.51.

Remark. The number θ_i is the *order of the element* y_i, $i \leq k$, *of* X^n *relative to the subgroup* U, that is, it is the least natural number θ such that $\theta y_i \in U$. In fact, if $\theta y_i \in U$, then

$$\theta y_i = \sum c_j \theta_j y_j$$

for some integers c_j. Then because of the linear independence of the elements y_i, $c_j = 0$, $j \neq i$, and $\theta = c_i \theta_i$.

Since $\theta_i y_i \in U$ $(1 \leq i \leq k)$, $y_i \in \hat{U}$, where $\hat{U}$ as usual denotes the division closure of the subgroup U (in X^n). On the other hand, let $x \in \hat{U}$ and let

$$x = c_1 y_1 + \cdots + c_k y_k + c_{k+1} y_{k+1} + \cdots + c_n y_n .$$

Then $cx \in U$ for some integer $c \neq 0$, i.e., there exist integers $a_1, \cdots, a_k$ such that

$$cc_1 y_1 + \cdots + cc_k y_k + \cdots + cc_n y_n = a_1 \theta_1 y_1 + \cdots + a_k \theta_k y_k .$$

Hence, since $y_1, \cdots, y_n$ are linearly independent,

$$cc_1 = a_1 \theta_1, \cdots, cc_k = a_k \theta_k, cc_{k+1} = \cdots = cc_n = 0,$$

so that $c_{k+1} = \cdots = c_n = 0$ and $x = c_1 y_1 + \cdots + c_k y_k$.

Therefore

2.511. *The elements* $y_1, \cdots, y_k$ *of a basis* $\{y_1, \cdots, y_n\}$ *form a basis for the group* $\hat{U}$.

We shall now introduce new notation which will be more convenient for the applications of Theorems 2.51 and 2.511. The elements y_i of a basis $y_1, \cdots, y_n$ for which $i > k$ will be denoted by $z_1, \cdots, z_\pi$; the elements y_i, $i \leq k$, for which $\theta_i > 1$, as well as the θ_i corresponding to these elements, will be renumbered in the converse order and denoted by $u_1, \cdots, u_\tau$ [we would then have $\theta_i \equiv 0 \pmod{\theta_{i+1}}$]; the corresponding θ_i's would then be $\theta_1, \cdots, \theta_\tau$, $\theta_i \equiv 0 \pmod{\theta_{i+1}}$. Finally we denote those y_i for which $\theta_i = 1$, by $v_1, \cdots, v_\sigma$. We may now restate the results obtained above:

2.52. *Let* X *be a free Abelian group and let* U *be an arbitrary subgroup of* X. *The group* X *has a basis*

$$z_1, \cdots, z_\pi; \quad u_1, \cdots, u_\tau; \quad v_1, \cdots, v_\sigma, \tag{2.52}$$

with the following properties:

a) $v_1, \cdots, v_\sigma$ *are elements of* U; *the order of* u_i *relative to* U *is a natural number* $\theta_i > 1$, *where* $\theta_i \equiv 0 \pmod{\theta_{i+1}}$;

b) *the elements* $\theta_1 u_1, \cdots, \theta_\tau u_\tau$ *and* $v_1, \cdots, v_\sigma$ *form a basis for* U;

c) *the elements* $u_1, \cdots, u_\tau; v_1, \cdots, v_\sigma$ *are a basis for* $\hat{U}$.

A basis (2.52) of X satisfying the conditions of Theorem 2.52 is called a *canonical basis* of the free group X (relative to the subgroup U).

§2.6. Some remarks on canonical bases. Let $U = \hat{U}$ be a division closed subgroup of a free Abelian group X. Let

$$\{z_1, \cdots, z_\pi; v_1, \cdots, v_\tau\}$$

be a canonical basis of X relative to U (in consequence of the fact that U is division closed, all the $\theta_i = 1$). Then

$$[v_1, \cdots, v_\tau] = U, \; X = [z_1, \cdots, z_\pi] + [v_1, \cdots, v_\tau],$$

and

$$X/U = [\mathfrak{z}_1, \cdots, \mathfrak{z}_\pi], \tag{2.61}$$

where $\mathfrak{z}_i \in X/U$ is the coset containing z_i. Hence

2.61. *The factor group* X/U, *where* U *is a division closed subgroup of* X, *is a free group*; *and* $X = U + V$, *where* V *is a subgroup of* X *isomorphic to* X/U.

Remark. The subgroup V is, in general, not uniquely determined by the condition $X = U + V$; but all V satisfying this condition are isomorphic (since they are all isomorphic to X/U). In our notation we may take as the group V the group $[z_1, \cdots, z_\pi]$.

Definition 2.62. Let $U = \hat{U}$ be a division closed subgroup of a free group X of rank n. The elements $x_1, \cdots, x_p$ of X form, by definition, a *U-basis for X* if their cosets $\mathfrak{x}_i \in X/U$ are a basis for the free group X/U.

It is clear that

2.621. The elements $x_1, \cdots, x_p$ of a group X are a U-basis if, and only if, no nontrivial linear combination of these elements is contained in U and, moreover, every $x \in X$ has a representation

$$x = u + \sum c_i x_i, \tag{2.621}$$

where $u \in U$ and the c_i are integers.

2.63. *If $\hat{U}$ is the division closure of the subgroup U of a free group X of rank n, then every $\hat{U}$-basis of X can be extended to a canonical basis of X relative to U.*

To prove this let $x_1, \cdots, x_p$ be a $\hat{U}$-basis of X and let

$$z_1, \cdots, z_\pi; \quad u_1, \cdots, u_\tau; \quad v_1, \cdots, v_\sigma \tag{2.630}$$

be a canonical basis of X relative to U. We shall show that

$$x_1, \cdots, x_p; \quad u_1, \cdots, u_\tau; \quad v_1, \cdots, v_\sigma \tag{2.63}$$

is a canonical basis of X (whence it follows, in particular, that $p = \pi$).

To show this, it is obviously enough to prove that

$$X = [x_1, \cdots, x_p] + [u_1, \cdots, u_\tau; v_1, \cdots, v_\sigma]. \tag{2.631}$$

The definition of a $\hat{U}$-basis implies that

$$[x_1, \cdots, x_p] \cap \hat{U} = 0$$

and that every $x \in X$ has a representation (2.621). Hence, by 1.31,

$$X = [x_1, \cdots, x_p] + \hat{U}.$$

Since $\hat{U} = [u_1, \cdots, u_\tau; v_1, \cdots, v_\sigma]$, it follows that (2.631) holds.

In conclusion, we shall prove the following important, but completely elementary proposition:

2.64. *Let L be a free group of rank n and let Z be a division closed subgroup of L. If $\mathfrak{r}_1, \cdots, \mathfrak{r}_p$ is a basis of the free group L/Z; $z_1, \cdots, z_q$ is a basis of Z; and x_i is an arbitrary element of the coset $\mathfrak{r}_i$, $i = 1, \cdots, p$, then*

$$\{x_1, \cdots, x_p; z_1, \cdots, z_q\}$$

is a basis for L.

We shall prove

a) every $x \in L$ has a representation

$$x = \sum_i a_i x_i + \sum_h c_h z_h;$$

b) the representation a) is unique; b) is obviously equivalent to

b′) the elements $x_1, \cdots, x_p; z_1, \cdots, z_q$ are linearly independent.

Proof of a). Let $\mathfrak{x}$ denote the coset of $x \in L$ relative to Z. Since $\mathfrak{x}_1, \cdots, \mathfrak{x}_p$ form a basis for L/Z, there exist $a_1, \cdots, a_p$ such that

$$\mathfrak{x} = \sum a_i \mathfrak{x}_i .$$

Then

$$x - \sum a_i x_i \in Z$$

and therefore there are integers c_h such that

$$x - \sum_i a_i x_i = \sum_h c_h z_h ,$$

that is,

$$x = \sum_i a_i x_i + \sum_h c_h z_h .$$

This proves a).

Proof of b′). Let $\sum a_i x_i + \sum c_h z_h = 0$. We must show that all the a_i and c_h are zero. We have

$$\sum a_i x_i = -\sum c_h z_h \in Z \tag{2.641}$$

or

$$\sum a_i \mathfrak{x}_i = 0.$$

Since the cosets $\mathfrak{x}_i$ (as elements of a basis of the free group L/Z) are linearly independent, all the a_i are zero and hence by (2.641),

$$\sum c_h z_h = 0.$$

Therefore, in view of the linear independence of the z_h, all the c_h are zero.

§2.7. Homomorphisms and endomorphisms of free groups. The matrix of a homomorphism (for prescribed bases). The trace of an endomorphism. Let f be a homomorphism of a free group X of rank p into a free group Y of rank q.

If $X = [x_1, \cdots, x_p]$, $Y = [y_1, \cdots, y_q]$, then

$$f(x_h) = \sum_{j=1}^{q} a_{hj} y_j , \qquad h = 1, \cdots, p.$$

The matrix

$$\left\| \begin{matrix} a_{11}, & \cdots, & a_{1j}, & \cdots, & a_{1q} \\ \cdot & \cdot & \cdot & \cdot & \cdot \\ a_{h1}, & \cdots, & a_{hj}, & \cdots, & a_{hq} \\ \cdot & \cdot & \cdot & \cdot & \cdot \\ a_{p1}, & \cdots, & a_{pj}, & \cdots, & a_{pq} \end{matrix} \right\|$$

is called the *matrix of the homomorphism* f of the group X into the group Y (relative to the bases $x_1, \cdots, x_p$; $y_1, \cdots, y_q$).

A homomorphism of a group into itself is called an endomorphism.

Let f be an endomorphism of a free group X of rank n and let $x_1, \cdots, x_n$ be a basis of X, with

$$f(x_h) = \sum_i a_{hi}x_i . \tag{2.71}$$

The number

$$\sum_h a_{hh}$$

is called the *trace of the endomorphism* f (relative to the basis $x_1, \cdots, x_n$).

THEOREM 2.7. *The trace of an endomorphism is independent of the choice of a basis.*

Proof. Let $y_1, \cdots, y_n$ be another basis of X, related to $x_1, \cdots, x_n$ by the equations

$$y_p = \sum_q u_{pq}x_q , \tag{2.72}$$

$$x_q = \sum_r v_{qr}y_r . \tag{2.73}$$

Suppose that the endomorphism f relative to the basis $y_1, \cdots, y_n$ is given by the equations

$$f(y_j) = \sum_k b_{jk}y_k . \tag{2.74}$$

We must show that

$$\sum_h a_{hh} = \sum_j b_{jj} . \tag{2.75}$$

If we substitute the expression (2.73) for x_q into (2.72), we get

$$y_p = \sum_r \sum_q u_{pq}v_{qr}y_r ;$$

whence, in consequence of the independence of y_r,

$$\sum_q u_{pq}v_{qr} = \delta_{pr} , \tag{2.76}$$

where

$$\delta_{pr} = \begin{cases} 0, & p \neq r, \\ 1, & p = r. \end{cases}$$

In the same way, substituting (2.72) for y_p with the subscripts p, q replaced by r, s into (2.73), we obtain as a consequence of the independence of the x_s :

$$\sum_r v_{qr}u_{rs} = \delta_{qs} . \tag{2.77}$$

Moreover, by (2.72) and (2.71),

$$f(y_j) = f(\sum_h u_{jh}x_h) = \sum_h u_{jh}f(x_h) = \sum_h \sum_q u_{jh}a_{hq}x_q ;$$

and from (2.74) and (2.72),

$$f(y_j) = \sum_k b_{jk} y_k = \sum_k \sum_q b_{jk} u_{kq} x_q .$$

Consequently

$$\sum_h \sum_q u_{jh} a_{hq} x_q = \sum_k \sum_q b_{jk} u_{kq} x_q ,$$

so that

$$\sum_h u_{jh} a_{hq} = \sum_k b_{jk} u_{kq}$$

since the x_q are independent. Multiplying this equation by v_{qr}, summing over q, and applying (2.76), we obtain

$$\sum_h \sum_q u_{jh} a_{hq} v_{qr} = \sum_k b_{jk} \delta_{kr} = b_{jr} .$$

Then (2.77) yields

$$\sum_j b_{jj} = \sum_j \sum_h \sum_q v_{qj} u_{jh} a_{hq} = \sum_q \sum_h \delta_{qh} a_{hq} = \sum_h a_{hh} .$$

This proves the assertion.

Accordingly, we may now speak of the *trace of an endomorphism* independently of any basis.

§2.8. Addition theorem for traces. Let f be an endomorphism of a free group X into itself and suppose that Y is a subgroup of X such that $f(Y) \subseteq Y$. Since Y is a free group (see 2.5), the trace SpY of the endomorphism f of Y is defined. We shall write the trace of the endomorphism f of the group X as SpX.

Since $f(Y) \subseteq Y$, the division closure $\hat{Y}$ of Y in X is also mapped into itself by f. For, if $x \in \hat{Y}$, there is a natural number m such that $mx \in Y$; then $mf(x) = f(mx) \in Y$ and $f(x) \in \hat{Y}$. Since f maps $\hat{Y}$ into itself, f induces (see 1.1) an endomorphism $\mathfrak{f}$ of the factor group $X/\hat{Y}$.

If $X/\hat{Y} \neq 0$, $X/\hat{Y}$ is a free group (by 2.61). Hence the trace $Sp(X/\hat{Y})$ of $\mathfrak{f}$ is defined. If $X/\hat{Y} = 0$, we shall set $Sp(X/\hat{Y}) = 0$, by definition.

THEOREM 2.8. *If an endomorphism f of a free group X maps a subgroup Y into itself, then*

$$SpX = SpY + Sp(X/\hat{Y}).$$

Proof. If $Y = 0$, the theorem is obvious. If $Y \neq 0$, choose a canonical basis $x_1, \cdots, x_n$ of X relative to Y. Then for some $r \leq n$,

$$\hat{Y} = [x_1, \cdots, x_r], \qquad Y = [\theta_1 x_1, \cdots, \theta_r x_r],$$

where some of the θ_i, and in fact all the θ_i starting with some $i = k \leq r$, may be equal to 1. If $\mathfrak{x}_{r+1}, \cdots, \mathfrak{x}_n$ are the cosets of X relative to $\hat{Y}$ which contain the elements $x_{r+1}, \cdots, x_n$, then by (2.61)

$$X/\hat{Y} = [\mathfrak{x}_{r+1}, \cdots, \mathfrak{x}_n].$$

Let

(2.81) $$f(x_h) = \sum_{i=1}^{n} a_{hi}x_i, \qquad h = 1, \cdots, n.$$

Since $\hat{Y}$ is mapped into itself,

(2.82) $$a_{hi} = 0 \begin{cases} h = 1, \cdots, r; \\ i = r+1, \cdots, n. \end{cases}$$

Since Y is also mapped into itself,

(2.83) $$f(\theta_h x_h) = \sum_{i=1}^{r} a'_{hi}\theta_i x_i, \qquad h = 1, \cdots, r.$$

On the other hand, (2.81) and (2.82) imply

(2.84) $$f(\theta_h x_h) = \theta_h \sum_{i=1}^{r} a_{hi}x_i, \qquad h = 1, \cdots, r.$$

Since $a'_{hi}\theta_i = \theta_h a_{hi}$ by (2.83) and (2.84), it follows that $a'_{hh} = a_{hh}$ for $i = h$, so that

(2.85) $$SpY = \sum_{h=1}^{r} a_{hh}.$$

This proves the theorem for $r = n$, i.e., $X/\hat{Y} = 0$. Suppose $r < n$. Applying (2.81) for $h = r + 1, \cdots, n$ and using the fact that $x_1, \cdots, x_r$ are elements of $\hat{Y}$ contained in the identity coset of $X/\hat{Y}$, we get

$$f(\mathfrak{x}_h) = \sum_{i=r+1}^{n} a_{hi}\mathfrak{x}_i, \qquad h = r+1, \cdots, n.$$

Hence

(2.86) $$Sp(X/\hat{Y}) = \sum_{h=r+1}^{n} a_{hh}.$$

The theorem now follows from (2.85) and (2.86).

Remark. If $X = \hat{Y}$, we get the special case

$$SpY = Sp\hat{Y}.$$

§3. Theorems on the rank of groups

3.1. *The rank of a group A is zero if, and only if, all the elements of A have finite order.*

Proof. If all the elements of A are of finite order, then

$$c_1x_1 + \cdots + c_nx_n = 0$$

for every set $x_1, \cdots, x_n \in A$ and c_i equal to the order of x_i. Hence every set $x_1, \cdots, x_n \in A$ is linearly dependent. Therefore A contains no nonempty set of linearly independent elements and the rank of A is zero.

Conversely, if $\rho A = 0$ and $x \in A$, there is a nonzero integer c such that $cx = 0$. Hence all the elements of A have finite order.

3.2. *If A is a group and A_0 is an arbitrary subgroup of A,*

(3.2) $$\rho A = \rho A_0 + \rho(A/A_0)$$

(the ranks of the above groups may be finite or infinite.)

We shall first prove the inequality

$$\rho A \geq \rho A_0 + \rho(A/A_0). \tag{3.21}$$

For this it is enough to prove the following proposition:

3.21. *If the elements* $x_1, \cdots, x_p$ *of* A_0, *and the elements* $\mathfrak{y}_1, \cdots, \mathfrak{y}_q$ *of* A/A_0 *are linearly independent, and if* $y_i \in \mathfrak{y}_i$, *then*

$$x_1, \cdots, x_p, \qquad y_1, \cdots, y_q$$

are linearly independent.

To prove this suppose that

$$\textstyle\sum_{i=1}^{p} a_i x_i + \sum_{j=1}^{q} b_j y_j = 0 \tag{3.211}$$

for integers a_i and b_j. Remembering that $\sum_{i=1}^{p} a_i x_i \in A_0$, (3.211) yields for the cosets:

$$\textstyle\sum_{j=1}^{q} b_j \mathfrak{y}_j = 0. \tag{3.212}$$

Hence, because of the linear independence of the $\mathfrak{y}_j$, $b_j = 0$ $(j = 1, \cdots, q)$ and (3.211) becomes

$$\textstyle\sum_{i=1}^{p} a_i x_i = 0. \tag{3.213}$$

The linear independence of the x_i then implies that $a_i = 0$ $(i = 1, \cdots, p)$. This proves (3.21).

Before we go on to the proof of the inequality

$$\rho A \leq \rho A_0 + \rho(A/A_0), \tag{3.22}$$

we note that (3.21) implies

3.210. *If* A_0 *is a subgroup of* A,

$$\rho A_0 \leq \rho A \quad \text{and} \quad \rho(A/A_0) \leq \rho A.$$

We shall now prove (3.22). In the proof we may obviously assume that the ranks of A_0 and A/A_0 are finite, since (3.22) is obvious in the contrary case.

Suppose $\rho A_0 = p$, $\rho(A/A_0) = q$, and let $x_1, \cdots, x_p$ be linearly independent elements of A_0, and $\mathfrak{y}_1, \cdots, \mathfrak{y}_q$ linearly independent elements of A/A_0, with $y_i \in \mathfrak{y}_i$.

LEMMA. *If* $z \in A$, *there exists a nonzero integer* c *and there exist integers* c_i, d_j *such that*

$$cz = \textstyle\sum_{i=1}^{p} c_i x_i + \sum_{j=1}^{q} d_j y_j.$$

For, if $z \in \mathfrak{z} \in A/A_0$, the elements $\mathfrak{z}, \mathfrak{y}_1, \cdots, \mathfrak{y}_q$ of A/A_0 are linearly dependent. Hence a combination $b\mathfrak{z} - \sum_{j=1}^{q} b_j \mathfrak{y}_j$ is zero:

$$b\mathfrak{z} - \textstyle\sum_{j=1}^{q} b_j \mathfrak{y}_j = 0. \tag{3.221}$$

Then, because of the linear independence of the $\mathfrak{y}_j$, $b \neq 0$.

(3.221) is equivalent to

$$z' = bz - \sum_{j=1}^{q} b_j y_j \in A_0 . \tag{3.222}$$

The elements $z', x_1, \cdots, x_p$ of A_0 are linearly dependent, so that a linear combination

$$az' - \sum_{i=1}^{p} a_i x_i = 0. \tag{3.223}$$

Hence, due to the linear independence of the x_i, $a \neq 0$.

From (3.222) and (3.223) we obtain

$$abz = az' + \sum_{j=1}^{q} ab_j y_j = \sum_{i=1}^{p} a_i x_i + \sum_{j=1}^{q} ab_j y_j , \tag{3.224}$$

so that, setting $c = ab \neq 0$, $c_i = a_i$, $d_j = ab_j$,

$$cz = \sum_{i=1}^{p} c_i x_i + \sum_{j=1}^{q} d_j y_j .$$

This proves the lemma.

Let us now write the elements $y_1, \cdots, y_q$ as $x_{p+1}, \cdots, x_{p+q}$. We must show that $\rho A \leq p + q$, that is, that every $p + q + 1$ elements

$$z_1, \cdots, z_{p+q+1}$$

of A are linearly dependent. To prove this we note first that by the Lemma there are integers $c_k \neq 0$ and c_{ik} such that

$$c_k z_k = \sum_{i=1}^{p+q} c_{ik} x_i , \qquad k = 1, \cdots, p + q + 1. \tag{3.225}$$

On the other hand, the theorem on linear homogeneous equations (see Theorem 2.3) implies that there exist integers $\xi_1, \cdots, \xi_{p+q+1}$, not all zero, with the property

$$\sum_{k=1}^{p+q+1} c_{ik} \xi_k = 0, \qquad i = 1, \cdots, p + q. \tag{3.226}$$

We then have by (3.225) and (3.226)

$$\sum_{k=1}^{p+q+1} \xi_k c_k z_k = \sum_{i=1}^{p+q} \Big(\sum_{k=1}^{p+q+1} c_{ik} \xi_k \Big) x_i = 0;$$

since not all the ξ_k are zero and all the c_k are different from zero, $\sum_{i=1}^{p+q+1} \xi_k c_k z_k$ is a nontrivial linear combination, and $z_1, \cdots, z_{p+q+1}$ are linearly dependent.

This completes the proof of Theorem 3.2.

§3.3. The rank of a group (**mod** m). An integral linear combination $\sum c_i x_i$ of elements x_i of a group A is said to be *nontrivial* (mod m) if not all the c_i are divisible by m. The elements x_i are said to be linearly dependent (mod m) if a nontrivial (mod m) linear combination of these elements is zero. The maximum integer r for which A contains r linearly independent (mod m) elements is called the *rank of* A (mod m) and is denoted by $\rho_m(A)$.

The following theorem may be proved by an almost verbatim repetition of the reasoning of 3.2 (but with the equalities replaced by the correpsonding congruences):

3.3. *If m is a prime, and A is a group with the property that mA is the null group (for m a prime this means that all the elements of A are of order m), then*

$$\rho_m A = \rho_m A_0 + \rho_m(A/A_0)$$

for an arbitrary subgroup A_0 of A.

REMARK. The inequality

$$\rho_m A \geq \rho_m A_0 + \rho_m(A/A_0) \tag{3.31}$$

can be proved without the assumption that m is prime. The condition $mA = 0$ appears in the argument in the following way. To prove (3.31), following the proof of (3.21), we deduce that $b_j \equiv 0 \pmod{m}$ from (3.212), whence it follows, by (3.211), that

$$\textstyle\sum_{i=1}^{p} a_i x_i = -\sum_{j=1}^{q} b_j y_j \in mA.$$

This in turn, since $mA = 0$, yields

$$\textstyle\sum_{i=1}^{p} a_i x_i = 0,$$

i.e., $a_i \equiv 0 \pmod{m}$.

To prove that

$$\rho_m A \leq \rho_m A_0 + \rho_m(A/A_0)$$

we must again use the condition $mA = 0$ to derive (3.222). The fact that m is prime is used to show that

$$a \not\equiv 0 \pmod{m}, \qquad b \not\equiv 0 \pmod{m}$$

implies $ab \not\equiv 0 \pmod{m}$. We note, finally, that we may still require of the numbers ξ_i that they do not have a common divisor (see remark at the end of 2.3).

§4. Groups with a finite number of generators

§4.1. Definition. A system $\mathfrak{S}$ of elements of a group X is called a *system of generators of X* (the elements themselves are called *generators*) if every element of X has at least one representation as a linear combination $\sum a_i x_i$, with $x_i \in \mathfrak{S}$ and the a_i integers.

We shall now consider groups whose systems of generators consist of a finite number of elements, that is, *groups with a finite number of generators.* The simplest examples of groups with a finite number of generators are the free groups of finite rank; every basis of a free group is a system of generators of the group. Moreover, every finite group (that is, a group having a finite number of elements) is a group with a finite number of generators, since the set of all elements of a group is itself a system of generators of the group.

REMARK. We shall say that a group is generated by n elements if it has a system of generators consisting of n elements.

THEOREM 4.11. *In order that a group be generated by n elements, it is necessary and sufficient that it be the homomorphic image of a free group of rank* $\leq n$.

Proof. Necessity. Let $x_1, \cdots, x_n$ be a system of generators of the group X and consider the free group $Y = [y_1, \cdots, y_n]$ of rank n. The mapping

$$f(\sum a_i y_i) = \sum a_i x_i$$

is a homomorphism of Y onto X.

Sufficiency. If f is a homomorphism of the free group $[y_1, \cdots, y_n]$ onto X, the elements $f(y_1), \cdots, f(y_n)$ are a system of generators of X.

Theorems 4.11 and 3.2 imply that $\rho X \leq n$, i.e.,

4.12. *The rank of a group with n generators is* $\leq n$.

§4.2. THEOREM 4.21. *If a group X is generated by n elements, the same is true for X/Z, where Z is any subgroup of X.*

Proof. The group X is homomorphic to a free group Y of rank $\leq n$ (Theorem 4.11); the group X/Z is homomorphic to X. Since the product of two homomorphisms is a homomorphism, X/Z is homomorphic to Y.

THEOREM 4.22. *If a group X is generated by n elements, so is any subgroup Z of X.*

Proof. By Theorem 4.11 there is a homomorphism f of a free group Y of rank n onto X. It is easy to see that $f^{-1}(Z)$ is a subgroup of Y and is therefore a free group of rank $\leq n$. According to Theorem 4.11 its image is generated by n elements.

THEOREM 4.23. *If Z is a subgroup of a group X, and the groups Z and X/Z both have a finite system of generators, then so does X.*

Proof. Let $z_1, \cdots, z_r$ $(\mathfrak{r}_1, \cdots, \mathfrak{r}_s)$ be a system of generators of Z (X/Z), and let $x_1, \cdots, x_s$ be arbitrary elements of the corresponding cosets $\mathfrak{r}_1, \cdots, \mathfrak{r}_s$. Then $z_1, \cdots, z_r, x_1, \cdots, x_s$ generate X. For, if $x \in X$ and x is contained in the coset $\sum b_j \mathfrak{r}_j$, then $x = z + \sum b_j x_j$, with $z \in Z$. Consequently $z = \sum a_i z_i$ and $x = \sum a_i z_i + \sum b_j x_j$, which was to be proved.

§4.3. Fundamental theorem for groups with finite systems of generators.

DEFINITION 4.31. A representation of a group A as a direct sum of a finite number of cyclic groups

$$A = \Theta_1 + \cdots + \Theta_\tau + Z_1 + \cdots + Z_\pi, \tag{4.31}$$

with the Θ_i finite and the Z_j infinite cyclic groups, will be referred to as a *cyclic decomposition* of A.

The cyclic decomposition (4.31) will be called *canonical* if the orders θ_i of the groups Θ_i have the additional property

$$\theta_i \equiv 0 \pmod{\theta_{i+1}}. \tag{4.310}$$

FUNDAMENTAL THEOREM 4.32. *Every Abelian group with a finite number of generators admits of a canonical cyclic decomposition.*

Proof. According to Theorem 4.11 every group A with a system of n generators is a homomorphic image of a free group X of rank n and there is, therefore, a subgroup U of X such that $A \approx X/U$.

Suppose that

$$z_1, \cdots, z_\pi; u_1, \cdots, u_\tau; v_1, \cdots, v_\sigma$$

is a canonical basis for X relative to U (see Theorem 2.52). Then

$$X = [z_1] + \cdots + [z_\pi] + [u_1] + \cdots + [u_\tau] + [v_1] + \cdots + [v_\sigma],$$

$$U = [\theta_1 u_1] + \cdots + [\theta_\tau u_\tau] + [v_1] + \cdots + [v_\sigma].$$

Hence, by Theorem 1.52,

$$X/U = [z_1] + \cdots + [z_\pi] + [u_1]/[\theta_1 u_1] + \cdots + [u_\tau]/[\theta_\tau u_\tau] + 0 + \cdots + 0,$$

that is,

$$A = Z_1 + \cdots Z_\pi + \Theta_1 + \cdots + \Theta_\tau,$$

where $Z_1, \cdots, Z_\pi$ are infinite cyclic, $\Theta_1, \cdots, \Theta_\tau$ are cyclic groups of order $\theta_1, \cdots, \theta_\tau$, and

$$\theta_i \equiv 0 \pmod{\theta_{i+1}}.$$

This completes the proof of the theorem.

REMARK. The numbers $\theta_1, \cdots, \theta_\tau$ are called the torsion numbers of the given canonical decomposition of A, or simply the *torsion numbers of* A. The following theorem justifies the use of the latter terminology:

4.320. *Every two canonical decompositions of a group A are isomorphic in the sense that they consist of the same number of cyclic groups, corresponding pairs of which are isomorphic, and have the same torsion numbers.*

We shall not prove this theorem (see Seifert-Threlfall [S–T, §§86–87]). We merely note that (3.2) and Theorem 3.1 immediately imply one of the assertions of Theorem 4.320:

4.33. *The number of infinite cyclic groups in an arbitrary cyclic decomposition of a group X is equal to the rank ρX of X.*

Moreover,

4.34. *In an arbitrary cyclic decomposition of a group A the direct sum of the finite groups of the decomposition is identical with the subgroup Θ of A consisting of the elements of A having finite order.*

Let

$$A = \Theta_1 + \cdots + \Theta_\tau + Z_1 + \cdots + Z_\pi$$

be a cyclic decomposition of A, with $\Theta_1, \cdots, \Theta_\tau$ finite and $Z_1, \cdots, Z_\pi$ infinite. It may happen that one of the numbers τ or π is zero (if $\tau = \pi = 0$, A is the null group).

It is obvious that every element of the direct sum $\Theta_1 + \cdots + \Theta_\tau$ is of finite order. Conversely, suppose that

$$u = a_1u_1 + \cdots a_\tau u_\tau + b_1z_1 + \cdots + b_\pi z_\pi \in \Theta,$$

where the u_i (z_h) are generators of the groups Θ_i (Z_h), is an element of A having a finite order. If $mu = 0$, that is,

$$ma_1u_1 + \cdots + ma_\tau u_\tau + mb_1z_1 + \cdots + mb_\pi z_\pi = 0,$$

and $m \neq 0$, then $b_1 = \cdots = b_\pi = 0$, so that $u \in \Theta_1 + \cdots + \Theta_\tau$.

Hence

$$\Theta_1 + \cdots + \Theta_\tau = \Theta.$$

A consequence of Theorems 4.33 and 4.34 is the following important

THEOREM 4.35. *Every group A with a finite number of generators may be written as a direct sum*

$$A = \Theta + Z,$$

where Θ is the subgroup of all the elements of finite order of A and Z is a free group of rank $\rho Z = \rho A$ (a free group of rank 0 is the null group).

COROLLARY 4.351. *The structure of a group with a finite number of generators completely determines its rank and the structure of its subgroup of elements of finite order.*

REMARK. Two groups have the same structure if they are isomorphic.

COROLLARY 4.352. A group with a finite number of generators which does not contain elements of finite order is a free group.

REMARK. 4.32, 4.320, and 4.33 imply

4.353. The structure of a group with a finite system of generators is completely determined by its rank and its torsion numbers.

§4.4. Further consequences.

THEOREM 4.41. *Suppose that X and Y are two groups of which one, say X, has a finite system of generators. If X', Y' are subgroups of X, Y, respectively, such that*

$$X \approx Y' \tag{4.41$_x$}$$

and

$$Y \approx X', \tag{4.41$_y$}$$

then

$$X \approx Y.$$

Proof. First, it follows from (4.41_y) and Theorem 4.22 that Y also has a finite number of generators. Further, (4.41_x) and (4.41_y) combined with (3.21) imply that

$$\rho X \leq \rho Y \quad \text{and} \quad \rho Y \leq \rho X,$$

so that

$$\rho X = \rho Y.$$

Because of 4.351 it merely remains to be proved that

$$\Theta_x \approx \Theta_y , \tag{4.42}$$

where Θ_x (Θ_y) is the subgroup of X (Y) containing its elements of finite order. For analogous definitions of the groups $\Theta_{x'}$, $\Theta_{y'}$, using (4.41_x) and (4.41_y) we obtain

$$\Theta_x \approx \Theta_{y'} \tag{4.43_x}$$

and

$$\Theta_y \approx \Theta_{x'} . \tag{4.43_y}$$

If the orders of the groups Θ_x , Θ_y , $\Theta_{x'}$, $\Theta_{y'}$ are a, b, a', and b', then

$$a = b' \tag{4.44_x}$$

and

$$b = a'. \tag{4.44_y}$$

Since $\Theta_{y'} \subseteq \Theta_y$, $b' \leq b$. Hence, by (4.44_x) and (4.44_y), $a \leq a'$. But $\Theta_{x'} \subseteq \Theta_x$, so that $a' \leq a$, i.e., $a = a'$. Consequently, $\Theta_{x'} = \Theta_x$ and (4.42) follows from (4.43_y).

THEOREM 4.42. *If X is a group with a finite number of generators and Z is a division closed subgroup of X, then X contains a free (or null) subgroup Y such that*

$$X = Y + Z.$$

Proof. The factor group X/Z contains no elements of finite order (Theorem 1.22) and has a finite number of generators (Theorem 4.21). Hence it is a free group (or the null group). The case $X/Z = 0$, i.e., $X = Z$, is trivial. Suppose

$$X/Z = [\mathfrak{y}_1 , \cdots , \mathfrak{y}_s],$$

where $\mathfrak{y}_1 , \cdots , \mathfrak{y}_s$ are cosets relative to Z. Choose elements $y_1 , \cdots , y_s$ from these cosets and denote by Y the group of all linear combinations

$\sum_{i=1}^{s} a_i y_i$. To prove the theorem it is enough to show: 1) Every $x \in X$ has a representation $x = y + z$, $y \in Y$, $z \in Z$, and 2) $Y \cap Z = 0$.

Proof of 1): if $x \in \sum_i a_i y_i$, then $x - \sum a_i y_i = z \in Z$. Hence

$$x = y + z,$$

where $y = \sum a_i y_i \in Y$.

Proof of 2): if $\sum a_i y_i \in Z$, then $\sum a_i \mathfrak{y}_i = 0$, and so $a_i = 0$ for all i.

§4.5. Remarks.

REMARK 1. Let

$$y_1, \cdots, y_\tau, z_1, \cdots, z_p \tag{4.51}$$

be a system of generators of a group A, where the elements $y_1, \cdots, y_\tau$ have finite orders $\theta_1, \cdots, \theta_\tau$, and the elements $z_1, \cdots, z_p$ have infinite order. The system of generators (4.51) is said to be a basis for A if

$$\sum_i a_i y_i + \sum_h c_h z_h = 0$$

is equivalent to $a_i \equiv 0 \pmod{\theta_i}$, $i = 1, \cdots, \tau$, and $c_h = 0$, $h = 1, \cdots, p$.

REMARK 2. We already know that the subgroup mX of a group X consists, by definition, of all the elements $mx \in X$, where x takes all values in X. Let us now denote by ${}_mX$ the subgroup of X consisting of all $x \in X$ for which $mx = 0$, and by X_m the factor group X/mX. It is easy to see that if

$$X = X' + X'' + \cdots + X^{(n)},$$

then

$${}_mX = {}_mX' + {}_mX'' + \cdots + {}_mX^{(n)}, \tag{4.52}$$

$$mX = mX' + mX'' + \cdots + mX^{(n)}, \tag{4.53}$$

and also (Theorem 1.52)

$$X_m = X'_m + X''_m + \cdots + X_m^{(n)}. \tag{4.54}$$

It is furthermore quite easy to see that for $m, m' \geq 2$,

$${}_{m'}(J_m) \approx (J_m)_{m'} \approx J_{(m,m')},$$

where J_m is the group of integers (mod m) and (m, m') is the greatest common divisor of m and m'. Since the same is, of course, true for any group isomorphic to J_m, i.e., for every finite *cyclic* group $\Theta \approx J_n$, ${}_m\Theta \approx J_{(n,m)} \approx (J_n)_m \approx \Theta_m$; and since every finite group is a direct sum of finite cyclic groups, we obtain the following theorem by using (4.52) and (4.54):

THEOREM 4.5. *If Θ is a finite group and $m \geq 2$,*

$${}_m\Theta \approx \Theta_m.$$

§5. Modules

§5.1. Scalar product in groups with a prescribed basis. Let $\{t_1, \cdots, t_n\}$ be a basis for a free Abelian group X^n of rank n. Every $x \in X^n$ has a unique representation $x = \sum a_i t_i$.

DEFINITION 5.11. If

$$x = \sum a_i t_i, \qquad y = \sum b_i t_i$$

are two elements of X^n, the integer

$$(x \cdot y) = \sum a_i b_i$$

will be called the *scalar product of $x, y \in X^n$ relative to the basis* $\{t_1, \cdots, t_n\}$.

Clearly,

$$(x \cdot y) = (y \cdot x),$$

$$((x_1 + x_2) \cdot y) = (x_1 \cdot y) + (x_2 \cdot y),$$

$$(x \cdot (y_1 + y_2)) = (x \cdot y_1) + (x \cdot y_2).$$

For arbitrary $x = \sum a_i t_i$, $(x \cdot t_i) = a_i$; and, in particular, $(t_i \cdot t_j) = \delta_{ij}$.

§5.2. Modules. The free Abelian group of rank n consisting of all the linear forms $\sum a_i t_i$ with integral coefficients in given variables $t_1, \cdots, t_n$, with the usual addition as the group operation, is called an *integral module* of rank n and is denoted by $[t_1, \cdots, t_n]$. (A module, in other words, is a free group one of whose bases is singled out once and for all.) The basis $t_1, \cdots, t_n$ of the group $[t_1, \cdots, t_n]$ (that is, the linear forms $\sum a_i t_i$, all of whose coefficients are zero except for a single one, equal to 1) is called the *initial basis* of the module. The scalar product relative to this basis is called simply the scalar product in the given module. Naturally, *every* basis of the group $[t_1, \cdots, t_n]$ is called a basis of the module.

If $\{x_1, \cdots, x_n\}$ is any basis of the module $[t_1, \cdots, t_n]$, then every element x of the module has a unique representation

$$x = \sum a^i x_i$$

and the coefficients a^i are referred to as the *contravariant components of x relative to the basis* $\{x_1, \cdots, x_n\}$. It is clear that every element of the module $[t_1, \cdots, t_n]$ is completely determined by its contravariant components relative to any basis of the module.

The numbers

$$a_1 = (x \cdot x_1), \cdots, a_n = (x \cdot x_n)$$

are called the *covariant components* of x relative to the basis $\{x_1, \cdots, x_n\}$. Clearly, relative to the initial basis $\{t_1, \cdots, t_n\}$, the contravariant and covariant components of every element of $[t_1, \cdots, t_n]$ coincide.

5.21. *Every element of a module is completely determined by its covariant components relative to any basis of the module.*

Proof. Let $x_i = \sum a_i^j t_j$, $i = 1, \cdots, n$, be a given basis and $a_i = (x \cdot x_i)$. It is required to find the coefficients c^i of the representation $x = \sum c^i t_i$.

We note first that since $x_1, \cdots, x_n$ form a basis, we may write

$$t_j = \sum b_j^i x_i, \qquad j = 1, \cdots, n,$$

where $\det a_i^j \cdot \det b_j^i = 1$. Since $\det a_i^j$ and $\det b_j^i$ are integers, each of them is equal to ± 1.

To determine c^i we have the equations

$$(x \cdot x_1) = ((c^1 t_1 + \cdots + c^n t_n) \cdot x_i) = a_i, \quad i = 1, \cdots, n. \tag{5.21}$$

Since $x_i = \sum a_i^j t_j$, (5.21) may be rewritten as

$$a_i^1 c^1 + \cdots + a_i^n c^n = a_i, \qquad i = 1, \cdots, n,$$

and solved by Kramer's rule (since $\det a_i^j = \pm 1$, the solutions are integers).

§5.3. Dual bases of a module. Two bases

$$X = \{x_1, \cdots, x_n\},$$

$$\bar{X} = \{\bar{x}_1, \cdots, \bar{x}_n\}$$

of a module $[t_1, \cdots, t_n]$ are said to be *dual* if

$$(x_i \cdot \bar{x}_j) = \delta_{ij}$$

for arbitrary $i, j = 1, \cdots, n$; x_i and $\bar{x}_i$ are said to be *corresponding* elements of the two dual bases.

5.3. *For every basis X of a module $[t_1, \cdots, t_n]$ there exists a unique dual basis X.*

Indeed, if

$$x_i = a_{i1} t_1 + \cdots + a_{in} t_n, \qquad i = 1, \cdots, n,$$

are the elements of the given basis X and

$$t_h = b_{h1} x_1 + \cdots + b_{hn} x_n, \qquad h = 1, \cdots, n,$$

then $\bar{x}_1, \cdots, \bar{x}_n$ are immediately defined by the conditions

$$(\bar{x}_i \cdot t_h) = (\bar{x}_i \cdot \textstyle\sum_j b_{hj} x_j) = \sum_j b_{hj} (\bar{x}_i \cdot x_j) = b_{hi}.$$

Since the matrix $\| b_{hi} \|$ is unimodular (as the inverse of the matrix $\| a_{ih} \|$), it follows that $\bar{x}_1, \cdots, \bar{x}_n$ form a basis for $[t_1, \cdots, t_n]$. This completes the proof.

§5.4. Dual homomorphisms. Let

$$A = [t_1^\alpha, \cdots, t_p^\alpha], \qquad B = [t_1^\beta, \cdots, t_q^\beta]$$

be modules. A homomorphism f of A into B and a homomorphism g of B into A are said to be *dual* if

$$(f(x) \cdot y) = (x \cdot g(y))$$

for arbitrary $x \in A$, $y \in B$.

We shall prove the following theorem on dual homomorphisms:

5.4. *Let*

$$X = \{x_1, \cdots, x_p\}, \qquad \bar{X} = \{\bar{x}_1, \cdots, \bar{x}_p\}$$

be dual bases of a module A, and let

$$Y = \{y_1, \cdots, y_q\}, \qquad \bar{Y} = \{\bar{y}_1, \cdots, \bar{y}_q\}$$

be dual bases of a module B. If f and g are dual homomorphisms (of A into B and B into A, respectively), and

$$f(x_i) = \sum_{j=1}^{q} c_{ij} y_j, \qquad i = 1, \cdots, p,$$

then

$$g(\bar{y}_j) = \sum_{i=1}^{p} c_{ij} \bar{x}_i, \qquad j = 1, \cdots, q.$$

For the proof it is enough to note that for arbitrary $j = 1, \cdots, q$ the elements $g(\bar{y}_j) \in A$ and $\sum_{i=1}^{b} c_{ij} \bar{x}_i \in A$ have the same covariant components relative to X. These components are

$$(g(\bar{y}_j) \cdot x_h) = (x_h \cdot g(\bar{y}_j)) = (f(x_h) \cdot \bar{y}_j)$$
$$= (\textstyle\sum_{k=1}^{q} c_{hk} y_k \cdot \bar{y}_j) = \sum_{k=1}^{q} c_{hk}(y_k \cdot \bar{y}_j) = c_{hj}$$

and

$$(\textstyle\sum_{i=1}^{p} c_{ij} \bar{x}_i \cdot x_h) = \sum_{i=1}^{p} c_{ij}(\bar{x}_i \cdot x_h) = c_{hj} .$$

This completes the proof.

Remark. The above formula

$$(g(\bar{y}_j) \cdot x_h) = c_{hj}$$

implies that c_{ij}, the jth contravariant component of $f(x_i)$ relative to Y, is also the ith covariant component of $g(\bar{y}_j)$ relative to X.

§5.5. Modules (mod m), m a prime.

Definition 5.5. We shall denote by $^m[t_1, \cdots t_n]$ the group whose elements are all the linear forms $\sum a_i t_i$, where the t_i are given variables and the coefficients are residues (mod m), m a prime. The group $^m[t_1, \cdots, t_n]$ is called a *module of rank n* (mod m). [The rank of the group $^m[t_1, \cdots, t_n]$ (mod m), m a prime, is indeed n (see 3.3).] The same variables $t_1, \cdots, t_n$

form an initial basis of the module (moreover, we shall identify t_i with the linear form $1 \cdot t_i$, where 1 is the unit of the field J_m).

The definitions of scalar product, contravariant and covariant components, dual bases and dual homomorphisms, and the related Theorems 5.3, 5.4 are to be retained (as well as the proofs of the theorems, which are even somewhat simpler, due to the fact that instead of the ring J we now have the field J_m).

LIST OF SYMBOLS

	Page
$\mathfrak{A}$	19
$\mathfrak{A}[\theta_k^{r-1}]$	109
$b^r(\Phi)$	125
$b_m^r(\Phi)$	125
$\mathcal{B}^r(\Phi)$	126
$\mathfrak{B}^r(\Phi)$	126
B_i^{r-1}	129
$\mathbf{B}_i^{r-1}$	129
β	137
Δ	25
$\Delta_{\mathfrak{K}}$	24
$\Delta^r(\mathfrak{K}, \mathfrak{A})$	50
$\Delta_0^r(\mathfrak{K}) = \Delta^r(\mathfrak{K}, J)$	50
$\Delta_m^r(\mathfrak{K}) = \Delta^r(\mathfrak{K}, J_m)$	50
$\Delta^{00}(K, \mathfrak{A})$	65
$\Delta_{00}(\mathfrak{K}, \mathfrak{A})$	69
∇	90
$\nabla^r(\mathfrak{K}, \mathfrak{A})$	94
δ	151
$\Delta_\Phi^r(\Gamma, \mathfrak{A})$	185
Δ_Φ^r	166
$\Delta_p^r(\Phi)$	199
$\| e_0, e_1, \cdots, e_n \|$	2
$(e_0 \cdots e_n)$	3
$\| e_0 \cdots e_n \|$	4
$-\| e_0 \cdots e_n \|$	4
$\pm\| e_0 \|$	4
$(e_0 t^{n-1})$	6
$\mathfrak{E}^r$	17
ε_{ij}	16
$E_{\mathfrak{K}} x^r$	24
$(\epsilon\text{-}\sim)$	148
$\eta(K)$	151
ϵ_C	163
$(\Gamma - \sim)$	184
γ	194
$H_\Delta^r(\mathfrak{K}, \mathfrak{A})$	30
$H_0^r(\mathfrak{K}) = H^r(\mathfrak{K}, J)$	30
$H_m^r(\mathfrak{K}) = H^r(\mathfrak{K}, J_m)$	30
$\hat{H}_0^r(\mathfrak{K})$	35
$\hat{H}^r(\mathfrak{K})$	69
$H_\nabla^r(\mathfrak{K}, \mathfrak{A})$	93
$H_\Phi^r(\Gamma, \mathfrak{A})$	185
J	20
J_m	20
$J(\theta_i^r)$	112
$\mathfrak{K}$	13
K	13
$\mathfrak{K}_0 x^r$	24
K	129
$K(R, \epsilon)$	148
$K(\Phi, \epsilon)$	178
$K(\Gamma, \epsilon)$	178
$L^{-1}(\mathfrak{K}, \mathfrak{A})$	20
$L_0^r(\mathfrak{K}) = L^r(\mathfrak{K}, J)$	20
$L_m^r(\mathfrak{K}) = L^r(\mathfrak{K}, J_m)$	20
$L^r(\mathfrak{K}, \mathfrak{A})$	20
lirh	31
$<oK>$	40
(ox^r)	40
(ot^r)	41
oK	41
Π	43
$\pi^r(\mathfrak{K})$	50
$\pi_m^r(\mathfrak{K})$	50
$\Psi\alpha$	181
$(p\text{-}\sim)$	199
$R^n, -R^n$	2, 3
$\| R^n \|$	2
$R^0, -R^0$	3
$R^n \| e_0, e_1, \cdots, e_n \|$	2
$R^0 \| o \|$	3
ρ^r	16
$\mathfrak{R}$	20
$\mathfrak{R}_1$	20
$S_\alpha^\beta R_\beta^n$	7

INDEX